AF361941

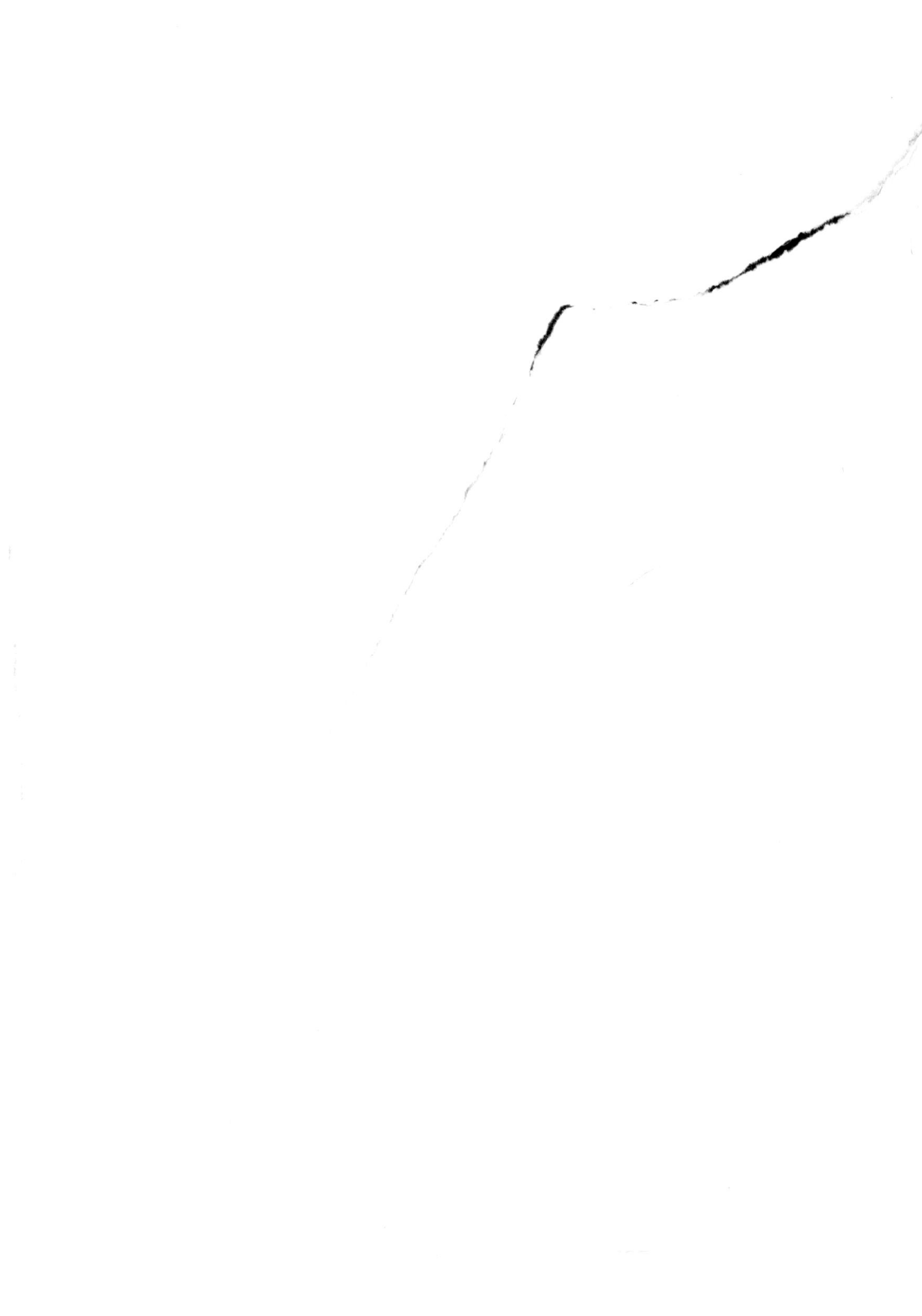

Science, Characterization and Technology of Joining and Welding

Science, Characterization and Technology of Joining and Welding

Special Issue Editor

Meysam Haghshenas

MDPI • Basel • Beijing • Wuhan • Barcelona • Belgrade

Special Issue Editor
Meysam Haghshenas
University of Toledo
USA

Editorial Office
MDPI
St. Alban-Anlage 66
4052 Basel, Switzerland

This is a reprint of articles from the Special Issue published online in the open access journal *Metals* (ISSN 2075-4701) from 2018 to 2019 (available at: https://www.mdpi.com/journal/metals/special_issues/joining_welding).

For citation purposes, cite each article independently as indicated on the article page online and as indicated below:

LastName, A.A.; LastName, B.B.; LastName, C.C. Article Title. *Journal Name* **Year**, *Article Number*, Page Range.

ISBN 978-3-03928-997-4 (Pbk)
ISBN 978-3-03928-998-1 (PDF)

Cover image courtesy of Sara Richmire and Meysam Haghshenas.

Dedication

I would like to dedicate this book to my wife, Mehrnoosh, and my son, Mehrsaam

Contents

About the Special Issue Editor

Meysam Haghshenas Prior to joining the University of Toledo in June 2019, Dr. Haghshenas was an Assistant Professor in the Department of Mechanical Engineering at the University of North Dakota, USA. Before that, he was a post-doctoral fellow (PDF) at the University of Waterloo (Canada), Center for Advanced Materials Joining (CAMJ), and a process engineer in Linamar Corporation, a Canadian manufacturing company that operates worldwide. His research background is in small-scale characterization of materials, fatigue and fracture mechanics, additive manufacturing, mechanical behavior of materials, as well as solid-state welding. He is a certified International Welding Engineer (IWE). He has taught various courses including Fatigue and Fracture, Manufacturing Processes, Advanced Manufacturing, Additive Manufacturing, and Fracture of Welded Structures.

Preface to "Science, Characterization and Technology of Joining and Welding"

As the Guest Editor of the Special Issue of *Metals* entitled "Science, Characterization, and Technology of Joining and Welding", I am pleased to have this book published by MDPI. Joining, including welding, soldering, brazing, adhesive bonding, and assembly, is an essential requirement in manufacturing processes and is classified as a secondary manufacturing process. Many welding processes are accomplished by heat alone, with no pressure applied; others by a combination of heat and pressure; and still others by pressure alone, with no external heat supplied. Among various joining methods, welding is the most popular and most commonly employed operation to join materials. Welding involves localized coalescence of two parts at their faying surfaces. The faying surfaces are the surfaces of the parts that are to be joined, and are in contact or close proximity. Welding is usually performed on parts made of the same metal, but some welding operations can be used to join dissimilar metals and/or non-metals. Though joining is an old process, it is still very much a live research field considering new technologies and advances in the characterization of welded joints. Some 50 different types of welding operations have been cataloged by the American Welding Society (AWS). However, the list of the joining processes is expanding continuously with the new processes developed around the globe.

This Special Issue of *Metals* includes technical and review papers on, but not limited to, different aspects of joining and welding, including welding technologies (i.e., fusion-based welding and solid-state welding), characterization, metallurgy and materials science, quality control, design, and numerical simulation. This Special Issue also includes the joining of different materials, including metal and non-metals (polymers and composites). This Special Issue includes 17 peer-reviewed papers from several researchers all around the globe (China, Germany, Brazil, South Korea, Slovakia, USA, Taiwan, Canada, and India). To date (April 2020), the papers in this Special Issue have been cited 47 times by other researchers, which reflects the high quality of the published papers in this issue.

This Special Issue includes a large diversity of various subjects in the field of joining: laser welding, friction stir welding, diffusion bonding, multipass welding, rotary friction welding, friction bit joining, adhesive bonding, weld bonding, simulation and experimentation, metal/FRP joints, welding simulation, plasma–TIG coupled arc welding, liquation cracking, soldering, resin bonding, microstructural characteristics, brazing, and friction stir butt and scarf welding. This large variety of papers covers different areas related to the science of joining and welding, in which significant research results are shared and may be used by anyone with an interest in these fields.

I would like to sincerely thank all the researchers, from all around the world, who contributed to this Special Issue for their high-quality research representing a great contribution to the field. I hope this Special Issue enhances our knowledge and understanding in the field of joining and assembly and serves as a reliable reference for students, researchers, and engineers working in this field.

Meysam Haghshenas
Special Issue Editor

Article

Mechanical and Microstructural Investigations of the Laser Welding of Different Zinc-Coated Steels

Eva Zdravecká [1],* and Ján Slota [2]

[1]   Department of Mechanical Technology and Materials, Faculty of Mechanical Engineering, Technical University of Košice, Mäsiarska 74, 040 01 Košice, Slovakia

[2]   Department of Computer Support of Technology, Faculty of Mechanical Engineering, Technical University of Košice, Mäsiarska 74, 040 01 Košice, Slovakia; jan.slota@tuke.sk

*   Correspondence: eva.zdravecka@tuke.sk; Tel.: +421-55-602-3516

Received: 9 December 2018; Accepted: 11 January 2019; Published: 16 January 2019

Abstract: Tailor welded blanks (TWB) represent an anisotropic and non-homogenous material. The knowledge of the mechanical properties and microstructure of the fusion zone and heat-affected zone (HAZ) obtained with laser welding is essential to ensure the reliability of the process. In this paper, laser-welded hot-dip Zn-coated low carbon microalloyed steels with different thickness and mechanical properties were used. The mechanical properties of the laser-welded blanks were determined by tensile tests and formability by Erichsen cupping tests. In addition, the pore formation during the laser welding process was analyzed. The microstructural analysis confirmed the formation of the favorable structure of the weld metal and the heat-affected zone without the presence of martensite. The obtained results showed that it is possible to produce TWBs with suitable mechanical properties by laser welding.

Keywords: tailor welded blanks; laser welding; Zn-coated low carbon microalloyed steels; microstructure; pores

---

1. Introduction

Laser beam welding is very flexible because welds can be made in continuous and complex shapes, can be performed at relatively high travel welding speeds, and can achieve deeply penetrated welds. Tailor welded blanks (TWB) are blanks that have been manufactured from sheets with similar or different thicknesses by a welding process. The differences in the material within a TWB can be in the thickness, grade, or coating of the material, e.g., galvanized versus ungalvanized. Weight and production cost-reducing automotive body design are achieved by using laser butt welded semi products such as tailored blanks made of steel or combined materials. The laser is preferred for welding owing to the high speed of the process, low distortion due to the small heat-affected zone, the manufacturing flexibility, and the ease of automation [1–5].

Designers are able to tailor the location in the blank where specific material properties are desired when creating a TWB. This trend of welding and forming of sheet-metal pieces allows notable flexibility in product design, structural stiffness, and crash behavior (crashworthiness).

The laser-welding of zinc-coated steel sheets is often used in the automotive industry. A major problem that arises when welding these materials is associated with the vaporization temperature of zinc (906 °C), which is much lower than the melting temperature of steel (1530 °C). The laser-welding of galvanized steel sheets in the thickness range of 0.6–1.5 mm is largely performed in the automotive industry for tailored blank welding applications [6–9].

Low carbon microalloyed steel sheets have long been a commonly-used material in consumer industries because it can be stamped into inexpensive parts with complicated shapes at very high production rates [10]. Interstitial-free (IF) steel sheets are the most frequently used sheet material for complex automotive applications due to their superior formability [11–15].

Previous studies and research have identified the factors that have an important effect on the strength of the laser beam-welded joints. These are the laser power (P), welding speed (S), and focal position (F), which were considered to be the factors influencing the Erichsen cupping test results and the microstructure of specimens [1,2,16].

Some important evaluated factors of TWBs are material property changes to welds made of different combinations of microalloyed steel sheets and changes of non-uniform deformation because of the differences in thickness, properties, and surface characteristics with respect to the application of the load.

This investigation addressed three types of galvanized automotive steel sheets, namely:

1. DX54D+Z (EN10346:2015)—interstitial-free (IF) steel sheets appropriate for galvanizing and annealing to produce specialized steel sheets required for automotive body manufacturing. IF steel sheets are free from carbide precipitates at the grain boundaries.

2. DX53D+Z (EN10346:2015)—low carbon steel sheets, where the microstructure is ferrite–pearlite with polyedric ferrite grains and sporadically precipitated deformed pearlite. Hot-dip galvanized sheets made of drawing grades are suitable for cold forming and deep-drawing. The sheets are used for the production of automobile parts, appliances, in the building industry, and for production of profiles, corrugated sheets, roof coverings, and engineering.

3. ZStE260Z (SEW 093/87)—the ferritic microstructure cementite is precipitated in ferrite grains. The structure is fine-grained.

The present work focuses on the characterization of the butt joint of low carbon galvanized steel sheets finished by laser welding. The influence of the laser welding on the microstructure, phase transformations, hardness, mechanical strength, and work hardening behavior is also investigated. The novelty of the paper concerns the evaluation of the special combinations of Zn-coated steel sheets (TWBs). The combination of materials labeled TWB1 and TWB2 has been requested by the manufacturer of tested materials for intended use in laser welding and has potential for applications in the automotive industry.

2. Materials and Methods

2.1. Materials

The experiments were conducted on zinc-coated steels. Three types of galvanized steel sheets in the thickness range of 0.8 to 1.75 mm with differential mechanical properties were used for tailored blank welding (TWB) applications and marked as TWB1 and TWB2 (see Tables 1 and 2). For the experiment, combinations of hot-dip zinc coated low carbon steel sheet DX53D+Z (deep drawing quality sheet steel for cold forming according to EN 10327) with DX54D+Z interstitial-free (IF) steel sheets marked as TWB1 were used. The second combination, marked as TWB2, consisted of low carbon microalloyed steel sheets DX53D+Z in combination with ZStE260Z. The chemical compositions and thicknesses of TWB1 and TWB2 are shown in Tables 1 and 2.

Table 1. Chemical composition and thickness of TWB1.

Material	Thickness (mm)	Chemical Composition TWB1 (%)								
		C_{max}	Mn_{max}	P_{max}	S_{max}	Si_{max}	Al	Ti	Nb_{max}	N_{max}
DX54D+Z (IF)	0.80	0.015	0.20	0.015	0.015	–	0.02	0.06–0.14	–	0.006
DX53D+Z	1.00	0.04	0.20	0.015	0.012	0.01	0.03–0.06	–	–	0.006

Table 2. Chemical composition and thickness of TWB2.

Material	Thickness (mm)	Chemical Composition TWB2 (%)								
		C_{max}	Mn_{max}	P_{max}	S_{max}	Si_{max}	Al	Ti	Nb_{max}	N_{max}
DX53D+Z	1.75	0.04	0.20	0.015	0.012	0.01	0.03–0.06	–	–	0.006
ZStE260Z	1.30	0.10	0.60	0.025	0.008	0.04	0.015	0.04	0.02–0.035	–

2.2. Process Parameters of Laser Welding

A CO_2 laser with a nominal power of 5 kW was used for laser welding. The CO_2 laser parameters were: wavelength—10.6 mm; power range—500 W to 5000 W; output stability—+2%; diameter of output beam—13 mm; mode—TEMoo; divergence—1.5 mrad; point stability—3.35 mrad; pulsing—up to 1 kHz; and horizontal beam polarization. In order to avoid accidental misalignment or movement during laser welding, a welding sample fixture was used to ensure stable and consistent experimental tests. Welding was carried out without gaps [17,18]. During welding, gas argon was used as the shielding gas. The butt weld was processed with full penetration on the sheet steels of uniform thickness. The welding parameters that were set are shown in Table 3.

Table 3. Welding parameters.

LBW Parameters	TWB1	TWB2
Laser power (P) (kW)	2.5	2.9
Welding speed (s) (mm·s^{-1})	45	45
Shielding gas pressure (P) (MPa)	0.2 Ar	0.2 Ar
Focal position (F) (mm)	0	0

2.3. Metallography and Hardness Testing

Samples for metallographic examination were cut from each of the TWBs and prepared following standard procedures. Polished specimens were etched using 3% Nital and the microstructures viewed under the Carl Zeiss Jena NEOPHOT 32 optical microscope. These analyses were focused to study the microstructure in the fusion zone (FZ), heat-affected zone (HAZ), and base metal (BM).

One of the most-commonly used techniques to determine the mechanical properties in the welding zone is microhardness measurement. The microhardness of the base materials and welding joints were analyzed by SHIMADZU–DUH 202, Japan, Indenter VICKERS (diamant), using a load of 98 mN (10 gf). The Vickers hardness test was performed according to ISO 6507. The hardness was measured 3–5 times in each area, and the average value was taken as the test result.

2.4. Tensile Tests and Formability

The dimensions of the tailor welded sheet metals were 250 mm × 250 mm. The shaped specimens were cut out perpendicular to the weld line from welded blanks. All the specimens were machined carefully with the centerlines of the weld zone and the specimen being coincident, as illustrated in Figure 1. The aim of the tensile tests was to evaluate the strength and plasticity of welding joints and/or examine the influence of welding defects on the joint performance. The tensile test was performed on a universal testing machine ZWICK 50 according to EN 895 and EN ISO 6892-1:2016. Standard tensile characteristics include yield stress (YS), ultimate tensile strength (UTS), and percent over elongation at break (ductility). The YS, UTS, and elongation to fracture of the BM as well as TWB were evaluated [19–23]. In the calculation of the stress-strain diagram of the TWB samples, the thickness of the thinner sheet was considered as the initial thickness. Cracking of the sample was not observed in the FZ or HAZ, but on the side of the material with lower mechanical properties or lower thickness [24].

The Erichsen cupping test, one of the conventional formability test methods, was used to evaluate the welded blank formability. Specimens for the Erichsen cupping test were prepared according to ISO 20482.

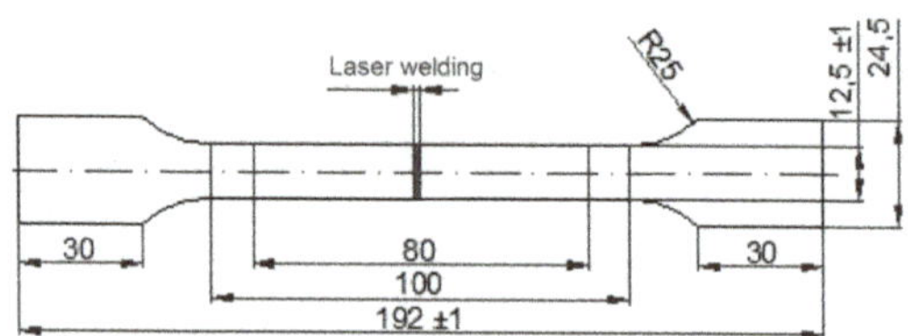

Figure 1. Shape and dimensions of the welded specimen (tension perpendicular to weld plane) (Unit: mm).

3. Results and Discussion

3.1. Microstructure Characterization of the Laser Welding Joints

The mechanical properties obtained in different welding zones are a function of the microstructure. To understand the welding behavior, it is helpful to examine the microstructure of the base materials and welded joints [25,26]. Analyses of the microstructure in the fusion zone (FZ), heat-affected zone (HAZ), and base metal (BM) for all specimens were performed. An overall cross-section photo of the microstructure of laser-welded steel sheets (DX54D+Z—0.8 mm/DX53D+Z—1.0 mm) marked as TWB1 is seen in Figure 2. The overall top width of the weld was approx. 0.7 mm, the bottom of the weld was 0.8 mm, the HAZ (DX54D+Z) was approx. 0.3 mm, and width of the HAZ (DX53D+Z) was approx. 0.4 mm (see Figure 2).

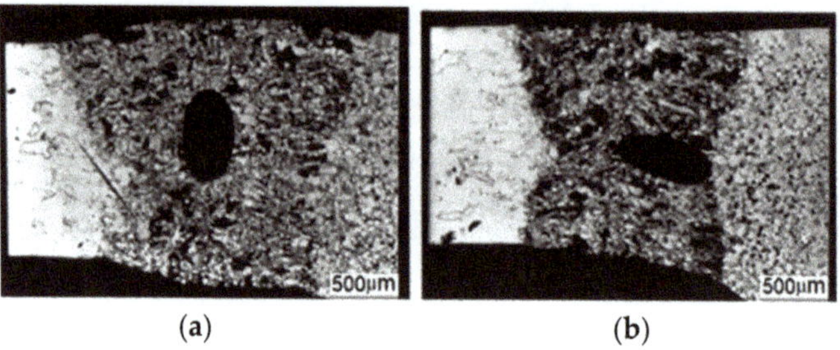

(a)   (b)

Figure 2. Micrographs of the TWB1 laser-welded specimens: (**a**) the pore in central area of FZ, (**b**) the pore near HAZ.

The transition zone FZ/BM was smooth, without micropores and notches. The weld metal contains multiple pores or cavities. The top and bottom of the weld have a smooth transition between sheets of different thicknesses. More detailed microstructures of the TWB1 laser-welded joint are shown in Figure 3a–e.

Figure 3a FZ: the ferritic-bainite microstructure, pearlite sporadically precipitated as untransformed pearlite (the presence of the cavity was observed).

Figure 3b HAZ: DX54D+Z (IF); ferritic-cementite microstructure is in the area of high temperature overheating of the base material, with a singularly formed bainite, and in the A_{C1}–A_{C3} temperature range the ferritic-cementite microstructure is with uniquely precipitated carbides.

Figure 3c HAZ: DX53D+Z; the ferritic-pearlitic microstructure in the area of high overheating of the base material with a singularly formed bainite; in the A_{C1}–A_{C3} temperature range the microstructure is ferritic-pearlitic with deformed pearlite, and ferrite has enlarged grains.

Figure 3d BM: DX54D+Z; the ferritic microstructure with a uniquely precipitated cementite.

Figure 3e BM: DX53D+Z; ferritic–pearlite microstructure with polyedric (quasi-equiaxed) ferrite grains and sporadically precipitated deformed pearlite.

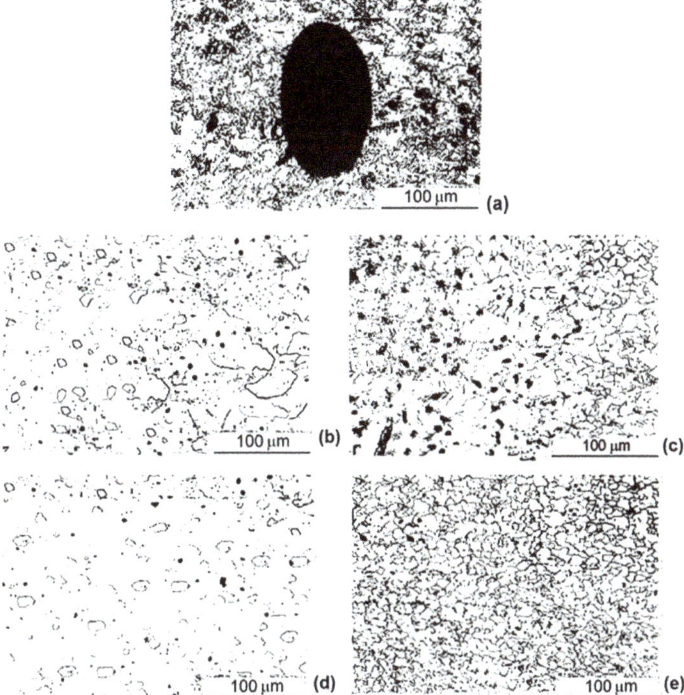

Figure 3. Microstructure of TWB1. (**a**) Fusion zone (FZ) (DX54D+Z/DX53D+Z); (**b**) Heat-affected zone (HAZ) (DX54D+Z); (**c**) HAZ (DX53D+Z); (**d**) Base metal (BM) (DX54D+Z); (**e**) BM (DX53D+Z).

Similar analyses for the TWB2 laser-welded specimens were performed. A photo of the microstructure of laser-welded steel sheets (ZStE260Z, 1.3 mm/DX53D+Z, 1.75 mm) marked as TWB2 is seen in Figure 4. The overall width of the top of the weld was approx. 1.0 mm, the bottom of the weld was 0.6 mm, and width of the HAZ (DX53D+Z) was approx. 0.6 mm, and the HAZ (ZStE260Z) was 0.6 mm (See in Figure 4).

Figure 4. Micrograph of the TWB2 specimen.

The microstructure of the TWB2 laser-welded joint are shown in Figure 5a–e.

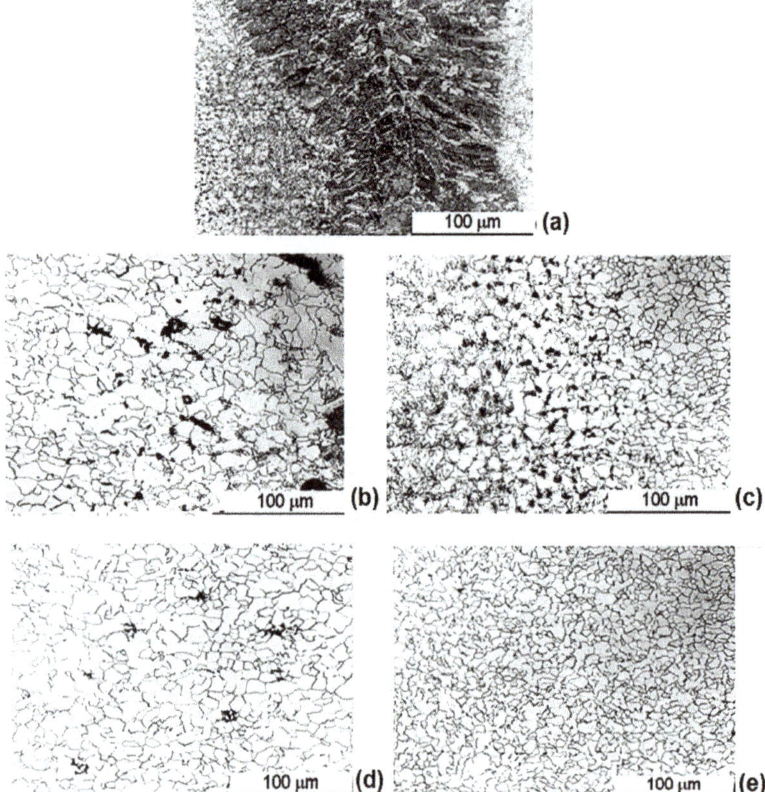

Figure 5. Microstructure of TWB2. (**a**) FZ (DX53D+Z/ZStE260Z); (**b**) HAZ (DX53D+Z); (**c**) HAZ (ZStE260Z); (**d**) BM (DX53D+Z); (**e**) BM (ZStE260Z).

Figure 5a FM: Weld metal; the bainitic-ferritic microstructure with acicular ferrite.

Figure 5b HAZ: DX53D; the ferritic-pearlite microstructure in the area of high temperature overheating of the base material with a singularly precipitated bainite; in the A_{C1}–A_{C3} temperature range is a ferritic-pearlite microstructure with deformed pearlite and enlarged ferrite grains.

Figure 5c HAZ: ZStE260Z; the ferritic-bainitic-pearlitic microstructure in the area of high temperature overheating of the base material; in the A_{C1}–A_{C3} temperature range is a ferritic-pearlite microstructure with enlarged ferrite grains and deformed pearlite.

Figure 5d BM: DX53D+Z; the ferritic-pearlite microstructure with polyedric ferrite grains and sporadically precipitated deformed pearlite.

Figure 5e BM: ZStE260Z; the ferritic microstructure, cementite is precipitated in ferrite grains. The structure is fine-grained.

The microstructure of TWB1 and TWB2 did not show any evidence of the excessive root penetration of the welds. In the HAZ, the grain thickness was slightly increased. For the microalloyed steel, pearlite and bainite were precipitated in the HAZ, and bainite was precipitated in the low carbon steel.

3.2. Analysis of the Microhardness

The microstructure hardness was measured in the base metal (BM), heat-affected zone (HAZ), and fusion zone (FZ). The microhardness distribution in the base metal, heat-affected zone, and fusion zone were measured for TWB1a and TWB2. In Figure 6 (TWB1) and Figure 7 (TWB2) marks represent the hardness testing position in the base metal zone, heat-affected zone, and fusion zone, respectively. The average values of HV0.5 microhardness and standard deviation (Stdev) for each material are shown in Figure 6.

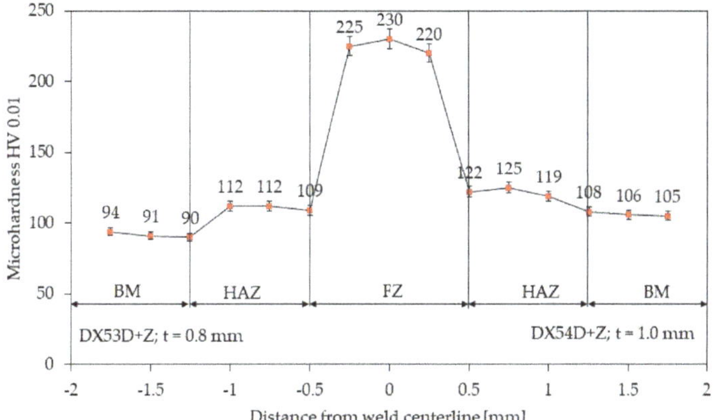

Figure 6. Microhardness distribution of TWB1.

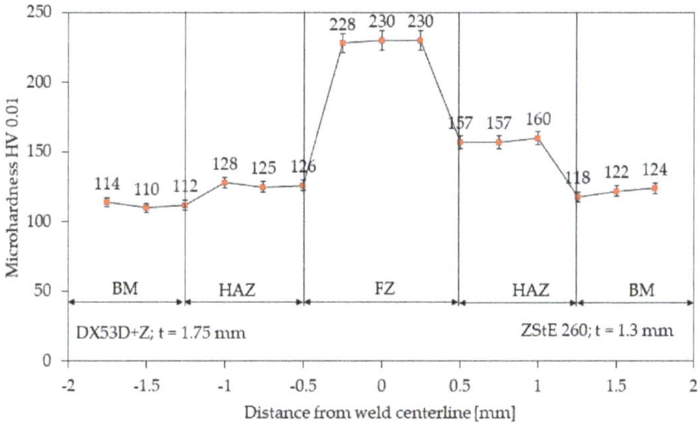

Figure 7. Microhardness distribution of TWB2.

The microhardness of the HAZ in DX53D+Z steel increased up to 112 HV and in the DX54Z+Z steel it increased up to 125 HV. The widths of the HAZ and FZ in the welded joints were similar. The higher microhardness of the FZ can be explained by the presence of bainite in the microstructure.

The ZStE260 material contains a larger amount of pearlite and a certain amount of cementite, which explains the higher microhardness in the HAZ up to 160 HV (Figure 7). Figures 6 and 7 show that the microhardness in the FZ of the welded joints was similar.

3.3. Results of the Tensile and Formability Tests

One of the aims of a tensile test is to evaluate the strength and plasticity of welding joints and examine the influence of welding defects on the joint performance [27]. Figures 8 and 9 show the results of the mechanical properties of the base materials (thinner blanks) and TWB welded joints.

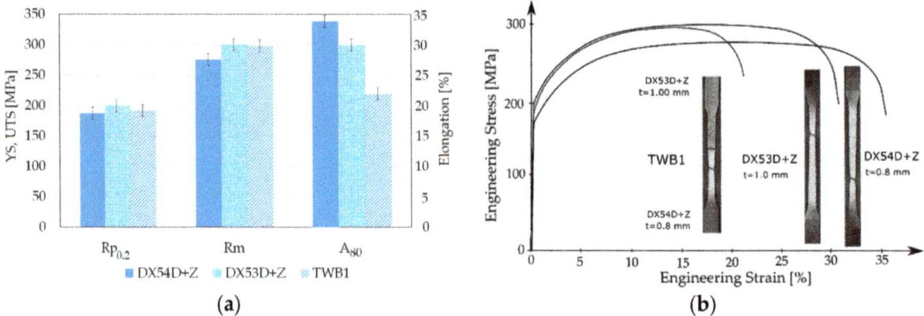

Figure 8. (**a**) The yield stress, tensile strength, and elongation of the base materials and TWB1; (**b**) the stress–strain curves of the base materials and TWB1.

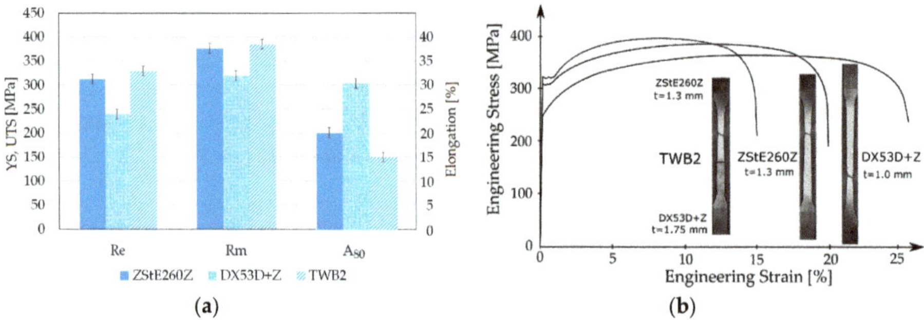

Figure 9. (**a**) The yield stress, tensile strength, and elongation of the base materials and TWB2; (**b**) the stress–strain curves of the base materials and TWB2.

The elongation of the welded joints decreased with the increased hardness of the welds. It was observed that the fracture occurred in the thinner/weaker material, but not in the weld, as shown in Figure 10.

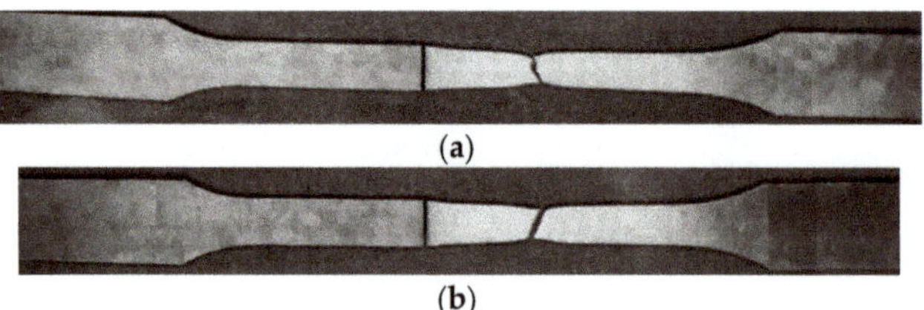

Figure 10. The TWB samples after tensile tests: (**a**) TWB1, (**b**) TWB2.

The failure occurred in the thinner sheet metal for both TWBs, which conforms with the literature [28–31]. A number of studies showed that increasing the thickness and/or strength ratios decreases the formability of the TWBs [29–31]. A larger thickness ratio forces more deformation into the weaker material and the strain is concentrated there, which results in premature failure. During deformation, the thinner material undergoes plastic deformation, whereas the thicker material undergoes primarily elastic deformation. An increase in the strength ratios has a similar effect on the failure mode as the thickness ratio, whereby the weaker material deforms more and fails first. In addition, due to non-uniform deformation, the weld line also tends to move towards the thicker/stronger materials.

The obtained results indicated that the microhardness of the weld zone was higher than the microhardness of the base sheets. This shows that the thinner blank is dominant in the overall deformation behavior. The thinner blank is the one that will decide the properties of the whole TWB. The thicker blank behaves as a rigid support to the thinner blank [32,33].

The Erichsen cupping test value of the base materials and TWB1 is shown in Figure 11. From the tested samples, the highest cup depth of 11.35 mm was obtained.

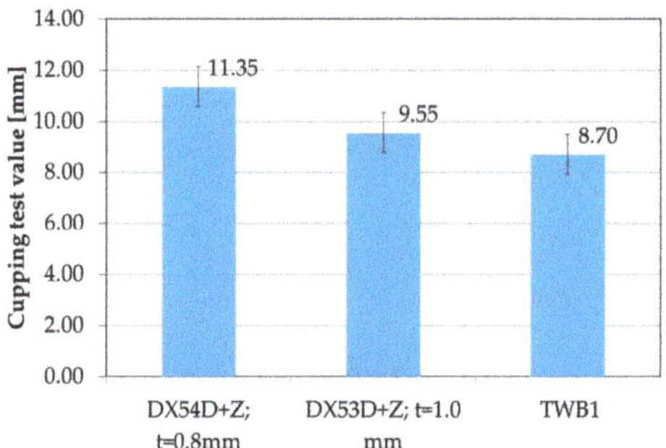

Figure 11. The Erichsen cupping test value of the base materials and TWB1.

The crack spread in a circular shape along the weld. On TWB1, the crack was initiated outside of the HAZ, despite internal defects in the weld (Figure 3). Figure 12a shows the initiation of the failure in base material DX54Z+Z and propagation transverse to the weld line into the base metal DX53D+Z. Figure 12b shows a ductile fracture of the base material DX54D. From the measured microhardness values, we can see that the laser welding caused material hardening.

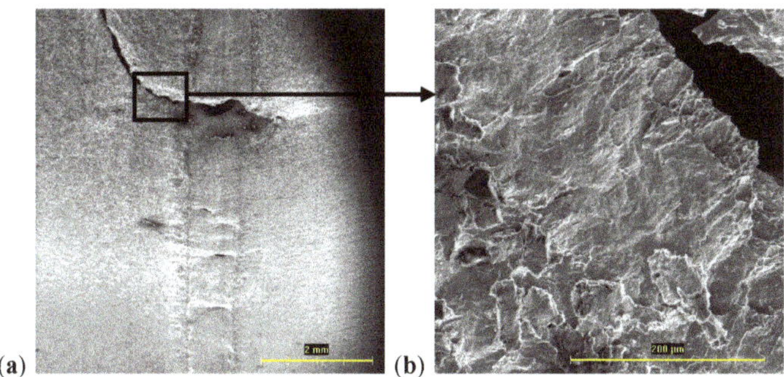

Figure 12. Failure of TWB1 after the Erichsen test. (**a**) The initiation and propagation of the failure. (**b**) Fracture surface.

The presence of defects in the welded joint could be affected by the improper machining of the welding edges and degreasing. The concentration of large amounts of energy during laser welding vaporizes the burrs after cutting, as well as impurities and grease on the surface of the sheet. The vapors from zinc coating as well as impurities get into the weld metal and remain blocked due to rapid cooling. The pore formation was identified on the bottom of weld for TWB1 (Figures 12 and 13a,b). The Energy Dispersive X-ray Analysis (EDX) analysis of the pore in TWB1 is shown in Figure 13c.

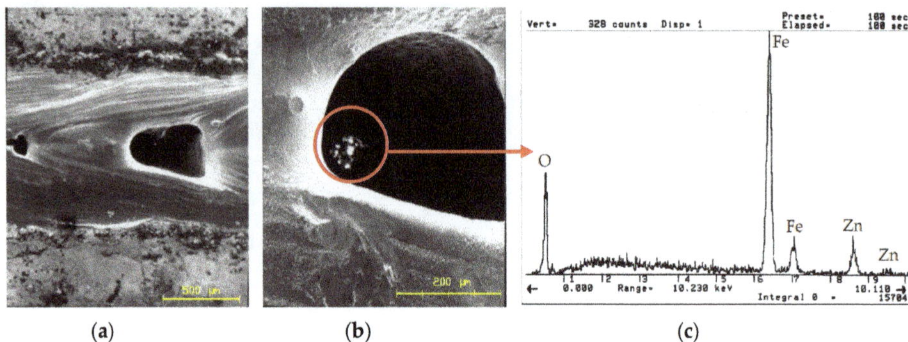

Figure 13. Laser weld line on TWB1 with a pore: (**a**) Pore formation in the weld, (**b**) A detailed view of the pore, (**c**) EDX analysis of the pore in the TWB1 sample.

It was found that thin sheets are more sensitive to the welding parameters, surface cleanness, and preparation of edges. The bubbles showed mixed behavior, indicating that the capillary is partly filled with zinc vapor and partly with ambient gas. Figure 14b shows the SEM of the cross–section of the TWB1 weld with the internal pore. Figure 14a,c shows the results of the EDX analysis. The advantage of the EDX analysis, in this case, is the exact identification of the chemical composition of each compound in the pore. This will contribute to the search for the source of the pore formation. The porosity in the TWBs could be due to various causes, e.g., keyhole instability, improper gas shielding, surface contamination, improper edge preparation, and the improper setting of welding parameters. The setting of the laser power affects the evaporation of the substrate and creates a good key effect. This creates an opening that can pump high-pressure zinc gas and reduce the risk of explosion.

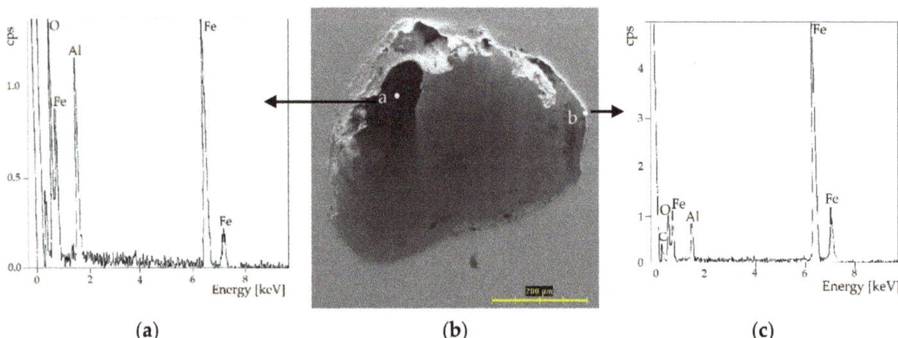

Figure 14. Cross-section through the TWB1pore. (**a**) EDX analysis of the pore; (**b**) SEM of the cross-section; (**c**) EDX analysis of the edge of the pore.

The Erichsen cupping test value of the base materials and TWB2 is shown in Figure 15. The base material ZStE260, with a thickness of 1.3 mm, showed a circular-shaped crack. In TWB2 specimens, the crack was of circular shape, and failure occurred a sufficient distance from the weld in the thinner sheet metal.

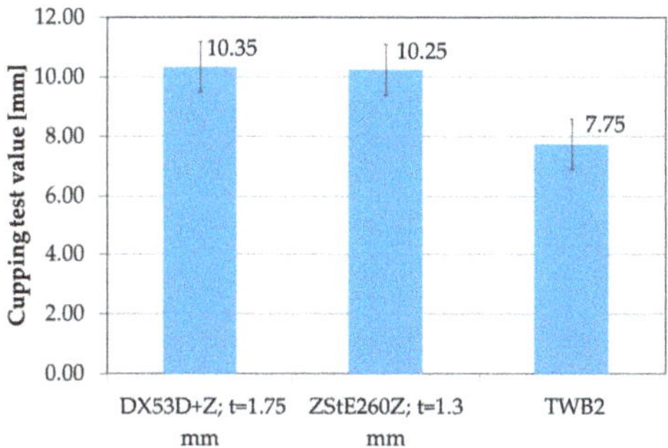

Figure 15. The Erichsen cupping test value of the base materials and TWB2.

4. Conclusions

The laser beam welding process and its effects of the on the quality of the welded joints were analyzed on microalloyed, high-strength, Zn-coated steels with different thicknesses (0.8 and 1.0 mm/TWB1 and 1.3 and 1. 75 mm/TWB2). The microstructure and mechanical properties of the laser-welded butt joints were investigated. The main results are summarized as follows:

i. The good weldability of microalloyed high-strength, Zn-coated steels was confirmed.

ii. In the microhardness test, hardness peaks were found in the weld metal. There was no evidence of martensite in the HAZ or the weld metal. For microalloyed steel, pearlite and bainite were precipitated in the HAZ, and bainite was precipitated in the low carbon steel. In all the TWBs, the FZ was predominantly formed by ferritic structures, with some grains of low-carbon bainite.

iii. A change in the mechanical properties of the welded joints and the base materials of the TWB was observed. A decrease of the ductility of both the TWB1 and TWB2 can be related to the heat-affected welding zone due to hardening. The thickness ratio and strength ratio had an effect on the failure, whereby the weaker material (low thickness, lower mechanical properties) deformed more and failed first. A larger thickness ratio forced more deformation into the weaker material and the strain was concentrated there, which resulted in premature failure.

iv. During the Erichsen test for TWB1, a failure was initiated in the DX54Z+Z base material and its propagation was perpendicular to the weld line. In the TWB2 specimens, the crack was of a circular shape, and the failure occurred a sufficient distance from the weld in the thinner sheet metal.

v. The possible causes of porosity in the laser welds of hot-dip, Zn-coated, low carbon, microalloyed steel sheets were keyhole instability, improper gas shielding, surface contamination, improper edge preparation, and the improper setting of welding parameters.

vi. The tensile tests showed lower sensitivity to the detected defect bands. The pores did not have a detrimental effect on the tensile properties of the welded joint, which may be due to the high strength of the fusion zone, which effectively protects the defective weld zone. The strong weld retained the defect and prevented it from spreading. The Erichsen test showed higher sensitivity in the presence of pores.

vii. The preparation of thinner sheets for welding required consistent weld surface finishing as well as sheet metal fitting. It was found that the thin sheets were more sensitive to welding parameters, surface cleanness, and preparation of edges.

Based on the experimental results, it can be stated that laser-welded blanks designed to produce parts exposed to plastic deformations require quality joints that do not complicate the stamping process. Laser welding is both an instrumentally and technically demanding method, so it is suitable for large-scale production. The problems of setting the right welding parameters and reproducible welding positioning affect the final weld quality.

Author Contributions: E.Z. conducted methodology, microhardness test, microscopical analyses, and analysis and interpretation of the results, and funding acquisition; J.S. conducted measurements of the tensile test and the Erichsen test, writing—original draft preparation, editing, visualization.

Funding: This research was funded by the Grant Agency of the Ministry of Education, Science, Research, and Sport of the Slovak Republic under grant number VEGA 1/0117/15 and VEGA 1/0259/19.

Acknowledgments: The authors are grateful for the support of experimental works by the Grant Agency of the Ministry of Education, Science, Research, and Sport of the Slovak Republic for the support of the project VEGA 1/0117/15 and VEGA 1/0259/19.

Conflicts of Interest: The authors declare no conflict of interest.

References

1. Kim, J.D.; Na, H.; Park, C.C. CO_2 Laser Welding of Zinc-Coated Steel Sheets. *J. Mech. Sci. Technol.* **1998**, *12*, 606–614. [CrossRef]
2. Lu, J.; Kujanpää, V. Review study on remote laser welding with fiber lasers. *J. Laser Appl.* **2013**, *25*, 052008. [CrossRef]
3. Popescu, A.C.; Delval, C.; Leparoux, M. Control of Porosity and Spatter in Laser Welding of Thick AlMg5 Parts Using High-Speed Imaging and Optical Microscopy. *Metals* **2017**, *7*, 452. [CrossRef]
4. Wu, Q.; Gong, J.; Chen, G.; Xu, L. Research on laser welding of vehicle body. *Opt. Laser Technol.* **2008**, *40*, 420–426. [CrossRef]
5. Merklein, M.; Johannes, M.; Lechner, M.; Kuppert, A. A review on tailored blanks—Production, applications and evaluation. *J. Mater. Process. Technol.* **2014**, *214*, 151–164. [CrossRef]
6. Pan, Y.; Richardson, I.M. Keyhole behaviour during laser welding of zinc-coated steel. *J. Phys. D Appl. Phys.* **2011**, *44*, 045502. [CrossRef]
7. Panda, S.K.; Kumar, D.R.; Kumar, H.; Nath, A.K. Characterization of tensile properties of tailor welded IF steel sheets and their formability in stretch forming. *J. Mater. Process. Technol.* **2007**, *183*, 321–332. [CrossRef]
8. Ayres, K.R.; Hilton, P.A. CO_2 laser butt welding of coated steels for the automotive industry. *Weld. Met. Fabr.* **1994**, *62*, 10–12.
9. Hamidinejad, S.M.; Hasanniya, M.H.; Salari, N.; Valizadeh, E. CO_2 laser welding of interstitial free galvanized steel sheets used in tailor welded blanks. *Int. J. Adv. Manuf. Technol.* **2013**, *64*, 195–206. [CrossRef]
10. Villalobos, J.C.; Del-Pozo, A.; Campillo, B.; Mayen, J.; Serna, S. Microalloyed Steels through History until 2018: Review of Chemical Composition, Processing and Hydrogen Service. *Metals* **2018**, *8*, 351. [CrossRef]
11. Shao, H.; Gould, J.; Albright, C. Laser blank welding of high strength steels. *Metall. Mater. Trans.* **2007**, *38*, 321–331. [CrossRef]
12. Feliu, S., Jr.; Perez-Revenga, M.L. Effect of alloying elements (Ti, Nb, Mn and P) and the water vapour content in the annealing atmosphere on the surface composition of interstitial free steels at the galvanising temperature. *Appl. Surf. Sci.* **2004**, *229*, 112–123. [CrossRef]
13. Abbasi, M.; Ketabchi, M.; Shakeri, H.R.; Shafaat, A. Obtaining high formability of IF-galvanized steel tailor welded blank by applying proper CO_2 laser welding parameters. *Int. J. Mater. Res.* **2011**, *102*, 1295–1302. [CrossRef]
14. Wang, Z.; Wang, X. A new technology to improve the $\bar{r}$-value of interstitial-free (IF) steel sheet. *J. Mater. Process. Technol.* **2001**, *113*, 659–661. [CrossRef]
15. Yan, B.; Gallagher, M. Strength and fatigue of laser butt welds for IF, HSLA and dual phase sheet steels. In: International symposium on advanced high strength steels for the ground transportation industry. *Mater. Sci. Technol.* **2006**, *2*, 87–101.
16. Chung, B.G.; Rhee, S.; Lee, C.H. The effect of shielding gas types on CO_2 laser tailored blank weldability of low carbon automotive galvanized steel. *Mater. Sci. Eng. A* **1999**, *272*, 357–362. [CrossRef]
17. Zdravecka, E.; Matta, M. Selected procedures for evaluation of "tailored welded blanks". In Proceedings of the 6th International conference FORM 2002, Brno, Czech Republic, 16–18 September 2002; pp. 115–120.
18. Zdravecka, E.; Kalmar, P. *The Analysis of Structural Changes of Joints on Samples Created by Laser Welding Technology*; Technical University of Kosice: Kosice, Slovak, 2003; 12p.

19. Xie, J.; Ma, Y.; Ou, M.; Xing, W.; Zhang, L.; Liu, K. Evaluating the Microstructures and Mechanical Properties of Dissimilar Metal Joints Between a New Cast Superalloy K4750 and Hastelloy X Alloy by Using Different Filler Materials. *Materials* **2018**, *11*, 2065. [CrossRef]
20. Podany, P.; Reardon, C.; Koukolikova, M.; Prochazka, R.; Franc, A. Microstructure, Mechanical Properties and Welding of Low Carbon, Medium Manganese TWIP/TRIP Steel. *Metals* **2018**, *8*, 263. [CrossRef]
21. Wu, Y.; Liu, G.; Liu, Z.-Q.; Tang, Z.-J.; Wang, B. Microstructure, mechanical properties and post-weld heat treatments of dissimilar laser-welded Ti2AlNb/Ti60 sheet. *Rare Met.* **2018**, 1–11. [CrossRef]
22. Xu, W.; Westerbaan, D.; Nayak, S.S.; Chen, D.L.; Goodwin, F.; Zhou, Y. Tensile and fatigue properties of fiber laser welded high strength low alloy and DP980 dual-phase steel joints. *Mater. Des.* **2013**, *43*, 373–383. [CrossRef]
23. Razmpoosh, M.H.; Macwan, A.; Biro, E.; Zhou, Y. Effect of coating weight on fiber laser welding of Galvanneal-coated 22MnB5 press hardening steel. *Surf. Coat. Technol.* **2018**, *337*, 536–543. [CrossRef]
24. Schrek, A.; Svec, P.; Gajdosova, V. Deformation properties of tailor-welded blank made of dual phase steels. *Acta Mech. Autom.* **2016**, *10*, 38–42. [CrossRef]
25. Yan, Q.; Cao, N.; Yu, N. Experimental study on formability of blanks after laser welding. *Appl. Laser* **2003**, *2*, 1–10.
26. Svec, P.; Schrek, A.; Domankova, M. Microstructural characteristics of fibre laser welded joint of dual phase steel with complex phase steel. *Kovove Mater.* **2018**, *56*, 29–40. [CrossRef]
27. Quan, Y.J.; Chen, Z.H.; Gong, X.S.; Yu, Z.H. Effects of heat input on microstructure and tensile properties of laser welded magnesium alloy AZ31. *Mater. Charact.* **2008**, *59*, 1491–1497. [CrossRef]
28. Li, J. *The Effect of Weld Design on the Formability of Laser Tailor Welded Blanks*; University of Waterloo: Waterloo, ON, Canada, 2010.
29. Chan, S.M.; Chan, L.C.; Lee, T.C. Tailor Welded Blanks of Different Thickness Ratios Effects on Forming Limit Diagrams. *J. Mater. Process. Technol.* **2003**, *132*, 95–101. [CrossRef]
30. Chan, L.C.; Cheng, C.H.; Chan, S.M.; Lee, T.C.; Chow, C.L. Formability Analysis of Tailor-Welded Blanks of Different Thickness Ratio. *J. Manuf. Sci. Eng.* **2005**, *127*, 743–751. [CrossRef]
31. Chan, L.C.; Chan, S.M.; Cheng, C.H.; Lee, T.C. Formability and Weld Zone Analysis of Tailor-Welded Blanks for Various Thickness Ratio. *J. Eng. Mater. Technol.* **2005**, *127*, 179–185. [CrossRef]
32. Babic, Z.; Aleksandrovic, S.; Stefanovic, M.; Sljivic, M. Determination of tailor welded blanks formability characteristics. *J. Technol. Plast.* **2008**, *33*, 39–48.
33. Riahi, M.; Amini, A. Effect of different combinations of tailor-welded blank. *Int. J. Adv. Manuf. Technol.* **2013**, *67*, 1937–1945. [CrossRef]

Article

Optimization of Process Parameters for Friction Stir Welding of Aluminum and Copper Using the Taguchi Method

Nima Eslami [1], Yannik Hischer [1], Alexander Harms [1], Dennis Lauterbach [1] and Stefan Böhm [2,*]

[1] Volkswagen Aktiengesellschaft Corporate Research, Berliner Ring 2, 38440 Wolfsburg, Germany; nima.eslami@volkswagen.de (N.E.); yannik.hischer@me.com (Y.H.); alexander.harms@volkswagen.de (A.H.); dennis.lauterbach@volkswagen.de (D.L.)

[2] Department for Cutting and Joining Manufacturing Processes, University of Kassel, Kurt Wolters Str. 3, 34125 Kassel, Germany

* Correspondence: s.boehm@uni-kassel.de; Tel.: +49-561-804-3141

Received: 10 December 2018; Accepted: 8 January 2019; Published: 10 January 2019

Abstract: Producing joints of aluminum and copper by means of fusion welding is a challenging task. However, the results of various studies have proven the potential of friction stir welding (FSW) for manufacturing aluminum-copper joints. Despite the proven feasibility, there is currently no series application in automotive industry to produce aluminum-copper joints for electrical contacts by means of FSW. To make FSW as efficient as possible for large-scale production, maximized welding speed is desired. Taking this into account, this paper presents results of a parametric investigation, the objective of which was to increase the welding speed for FSW of aluminum and copper in comparison to welding speeds that are considered to be state of the art. Taguchi method was used to design an experimental plan and target figures of the investigations were the resultant tensile strengths and electrical resistances. Dependencies between input parameters and target figures were determined systematically. The optimal welding parameters, at which joints failed in the weaker aluminum material, included a welding speed of 700 mm/min. Consequently, it could be shown that joints with a performance similar to those of the base materials can be obtained using significantly higher welding speeds than reported in the relevant literature.

Keywords: friction stir welding; aluminum; copper; dissimilar joints; design of experiments; taguchi design; mechanical properties; electrical properties

---

1. Introduction

Excellent electrical and thermal conductivity combined with high ductility, creep resistance and corrosion resistance are the reasons for copper materials being considered to be state of the art in current-carrying components for automotive applications. However, using copper is disadvantageous regarding the high procurement costs and the high material density. Taking this into account, dissimilar aluminum-copper joints represent a solution with great potential for weight and cost-optimized conductors [1,2]. In order to produce joints for electrical contacts, it is well-known that firmly bonded joining is preferred to interlocking and force-locking joining techniques, due to better electrical performance of the joint [3]. However, joining aluminum and copper is a challenging task by means of conventional fusion welding. Different melting temperatures of the base materials, the high thermal conductivities, and the low mutual solubility, which leads to the formation of brittle intermetallic phases, make it difficult to achieve sound welds [4]. Instead, joining processes in which the formation of a melt is avoided are receiving much interest [5]. Friction stir welding (FSW) also belongs to these so-called solid state joining techniques and various authors report on the suitability of this process for joining aluminum and copper materials [2,6–10].

FSW was developed and patented in 1991 by Thomas et al. [11]. In order to perform a firmly bonded joint, this process uses a non-consumable tool, which typically consists of a shoulder and a pin. This rotating tool is pressed into the joint gap and then traversed along the joint line. As a result of tool rotation and feed, the two joining partners are plasticized and stirred [12].

Most studies carried out in the field of FSW of aluminum and copper provide proof of feasibility and focus on the influence of tool and process parameters on the resulting mechanical and microstructural joint properties. Important findings have been obtained through the work of Xue et al. [9] and Akinlabi [7]. These authors inform unanimously on the importance of positioning the harder copper material on the advancing side (AS) and the softer aluminum workpiece on the retreating side (RS) in order to manufacture sound welds free of defects. Moreover, a lateral offset towards the softer aluminum material is recommended to improve the material flow, and thus, the weld quality. Further publications on FSW of aluminum and copper are summarized in Table 1. All of these literature references report on the successful joining of aluminum and copper using FSW.

Table 1. Overview of previous studies on dissimilar FSW of aluminum and copper.

Reference	Sheet Metals	Thickness (mm)	Test Parameters	Target Figure/Object of Investigation	Recommended Traverse Speed (mm/min), Tool Rotation Speed (rpm), n/v-Ratio (1/mm)
[2]	AW1350/Cu	3	-	Joint strength Hardness Microstructure	80, 1000, 12.5
[9]	AW1060/Cu	5	Rotation speed Positioning AS/RS Offset	Defect-free welds Joint strength	100, 1000, 10
[13]	AW1050/Cu	3	Rotation speed Traverse speed	Joint strength Hardness Microstructure	50–100, 1200–1400, 12–28
[14]	AW5083/CW024	1	Tool shoulder type	Microstructure	160–250, 760–1000, 3–6.25
[15]	AW1100/Cu	6	Traverse speed	Joint strength Microstructure	80, 1075, 13.43
[16]	AW6061/Cu	12.7	Traverse speed Rotation speed	Temperature distribution Microstructure	95, 914, 9.62
[17]	AW1060/Cu	3	Traverse speed Rotation speed	Joint strength Hardness Microstructure	30, 1050, 35

Despite the proven potential, as far as the authors of this study know, there is currently no series application in the automotive industry to produce aluminum-copper joints for electrical contacts by means of FSW. In order to achieve this, there are several aspects that require further investigation. This study addresses a research questions that is of particular relevance to the use of FSW for the production of aluminum-copper joints in the automotive industry. As can be seen in Table 1, FSW in previous research studies has been conducted at relatively low welding speeds. The objective of this work is to determine a significantly higher welding speed than in published studies, at which butt welds with excellent mechanical and electrical performance can be manufactured in order to make the FSW process as efficient as possible for large-scale production.

2. Materials and Methods

The applied materials in this study were EN AW-1050A and EN CW004A. Table 2 shows the chemical compositions of both materials, which were taken from the material supplier. The dimensions of the blanks were 160 mm, 100 mm, and 3 mm (length, width, thickness). The FSW experiments were performed on a PTG Powerstir portal system (PTG Heavy Industries Ltd, West Yorkshire, UK) in position-controlled operation. The clamping setup used for fixation of the blanks is shown in Figure 1. The FSW tool used for the welding tests was made of heat treated steel (X40CrMoV5-1) and consisted of a flat shoulder with a diameter of 18 mm and an unthreaded pin with a diameter of 6 mm. The length of the variably adjustable pin was set to 2.9 mm. All the friction stir welds produced within this study had a length of 120 mm.

Table 2. Chemical compositions of the applied base materials [18,19].

Materials	Al	Fe	Si	Mn	Mg	Zn	Ti	Pb	O	Bi	Cu
EN AW-1050A	≥99.50	≤0.40	≤0.25	≤0.05	≤0.05	≤0.07	≤0.05	-	-	-	≤0.05
EN CW004a	-	-	-	-	-	-	-	≤0.005	≤0.04	≤0.0005	≥99.90

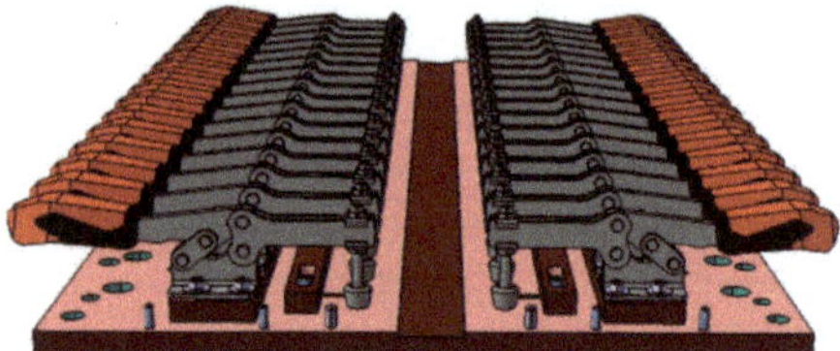

Figure 1. Clamping setup used for FSW experiments.

In order to increase the welding speed in comparison to published studies in the field of Al-Cu FSW, it was necessary to consider a wider parameter window for the parametric investigation. Design of experiments (DoE) was used to ensure an efficient procedure in terms of test effort and quality of results. Using the statistics software Minitab 18 (Minitab GmbH, Munich, Germany), an experimental plan was created. This was a fractional factorial Taguchi L25 design with three factors and five levels. Taguchi orthogonal plans are known to be suitable for parameter optimization purposes. The process parameters that were kept constant during the welding tests are listed in Table 3. The plunge depth and the tool tilt angle were determined based on preliminary tests and were not varied during the welding tests in order to achieve a complete penetration depth. As recommended by Xue et al. [9] and Akinlabi [7], the copper workpiece was positioned on the AS throughout the investigations.

Table 3. Constant process parameters for the welding experiments.

Process Parameter	
Plunge depth (mm)	2.98
Tool tilt angle (°)	2
Plunge speed (mm/s)	70
Dwell time (s)	2

The process parameters, hereinafter also referred to as factors, which were varied equidistantly during the parametric investigation, are the traverse speed (factor 1), the tool rotation speed (factor 2), and the offset towards the aluminum side (factor 3). The structure of the Taguchi L25 design with 25 individual experiments is shown in Table 4, and the levels for each process parameter are listed in Table 5. For statistical purposes, three samples were welded for each factor-level combination.

Table 4. Structure of Taguchi L25 design with three factors and five levels.

Experiment	1	2	3	4	5	6	7	8	9	10	11	12	13	14	15	16	17	18	19	20	21	22	23	24	25
													Level												
Factor 1	1	1	1	1	1	2	2	2	2	2	3	3	3	3	3	4	4	4	4	4	5	5	5	5	5
Factor 2	1	2	3	4	5	1	2	3	4	5	1	2	3	4	5	1	2	3	4	5	1	2	3	4	5
Factor 3	1	2	3	4	5	2	3	4	5	1	3	4	5	1	2	4	5	1	2	3	5	1	2	3	4

Table 5. Factors and their levels.

Factor	Level				
	1	2	3	4	5
Traverse speed (mm/min)	500	700	900	1100	1300
Tool rotation speed (rpm)	200	300	400	500	600
Offset (mm)	1.4	1.8	2.2	2.6	3.0

The evaluation of joint quality for the individual factor-level combinations was carried out by means of tensile testing and electrical resistance measurement. Moreover, hardness tests and metallographic analyses were performed on selected samples by digital microscopy and scanning electron microscopy (SEM) to assess the quality of the welds.

The tensile tests were conducted according to DIN EN ISO 25239-5 [20] by the test machine Zwick Z100 (Zwick GmbH & Co. KG, Ulm, Germany) at an operating speed of 10 mm/min. Transversal sections of the friction stir welds were detached by water jet cutting for evaluation of the mechanical joint properties. In order to avoid excessive material consumption, a distance of 20 mm from the plunging spot has been set for detaching the samples. This distance deviates from the 50 mm specified in DIN EN ISO 25239-5 [20]. The shape of the samples for tensile testing accorded with DIN EN ISO 4136 [21]. In addition to the friction stir welds, five samples each of the respective base materials were tensile tested.

For analyzing the electrical joint properties, the four point resistance measurement setup that is shown in Figure 2 was applied. This setup consists of the Micro-Ohmmeter MR5-600 (Schuetz Messtechnik GmbH, Teltow, Germany) and a clamping device that was designed for the rectangular samples with widths of 40 mm and lengths of 190 mm. A test current of 200 A was chosen and the measuring tips had a distance of 30 mm. The used setup allowed the measurement of the electrical resistance of the weld seam and the respective base materials simultaneously. The electrical resistance of the copper base material was measured via measuring tips 1 and 2, and the aluminum base material was analyzed via measuring tips 3 and 4. The welded area was positioned between tips 2 and 3. For each of the three areas ten values were recorded that were averaged subsequently.

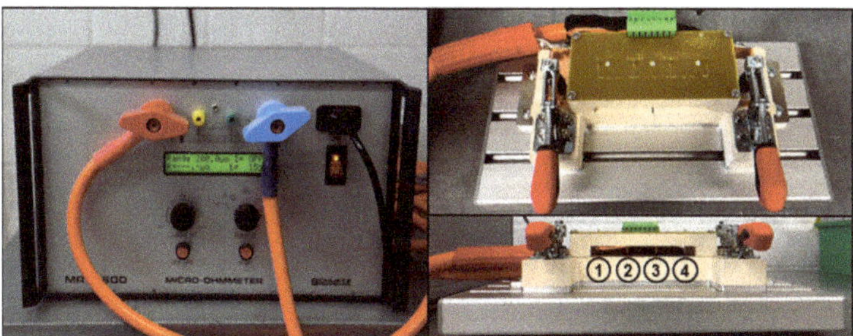

Figure 2. Four point resistance measurement setup.

The samples for digital microscopy were prepared using the standard metallographic procedures. After mounting, the samples were ground using 1200 SiC abrasive paper and then polished using 1 μm aluminum oxide suspension and 50 nm colloidal silica suspension. Grinding and polishing were done manually to avoid the shifting of aluminum particles into the copper side and vice versa. A digital microscope VHX-2000 (Keyence Deutschland GmbH, Neu-Isenburg, Germany) was used to analyze the metallographic features of the friction stir welds.

The samples for scanning electron microscopy were mounted, ground with 1200 and 2400 SiC abrasive papers, and then polished with 6 μm, 3 μm, and 1 μm diamond suspension. This procedure prevented topographical differences at the Al-Cu interfaces, so that the relevant areas could be analyzed properly. Scanning electron microscope model Scios (Field Electron and Ion Company, Hillsboro, OH, USA) was used for further analysis of the Al-Cu interfaces by means of backscattered electrons (BSE).

Vickers hardness tests were carried out using the Leco AMH-43 test device (Leco Corporation, Saint Joseph, MO, USA) with a test load of 0.1 kp.

Once the optimal FSW parameters had been determined, the scalability of the results was tested. Since the ratio of tool rotation speed to traverse speed is a key figure for the heat input in FSW, these parameters were scaled up, while the optimal ratio that was determined through the parametric investigation was kept constant. The motivation for these experiments was a further increase in welding speed.

In order to be able to compare the properties of the friction stir welds to those of the respective base materials, the base materials are characterized at first. Five tensile specimens were tested per base material. Moreover, the electrical properties of the base materials were analyzed using the four point resistance measurement method. The measured values of the 75 samples from the parametric investigation were used for both base materials.

The last part of this section describes the labelling of the samples. Table 6 includes all the different variants.

Table 6. Tabular list of material and specimen labeling.

Label	Description
Al	Aluminum base material EN AW-1050A
Cu	Copper base material EN CW004A
AlCu	Friction stir welds produced as part of the Taguchi experimental plan
AlCu$_{opt}$	Friction stir welds produced using welding parameters with the optimal ratio of tool rotation speed to traverse speed

3. Results and Discussion

3.1. Mechanical and Electrical Properties of the Base Materials

Table 7 provides an overview on the mean values and standard deviations of the tensile strengths and the electrical resistances of the base materials used.

Table 7. Tensile strength and electrical resistance of the base materials used.

Base Material	Tensile Strength (Mpa)	Electrical Resistance ($\mu\Omega$)
Al	120.69 $\pm$ 0.37	7.18 $\pm$ 0.12
Cu	238.41 $\pm$ 0.24	4.08 $\pm$ 0.09

3.2. Mechanical and Electrical Properties of the Al-Cu Friction Stir Welds

The evaluation of the welding experiments for parameter optimization starts with analyzing the mechanical properties of the friction stir welds for the different factor-level combinations (Figure 3). The diagram shows that the averaged tensile strength for parameter settings 1, 6, 22, and 23 is at the level of the aluminum base material. For three of the four parameter combinations mentioned, failure in all tensile specimens occurred in the weaker aluminum base material, which is always the objective when welding dissimilar joints. However, it was observed that most specimens failed in the area of the weld seam. This leads to the conclusion that the parameter window for the production of welds with optimal tensile strength is relatively small.

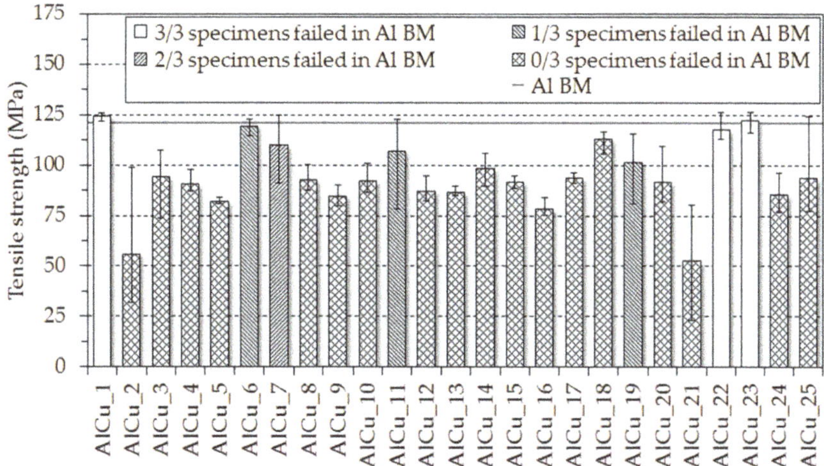

Figure 3. Tensile strengths of the friction stir welds from the Taguchi experimental plan.

In order to be able to compare the heat input between different welds, the ratio between tool rotation speed and traverse speed (n/v-ratio) will be used in the following. This n/v-ratio indicates the number of revolutions per mm feed, and thus, allows a rough estimation of the heat input [22]. Low heat input is represented by a low n/v-ratio and a high n/v-ratio stands for high heat input into the workpieces. Throughout the welding experiments, the n/v-ratio was in a range from 0.15 1/mm to 1.2 1/mm. Taking into account that the n/v-ratios for the parameter settings that lead to the highest tensile strengths are comparatively low, with values of 0.4 1/mm (AlCu_1), 0.29 1/mm (AlCu_6), 0.23 1/mm (AlCu_22), and 0.31 1/mm (AlCu_23), it can be concluded that cold welding tends to lead to better mechanical properties.

Figure 3 also shows that the tensile strengths of samples that were produced with parameter settings AlCu_5, AlCu_9, AlCu_13, AlCu_17, and AlCu_21 are amongst the lowest values. All of these parameter settings included an offset of 3.0 mm into the aluminum side. Since the tool pin has a diameter of 6 mm, no scratching of the copper workpiece should have taken place, leading to an insufficient material mixing. Consequently, the joint strength can only be attributed to an adhesive bonding of the base materials. In order to follow up this consideration, further examination is given in Section 3.5 by means of metallographic analyses.

The results of the electrical resistance measurements are given in Figure 4. The diagram shows that the averaged electrical resistances for the 25 parameter combinations are at a level of approximately 5.7 $\mu\Omega$. Since this value corresponds to the resistance average of both base materials, it can be concluded that the mass proportions of aluminum and copper in the joining area are balanced and that welds with excellent current-carrying behavior have been produced. This observation confirms a good choice of the considered parameter window for the experimental design.

A comparison of the results for tensile testing with electrical resistance measurements shows that the electrical resistances are subject to significantly lower deviations than the resultant tensile strength. Consequently, it is evident that the target figure electrical resistance is more robust against parameter changes than the tensile strengths of the friction stir welds.

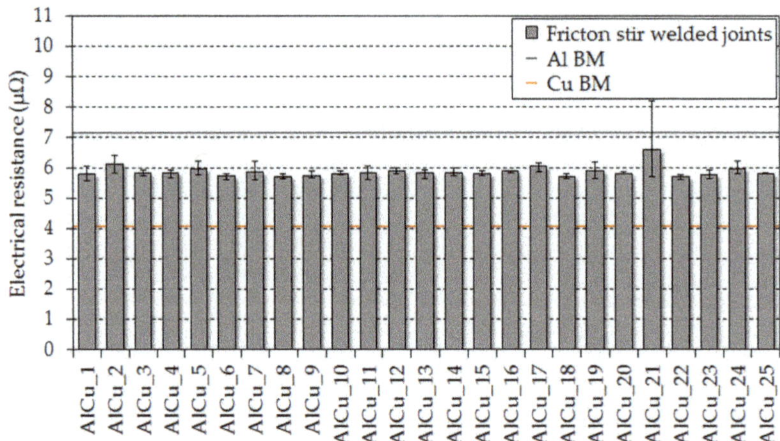

Figure 4. Electrical resistances of the friction stir welds from the Taguchi experimental plan.

3.3. Analysis of the Taguchi Experimental Plan

After the tensile strengths and the electrical resistances of the friction stir welds from the Taguchi experimental plan have been compared with each other and initial dependencies have been identified, the influence of each factor on the respective target figure is presented by the main effect plots in Figure 5.

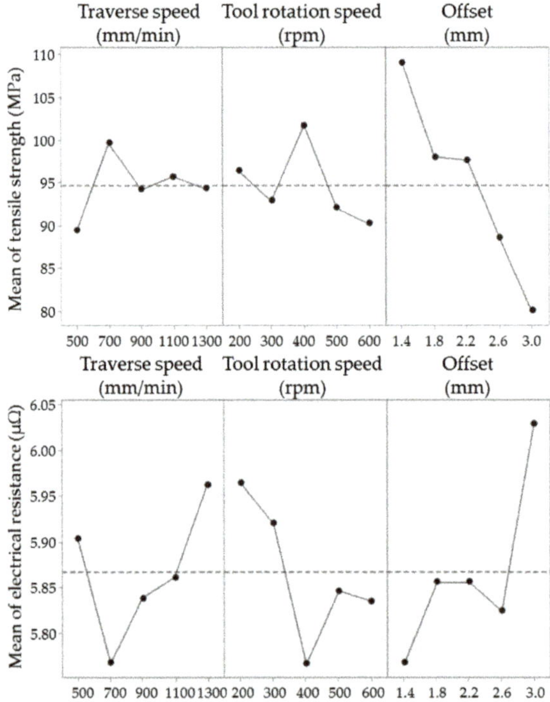

Figure 5. Main effect plots for mean of tensile strength and mean of electrical resistance.

From the main effect plots for the target figure tensile strength it can be seen that only the offset has a steady influence, whereby the tensile strength decreases with larger offsets. In contrast, the traverse speed and the tool rotation speed do not have a steady effect on the tensile strength. Due to the fact that the joint properties, and thus, also the tensile strength depend essentially on the heat input and the associated n/v-ratio during FSW, the effect of the factors traverse speed and tool rotation speed are difficult to separate from each other clearly. Instead, the interaction of these two factors, which is expressed by the n/v-ratio, is crucial for the joint quality. Consequently, no clear correlation between traverse speed and tensile strength or tool rotation speed and tensile strength results from the main effect plots. However, it should be noted that the most powerful levels for these two parameters (traverse speed 700 mm/min and tool rotation speed 400 rpm) result in a n/v-ratio of 0.57 1/mm. This value is relatively low compared to the highest n/v-ratio from the Taguchi experimental plan (1.2 1/mm). Hence, the observation that relatively cold welds achieve better tensile strengths could be confirmed by the main effect plots.

In accordance with the main effect plots for the target figure tensile strength, the main effect plots for the mean of electrical resistance also show a steady influence from the offset and an unsteady influence from the factors traverse speed and tool rotation speed. In addition, it can be seen that for all three factors the courses for the tensile strength are nearly contrary to those for the electrical resistance. Since each tensile strength maximum results in a minimum electrical resistance, the following optimal welding parameters can be considered to maximize the tensile strength, and at the same time minimize the electrical resistance of the friction stir welds.

- Traverse speed: 700 mm/min
- Tool rotation speed: 400 rpm
- Offset 1.4 mm:

3.4. Scaling of Optimal Welding Parameters

In order to verify the optimal welding parameters to maximize the tensile strength and minimize the electrical resistance, which were determined by the analysis of the Taguchi experimental plan, welding tests were carried out using these parameter settings. Since the aim of the parametric investigation is to maximize the welding speed, further welding experiments were performed. Therefore, the factors traverse speed and tool rotation speed were scaled up, while maintaining the n/v-ratio of 0.57 1/mm and using a constant offset of 1.4 mm. Table 8 gives an overview on the parameter combinations used. Three welds were produced per parameter setting.

Table 8. Parameter combinations for welding experiments with optimal n/v-ratio.

Labelling	Traverse Speed (mm/min)	Tool Rotation Speed (rpm)	n/v-ratio (1/mm)	Offset (mm)
AlCu$_{opt_}$1	700	400	0.57	1.4
AlCu$_{opt_}$2	1000	570	0.57	1.4
AlCu$_{opt_}$3	1300	741	0.57	1.4
AlCu$_{opt_}$4	1600	912	0.57	1.4
AlCu$_{opt_}$5	1900	1083	0.57	1.4
AlCu$_{opt_}$6	2200	1254	0.57	1.4
AlCu$_{opt_}$7	2500	1425	0.57	1.4

The tensile strengths and electrical resistances resulting from these parameter settings are presented in Figure 6. The diagram shows that only parameter combination AlCu$_{opt_}$1 leads to tensile strengths on the level of the aluminum base material. The average tensile strength for this parameter set is even higher, by 1.98 Mpa, than that for the most effective factor-level-combination from the Taguchi experimental design (AlCu_1). Also, the resulting electrical resistance for parameter combination AlCu$_{opt_}$1 is by 0.03 $\mu\Omega$ lower than that for the most low-resistant parameter combination from the Taguchi experimental design (AlCu_22). The conclusion is that the optimal welding parameters determined by the main effect diagrams could be verified.

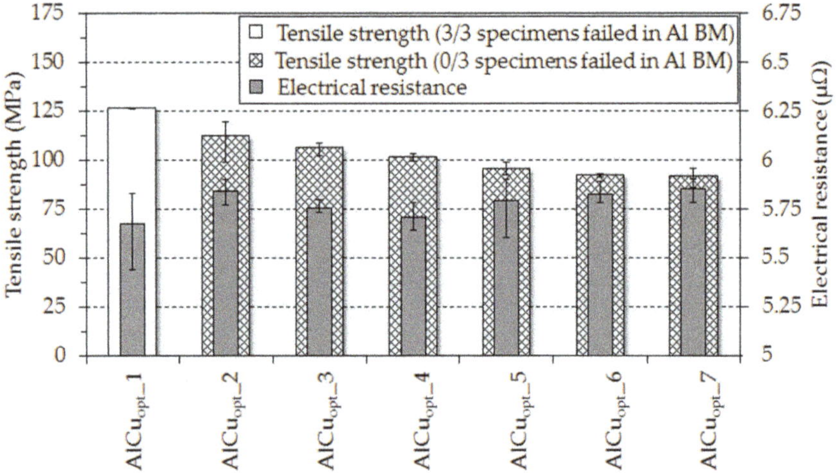

Figure 6. Tensile strengths and electrical resistances for welding experiments with optimal n/v-ratio.

On the other side, it can be seen that scaling up the traverse speed and the tool rotation speed leads to an almost linear decrease in tensile strength and to an increase in electrical resistance. Therefore, it is evident that scaling up these parameters is not feasible without a loss in joint quality.

However, based on the welding experiments carried out within this study, it was proved that significantly higher welding speeds than those specified in the state of the art can be achieved.

3.5. Metallographic Analysis of the Al-Cu Friction Stir Welds and Hardness Testing

In order to be able to understand the observations made in the previous subsections, metallographic analyses were carried out on selected specimens.

The first objective within this subsection is to explain why welds with lower offset lead to higher weld quality. Then, it is to be shown why parameter settings that represent lower heat input achieve friction stir welds with better tensile strengths. In addition, the reduced joint qualities when scaling up the factors traverse speed and tool rotation speed while maintaining the optimal n/v-ratio will be discussed.

As could be determined during the evaluation of the mechanical and electrical joint properties and the analysis of the Taguchi experimental plan, both the tensile strength and the electrical resistance are clearly dependent on the choice of the offset. Considering the macrostructures shown in Figure 7, it can be seen that the quantity as well as the size of copper particles stirred into the aluminum side vary depending on the chosen offset. Furthermore, it can be seen from the figure that with an offset of 3 mm there was no scratching of the copper through the tool pin. As a result, no copper particles were stirred into the aluminum side. These findings lead to the conclusion that more intense material mixing, which is achieved by smaller offsets, leads to better electrical and mechanical properties. However, it should be said that as shown by Xue et al. [9] and Akinlabi [7], the offset should not be too small to ensure a beneficial material flow.

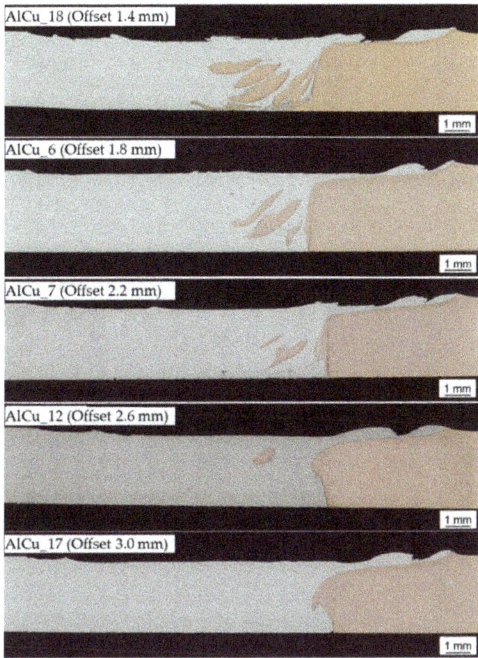

Figure 7. Cross-sectional macrostructures of Al-Cu friction stir welds produced using parameter settings AlCu_18, AlCu_6, AlCu_7, AlCu_12, AlCu_17.

In order to explain why parameter settings representing lower heat input tend to achieve higher tensile strengths than hot welds, the formation of intermetallic compounds (IMC) was investigated by scanning electron microscopy. Backscattered electron (BSE) images from the stir zone were taken for welds that were performed using parameter combinations AlCu_1 (n/v-ratio 0.4 1/mm), AlCu$_{opt}$_1 (n/v-ratio 0.57 1/mm), and AlCu_10 (n/v-ratio 0.86 1/mm). These three parameter sets include an offset of 1.4 mm, and thus, differ only by the heat input. Figure 8 shows that IMC could not be detected using parameter combination AlCu_1, neither at the Al-Cu interface nor at the copper particle stirred into the aluminum side. From this it can be concluded that no IMC were formed or that these phases are too small to be detected by the SEM. Taking into account the BSE images in Figure 9 for parameter combinations AlCu$_{opt}$_1 and AlCu_10, it can be seen that at both welds a continuous layer of IMC was formed at the transition between the examined copper particle to the aluminum matrix. The average thickness of this layer is 150 nm for the specimen that was welded according to parameter combination AlCu$_{opt}$_1 (n/v-ratio 0.57 1/mm) and 265 nm for parameter setting AlCu_10 (n/v-ratio 0.86 1/mm). As a result, a correlation between heat input and resulting intermetallic compound formation could be observed. This effect was also shown in previous work by Galvão et al. [14] and Khodir et al. [23]. However, the thickness of the determined IMC layers is so small that an effect of the IMC formation on the resultant tensile strengths is to be excluded, according to publications by Xue et al. [10], Khodir et al. [23], and Schmidt [24]. Due to the low thickness of the respective layers formed, it was not possible to determine an exact composition of the IMC by means of energy dispersive X-ray spectroscopy.

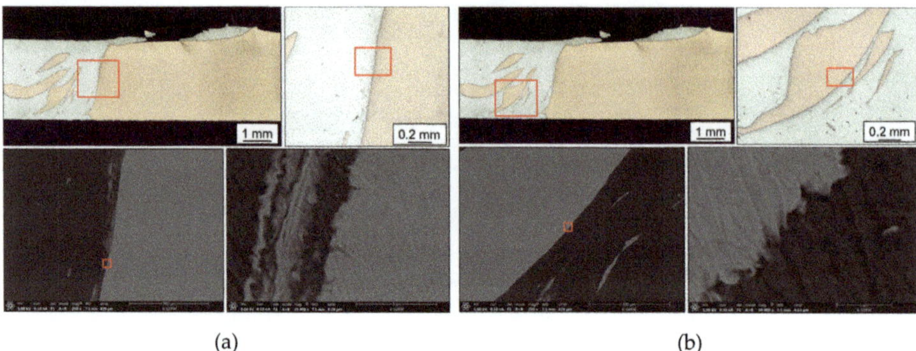

(a)   (b)

Figure 8. BSE images of Al-Cu friction stir weld produced with parameter setting AlCu_1:(**a**) (Al-Cu interface), (**b**) Cu-particle.

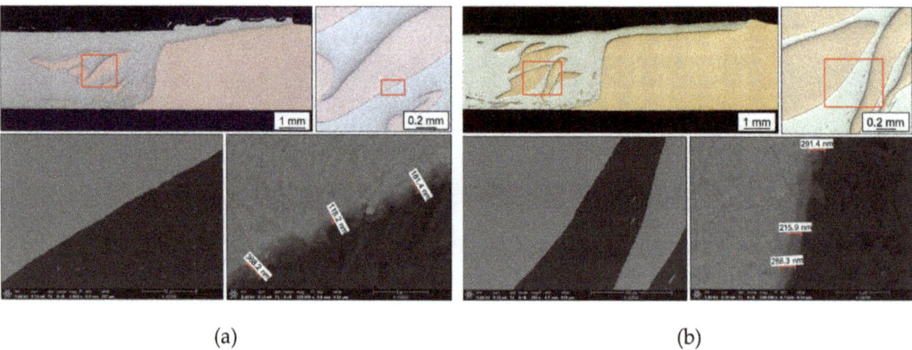

(a)   (b)

Figure 9. (**a**) BSE images of Al-Cu friction stir welds produced with parameter setting AlCuopt_1 (Cu particle). (**b**) BSE images of Al-Cu friction stir weld produced with parameter setting AlCu_10 (Cu particle).

For further investigation of the tensile strength differences between parameter sets representing low or high heat input, Figure 10 shows hardness profiles on cross-sections of welds that were obtained using parameter settings AlCu_1 (n/v-ratio 0.4 1/mm) and AlCu_10 (n/v-ratio 0.86 1/mm). By means of hardness testing, process-related hardening or softening of the examined welds can be detected, so that any occurred strength-reducing microstructural features can be localized. Vickers hardness of the respective base material was found to be 37.7 HV 0.1 for the aluminum base material and 80.1 HV 0.1 for the copper base material. As shown for parameter setting AlCu_1 (n/v-ratio 0.4 1/mm) in Figure 10a, both in the stir zone (SZ) and on both sides in the thermo-mechanically affected zone (TMAZ), there is a significant increase in hardness compared to the respective base materials, with a hardness peak of 122 HV 0.1 in the SZ. This increase in hardness is to be explained by the effect of work hardening due to the cold welding parameters. On the other side, for the weld that was obtained using the parameter combination AlCu_10 (n/v-ratio 0.86 1/mm), the peak hardness values are significantly lower. Furthermore, a decrease in hardness can be seen in aluminum-sided in the SZ, and the plateau, on which the copper bulk material undergoes cold hardening, is clearly smaller. Therefore, the effect of recrystallization seems to dominate here.

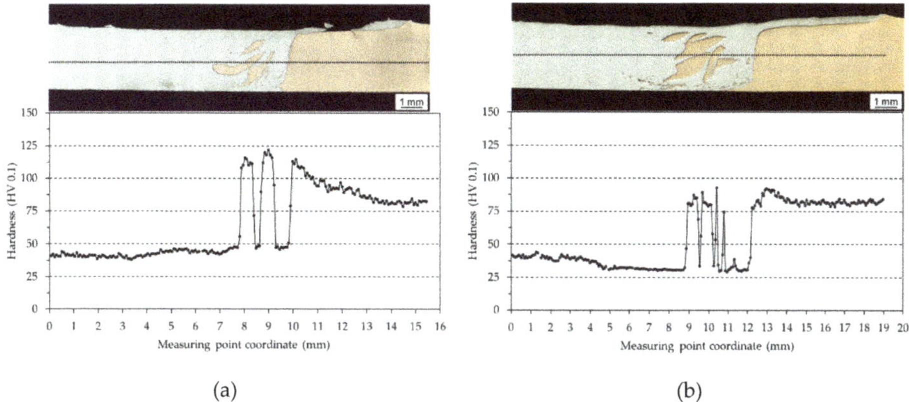

(a) (b)

Figure 10. Hardness profiles on cross-sections of Al-Cu joints: (**a**) Produced with parameter setting AlCu_1; (**b**) weld produced with parameter setting AlCu_10.

In order to understand how the mechanisms of work hardening and recrystallization affect a sample produced using the determined optimal parameter combination, the hardness profile shown in Figure 11 was analyzed. It can be seen that the aluminum material in the TMAZ as well as in the SZ is slightly hardened compared to the aluminum base material. A hardening of the copper particles introduced into the aluminum matrix cannot be detected, whereas the plateau, on which the copper bulk material undergoes cold hardening, is slightly wider than for AlCu_10. Taking into account parameter combination AlCuopt_1 achieving the highest tensile strength and the lowest electrical resistance, it is to be concluded that using these parameter settings, the ideal window for sufficient plasticization of the copper and for avoiding excessively high recrystallization in the SZ was determined.

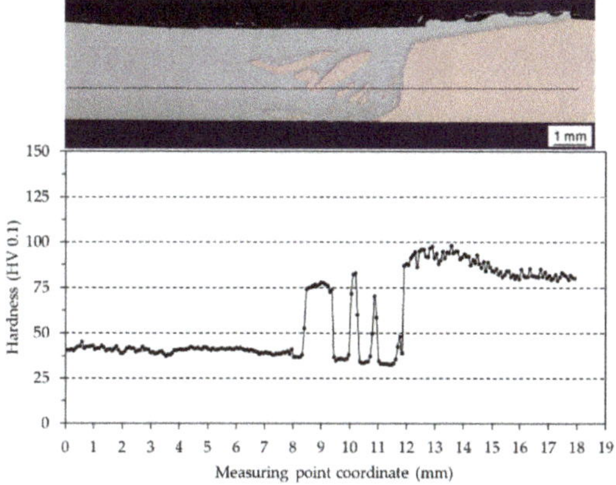

Figure 11. Hardness profile on cross-section of Al-Cu joint produced with parameter setting AlCuopt_1.

At the end of this subsection, it is aimed to understand why scaling up of the optimal welding parameters while maintaining the n/v-ratio 0.57 1/mm could not be realized without losses in mechanical and electrical properties. An explanation for this is provided by the cross-sectional macrostructures in Figure 12. From the macrostructures, it becomes clear that the number and size of defects in the welded area increases with increasing traverse speed. While parameter combination $AlCu_{opt}\_1$ shows a homogeneous distribution of the copper particles without the occurrence of cavities or any other defects, parameter setting AlCu$_{opt}$_4 leads to areas with insufficient bonding and strength-reducing tunnel defects in the root of the SZ. The parameter set AlCu$_{opt}$_7 finally leads to a completely open seam root. From this, it can be concluded that although the tool rotation speed has been adjusted according to the feed speed, the material transport in the vertical direction has been reduced with increasing welding speeds. Thus, the plasticized material does not have enough time to be stirred behind the tool pin and sufficiently compacted by the tool shoulder. The shorter the time for plasticizing and stirring the materials is, the more the inertia of the joining partners promotes the formation of defects in the weld. Moreover, as shown by two publications from Lambiase et al. [25,26], the heat exchange mechanisms during the friction stir welding process need to be considered. The authors have found that the heat dissipation into the clamping device and the preheating of material in front of the welding tool vary depending on the traverse speed and the tool rotation speed. Actually, it is stated that the parameters "traverse speed" and "tool rotation speed" have a different influence on the heat exchange mechanisms, and thus, on the resulting temperature in the welding area. Consequently, it is to say that using the n/v-ratio as a heat index allows only a rough comparison of the heat input between different parameter settings in a limited range of process parameters.

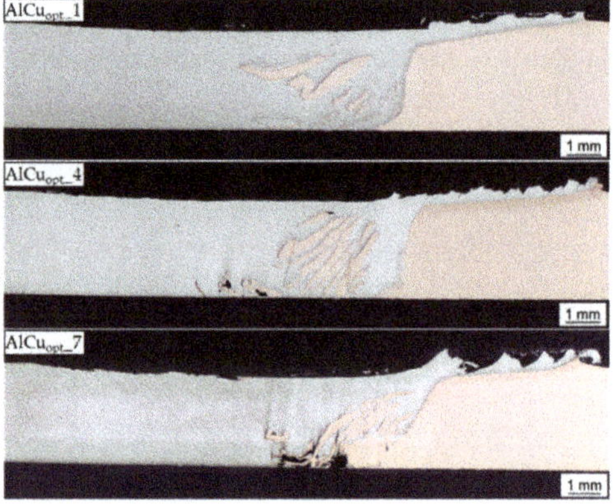

Figure 12. Cross-sectional macrostructures of Al-Cu friction stir welds produced using parameter settings AlCu$_{opt}$_1, AlCu$_{opt}$_4, AlCu$_{opt}$_7.

4. Conclusions

In this study, a parametric investigation on dissimilar friction stir butt welding of 3 mm thick aluminum EN AW-1050A and copper EN CW004A was performed, with the objective to maximize the welding speed at which joints with excellent mechanical and electrical performance can be produced. After designing a Taguchi experimental plan, welding tests were carried out and dependencies between input parameters and the target figure tensile strength and electrical resistance were determined.

1. The target figure electrical resistance is more robust against parameter changes than the tensile strengths of the friction stir welds.
2. It was found that the lowest offset in the considered parameter window (1.4 mm) led to the best mechanical and electrical properties. Cross-sectional macrostructures have proved that more intense material mixing when using low offsets improved the performance of the joint.
3. The main effect plots did not show a steady effect of the factors traverse speed and tool rotation speed on the resultant tensile strength and electrical resistance. Instead, it was shown that the interaction of these two factors, which was expressed by the n/v-ratio, is crucial for the quality of the friction stir welds.

4. It was recognized that cold welds, which were represented by a low n/v-ratio, tended to lead to better mechanical and electrical properties. This observation could be confirmed by the analysis of the Taguchi experimental plan.

5. The effect of IMC on the resultant joint properties could be excluded. Instead, the varying tensile strength when welds were obtained with low or high heat input could be explained by the results of Vickers hardness testing.

6. It was found that the optimal welding parameters for sufficient plasticization of the copper and for avoiding excessively high recrystallization in the SZ were traverse speed 700 mm/min, tool rotation speed 400 rpm, and offset 1.4. Friction stir welds that were manufactured using this parameter combination failed in the weaker aluminum base material during tensile testing and achieved an electrical resistance that was exactly between the resistances of the respective base materials. Scaling up the traverse speed and the tool rotation speed while maintaining the optimal n/v-ratio of 0.57 1/mm could not be realized without losses in mechanical and electrical joint properties. However, it could be shown by the investigations carried out that joints with a performance similar to those of the base materials used can be obtained using significantly higher welding speeds than reported in the relevant literature.

Author Contributions: Conceptualization, N.E., Y.H. and A.H.; methodology, N.E. and Y.H.; formal analysis, N.E., A.H., and D.L.; investigation, N.E. and Y.H.; writing—original draft preparation, N.E.; writing—review and editing, A.H., D.L., and S.B.; visualization, N.E. and Y.H.; supervision, A.H. and S.B.; project administration, A.H.

Funding: This research received no external funding.

Conflicts of Interest: The authors declare no conflict of interest.

Abbreviations

FSW	Friction stir welding
Al	Aluminum
Cu	Copper
SEM	Scanning electron microscopy
BSE	Backscattered electrons
IMC	Intermetallic compounds

References

1. Bargel, H.-J.; Schulze, G. *Werkstoffkunde*; Springer: Berlin, Germany, 2016.
2. Li, X.-W.; Zhang, D.-T.; Qiu, C.; Zhang, W. Microstructure and mechanical properties of dissimilar pure copper/1350 aluminum alloy butt joints by friction stir welding. *Trans. Nonferrous Met. Soc. China* **2012**, *22*, 1298–1306. [CrossRef]
3. Braunovic, M.; Myshkin, N.K.; Konchits, V.V. *Electrical Contacts. Fundamentals, Applications and Technology*; Taylor & Francis Distributor: Boca Raton, FL, USA, 2007.
4. Eslami, N.; Harms, A.; Deringer, J.; Fricke, A.; Böhm, S. Dissimilar friction stir butt welding of aluminum and copper with cross-section adjustment for current-carrying components. *Metals* **2018**, *8*, 661. [CrossRef]
5. Carlone, P.; Astarita, A.; Palazzo, G.S.; Paradiso, V.; Squillace, A. Microstructural aspects in Al–Cu dissimilar joining by FSW. *Int. J. Adv. Manuf. Technol.* **2015**, *79*, 1109–1116. [CrossRef]
6. Celik, S.; Cakir, R. Effect of friction stir welding parameters on the mechanical and microstructure properties of the Al-Cu butt joint. *Metals* **2016**, *6*, 133. [CrossRef]
7. Akinlabi, E.T. Characterisation of Dissimilar Friction Stir Welds between 5754 Aluminium Alloy and C11000 Copper. Ph.D. Thesis, Nelson Mandela Metropolitan University, Port Elizabeth, South Africa, 2010.
8. Al-Roubaiy, A.O.; Nabat, S.M.; Batako, A.D.L. Experimental and theoretical analysis of friction stir welding of Al–Cu joints. *Int. J. Adv. Manuf. Technol.* **2014**, *71*, 1631–1642. [CrossRef]
9. Xue, P.; Ni, D.R.; Wang, D.; Xiao, B.L.; Ma, Z.Y. Effect of friction stir welding parameters on the microstructure and mechanical properties of the dissimilar Al–Cu joints. *Mater. Sci. Eng. A* **2011**, *528*, 4683–4689. [CrossRef]

10. Xue, P.; Xiao, B.L.; Ni, D.R.; Ma, Z.Y. Enhanced mechanical properties of friction stir welded dissimilar Al–Cu joint by intermetallic compounds. *Mater. Sci. Eng. A* **2010**, *527*, 5723–5727. [CrossRef]

11. Thomas, W.M.; Nicholas, E.D.; Needham, J.C.; Murch, M.G.; Templesmith, P.; Dawes, C.J. Improvements Relating to Friction Welding. PCT World Patent Application WO 93/10935, 10 June 1933.

12. Mishra, R.S.; De, P.S.; Kumar, N. *Friction Stir Welding and Processing. Science and Engineering*; Springer International Publishing: Basel, Switzerland, 2014.

13. Barekatain, H.; Kazeminezhad, M.; Kokabi, A.H. Microstructure and mechanical properties in dissimilar butt friction stir welding of severely plastic deformed aluminum AA 1050 and commercially pure copper sheets. *J. Mater. Sci. Technol.* **2014**, *30*, 826–834. [CrossRef]

14. Galvão, I.; Oliveira, J.C.; Loureiro, A.; Rodrigues, D.M. Formation and distribution of brittle structures in friction stir welding of aluminium and copper: Influence of process parameters. *Sci. Technol. Weld. Join.* **2011**, *16*, 681–689. [CrossRef]

15. Muthu, M.F.X.; Jayabalan, V. Tool travel speed effects on the microstructure of friction stir welded aluminum–copper joints. *J. Mater. Process. Technol.* **2015**, *217*, 105–113. [CrossRef]

16. Ouyang, J.; Yarrapareddy, E.; Kovacevic, R. Microstructural evolution in the friction stir welded 6061 aluminum alloy (T6-temper condition) to copper. *J. Mater. Process. Technol.* **2006**, *172*, 110–122. [CrossRef]

17. Zhang, Q.-Z.; Gong, W.-B.; Liu, W. Microstructure and mechanical properties of dissimilar Al-Cu joints by friction stir welding. *Trans. Nonferrous Met. Soc. China* **2015**, *25*, 1779–1786. [CrossRef]

18. Deutsches Institut für Normung e. V. *Aluminium und Aluminiumlegierungen—Chemische Zusammensetzung und Form von Halbzeug—Teil 3: Chemische Zusammensetzung und Erzeugnisformen*; DIN EN 573-3; Beuth Verlag GmbH: Berlin, Germany, 2013.

19. Deutsches Institut für Normung e. V. *Kupfer und Kupferlegierungen—Platten, Bleche und Bänder aus Kupfer für die Anwendung in der Elektrotechnik*; DIN EN 13599; Beuth Verlag GmbH: Berlin, Germany, 2014.

20. Deutsches Institut für Normung e. V. *Rührreibschweißen—Aluminium—Teil 5: Qualitäts- und Prüfungsanforderungen. DIN EN ISO 25239-5*; Beuth Verlag GmbH: Berlin, Germany, 2012.

21. Deutsches Institut für Normung e. V. *Zerstörende Prüfung von Schweißverbindungen an Metallischen Werkstoffen—Querzugversuch*; DIN EN ISO 4136:2012; Beuth Verlag GmbH: Berlin, Germany, 2013.

22. Kleih, L.G. Theoretische und experimentelle Analyse des Bauteilverhaltens rührreibgeschweißter Überlappverbindungen. Ph.D. Thesis, Universitätsbibliothek der Universität Stuttgart, Stuttgart, Germany, 2014.

23. Khodir, S.A.; Ahmed, M.M.Z.; Ahmed, E.; Mohamed, S.M.R.; Abdel-Aleem, H. Effect of intermetallic compound phases on the mechanical properties of the dissimilar Al/Cu friction stir welded joints. *J. Mater. Eng. Perform.* **2016**, *25*, 4637–4648. [CrossRef]

24. Schmidt, P.A. *Laserstrahlschweissen Elektrischer Kontakte von Lithium-Ionen-Batterien in Elektro- und Hybridfahrzeugen*; Herbert Utz Verlag: Munich, Germany, 2015.

25. Lambiase, F.; Paoletti, A.; Di Ilio, A. Forces and temperature variation during friction stir welding of aluminum alloy AA6082-T6. *Int. J. Adv. Manuf. Technol.* **2018**, *99*, 337–346. [CrossRef]

26. Lambiase, F.; Paoletti, A.; Grossi, V.; Di Ilio, A. Analysis of loads, temperatures and welds morphology in FSW of polycarbonate. *J. Mater. Process. Technol.* **2018**, *266*, 639–650. [CrossRef]

Article

Diffusion Bonding of Ti$_2$AlNb Alloy and High-Nb-Containing TiAl Alloy: Interfacial Microstructure and Mechanical Properties

Hong Bian [1,2], Yuzhen Lei [1,2,*], Wei Fu [1,2], Shengpeng Hu [1,2], Xiaoguo Song [1,2,*] and Jicai Feng [1,2]

1   State Key Laboratory of Advanced Welding and Joining, Harbin Institute of Technology, Harbin 150001, China; whenstreaming163@163.com (H.B.); fuwei_drmf@126.com (W.F.); husp@hitwh.edu.cn (S.H.); fengjc@hit.edu.cn (J.F.)
2   Shandong Provincial Key Lab of Special Welding Technology, Harbin Institute of Technology at Weihai, Weihai 264209, China
*   Correspondence: lei.yuzhen@163.com (Y.L.); xgsong@hitwh.edu.cn (X.S.); Tel.: +86-631-5678474 (Y.L. & X.S.); Fax: 86-631-5678454(Y.L. & X.S.)

Received: 22 November 2018; Accepted: 11 December 2018; Published: 14 December 2018

Abstract: In this study, reliable Ti$_2$AlNb/high-Nb-containing TiAl alloy (TAN) joints were achieved by diffusion bonding. The effects of bonding temperature and holding time on the interfacial microstructure and mechanical properties were fully investigated. The interfacial structure of joints bonded at various temperatures and holding times was characterized by scanning electron microscopy (SEM), energy dispersive spectrometer (EDS) and X-ray diffraction (XRD). The results show that the typical microstructure of the Ti$_2$AlNb substrate/O phase/Al(Nb,Ti)$_2$ + Ti$_3$Al/Ti$_3$Al/TAN substrate was obtained at 970 °C for 60 min under a pressure of 5 MPa. The formation of the O phase was earlier than the Al(Nb,Ti)$_2$ phase when bonding temperature was relatively low. When bonding temperature was high enough, the Al(Nb,Ti)$_2$ phase appeared earlier than the O phase. With the increase of bonding temperature and holding time, the Al(Nb,Ti)$_2$ phase decomposed gradually. As the same time, continuous O phase layers became discontinuous and the Ti$_3$Al phase coarsened. The maximum bonding strength of 66.1 MPa was achieved at 970 °C for 120 min.

Keywords: Ti$_2$AlNb alloy; TAN alloy; direct diffusion bonding; interfacial microstructure; mechanical properties

---

1. Introduction

As new lightweight and high-temperature structure materials, Ti$_2$AlNb alloy and high-Nb-containing TiAl alloys (abbreviated as TAN alloys) have received significant attention in the aerospace field [1–3]. The addition of the Nb element gives them excellent properties, such as low density, high specific strength, favorable oxidation, creep resistance [4], and especially high temperature properties. The serving temperatures of Ti$_2$AlNb alloys and TAN alloys are 650–700 °C and up to 900 °C, respectively [5,6]. Based on the above advantages and their different serving temperatures, the two alloys have become the most promising candidates for hot-section components with temperature gradients in aero-engines [7–9]. However, it is difficult to make them into large-scale and complex components, due to their poor workability [10,11]. Therefore, it is significant and necessary to join them together, in order to make full use of their respective advantages and extend their applications.

Up to now, fusion welding, friction welding, and brazing have been employed for joining Ti$_3$Al based alloys or TiAl based alloys. However, defects such as cracks and voids usually appear in the joints obtained by fusion or friction welding, which can lead to hidden dangers in practical applications [12–14]. In brazing, the addition of filler can easily introduce impurities [15,16].

Diffusion bonding, an advanced and efficient method, has been proven to overcome the problems encountered in the above-mentioned welding techniques [17–19]. Some studies on diffusion bonding TiAl based alloys using interlayers such as Ni and Ti/Al have been reported [20,21]. In addition, several researchers have attempted to join Ti_2AlNb based alloys by direct diffusion bonding [22,23]. Sound joints were obtained, and the joint strength was almost equal to the substrate strength using this technique. However, these previous studies mainly focused on joining the same, rather than different kinds of, Ti-Al based alloys. To date, reports on joining different kinds of Ti-Al based alloys are rare. Zou et al. [24] joined Ti-22Al-23Nb-2Ta alloy and Ti-46.2Al-2Cr-2Nb-0.15B alloy by direct diffusion bonding. At low temperature, the $Al(Nb,Ti)_2$ phase was formed in the interface region adjacent to the O phase. When bonding temperature was relatively high, the $Al(Nb,Ti)_2$ phase did not form, but a B2-enriched zone formed instead after long bonding time.

In our work, direct diffusion bonding was applied to join a Ti_2AlNb (Ti_3Al-based) alloy and a TAN (TiAl-based) alloy. The interfacial microstructure of bonded joints was characterized, and its evolution with changing bonding temperature and holding time was investigated. The shear strength of joints was tested, and surface analyses of the fracture were conducted to understand the mechanism of the fracture.

2. Experimental Section

In this study, a Ti_2AlNb alloy with a nominal composition of Ti-22Al-25Nb (atom %) was provided by BaoTai Group, Shanxi, China. TAN alloy (Ti-45Al-8.5Nb (W, B, Y) (atom %)) was provided by the State Key Laboratory for Advanced Metals and Materials, Beijing, China. Figure 1 shows backscattered electron (BSE) images and X-ray diffraction (XRD) patterns of the Ti_2AlNb and TAN alloys. It reveals that the Ti_2AlNb alloy was composed of black α_2-Ti_3Al phase, grey O-Ti_2AlNb phase, and little white B2 phase around the α_2-Ti_3Al phase (Figure 1a), and the TAN alloy consisted of γ-TiAl phase, α_2-Ti_3Al phase (in lamellar colony structure [10]), and a little Nb-rich phase (Figure 1c).

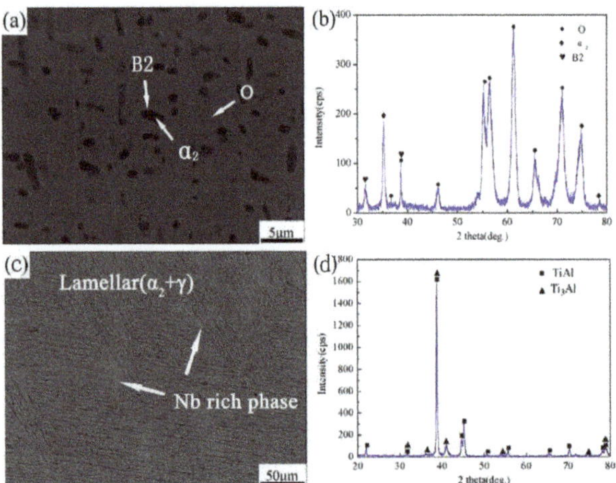

Figure 1. Microstructure and X-ray diffraction (XRD) patterns of substrates. (**a,b**) Ti_2AlNb, (**c,d**) TAN.

Before the bonding experiment, both Ti_2AlNb alloy and TAN alloy were cut into two kinds of rectangular specimens, with dimensions of 5 mm × 5 mm × 3 mm and 20 mm × 10 mm × 3 mm, by linear cutting. The joining surfaces of all specimens were ground by SiC grit papers down to 2000 grit, then cleared ultrasonically in acetone solution for ~20 min and dried by air blowing. Ti_2AlNb alloy was well overlapped on TAN alloy under a pressure of 5 MPa; the schematic diagram of assembling parts is shown in Figure 2a. Bonding took place in a furnace under a vacuum of 1.3–2.0 × 10^{-3} Pa. First of all, the bonding specimens were heated to bonding temperature (930 °C–1010 °C) at a rate of 20 °C/min. Afterwards, they were held for a certain time (60 min to 150 min), and then cooled down to 600 °C at a rate of 10 °C/min. Eventually, the specimens were cooled down to room temperature spontaneously in the furnace.

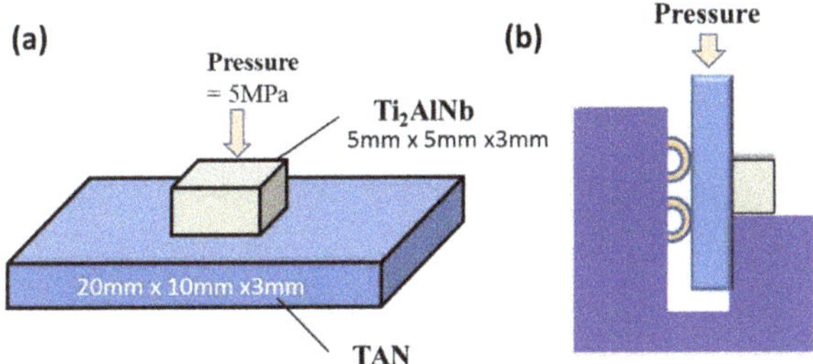

Figure 2. Schematic diagrams of (**a**) assembling diffusion bonding parts and (**b**) shear test experiment.

After direct diffusion bonding, the cross-sections of the Ti$_2$AlNb/TAN bonded joints were characterized by scanning electron microscopy (SEM, MERLIN Compact, Zeiss) (Stuttgart, Germany), and the composition of each phase in the joints was analyzed by energy dispersive spectrometer (EDS, OCTANE PLUS, EDAX) (Mahwah, NJ, USA). The shear tests were conducted at a constant speed of 0.5 mm/min by a universal testing machine (5967, Instron, Boston, MA, USA), and the schematic diagram of the shear test experiment is shown in Figure 2b. At least five samples obtained in same bonding condition were used to determine the average shear strength. SEM, EDS, and X-ray diffraction spectrometer equipped with Cu-Kα (XRD, DX-2700, Dandong Haoyuan Instrument Co., Ltd., Dandong, China) were applied to identify the fracture locations and phases on the fracture surface.

3. Results and Discussion

3.1. Typical Interfacial Microstructure of Ti$_2$AlNb/TAN Bonded Joints

Figure 3a illustrates the typical interfacial microstructure of Ti$_2$AlNb/TAN joints bonded at 970 °C for 60 min under a pressure of 5 MPa. It shows that reliable bonded joints without any pores or microcracks were achieved. In order to further investigate the interfacial microstructure of the Ti$_2$AlNb/TAN bonded joints, a high-magnification image is given in Figure 3b. For a better description, the joint can be divided into three zones according to the different morphology. Zone I, adjacent to the Ti$_2$AlNb substrate, was a continuous light gray layer (marked as D). Zone II mainly consisted of alternating black and white particles (marked as E and F, respectively). Zone III, adjacent to the TAN substrate, was alternating dark grey and black particles (marked as G and H, respectively). The corresponding EDS results of each phase in the joints are listed in Table 1. The EDS result of spot D in zone I suggests the presence of Ti, Al, and Nb in ratio of 2:1:1, which corresponds to O phase. Spots E in zone II, and G and H in zone III, mainly contained Ti and Al. The content of Ti was higher than that of Al. Combined with their morphology, similar to that of the α$_2$-Ti$_3$Al phase in the Ti$_2$AlNb substrate, the phases could be regarded as α$_2$-Ti$_3$Al phases with different content of Nb. In addition, according to the experimental isotherm section of Ti-Al-Nb at 1000 °C [25] and in Reference [24], the white phase (marked as spot F) in zone II was an Al(Nb,Ti)$_2$ phase. The components of the Ti$_2$AlNb substrate adjacent to the joint are also given in Table 1. Obviously, the microstructure of theTi$_2$AlNb substrate also consisted of grey O phase, black α$_2$-Ti$_3$Al phase, and white B2 phase, but it is worth mentioning that the quantity of B2 phase in the Ti$_2$AlNb substrate adjacent to the joint was higher than in the area far from the joint. Therefore, the typical interfacial microstructure of Ti$_2$AlNb/TAN joint bonded at 970 °C for 60 min was Ti$_2$AlNb substrate/O phase/Al(Nb,Ti)$_2$ + Ti$_3$Al/Ti$_3$Al/TAN substrate.

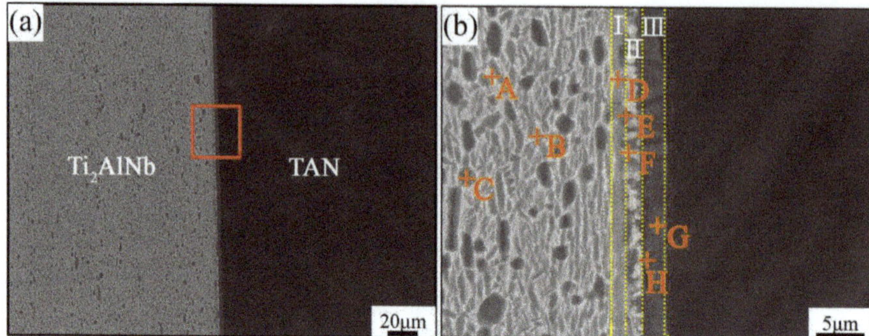

Figure 3. Typical interfacial microstructure of bonded joints at 970 °C/60 min/5 MPa. (**a**) Backscattered electron (BSE) image and (**b**) high-magnified image.

Table 1. Chemical compositions and possible phases of each spot marked in Figure 3b (atom %).

Spot	Ti	Al	Nb	Possible Phase
A	54.84	21.78	23.37	B2
B	55.24	24.09	20.67	O
C	59.85	24.91	15.24	α_2-Ti$_3$Al
D	52.25	27.83	19.93	O
E	56.28	26.87	16.85	α_2-Ti$_3$Al
F	42.55	36.69	20.75	Al(Nb,Ti)$_2$
G	54.53	35.25	10.22	α_2-Ti$_3$Al
H	55.81	27.84	16.35	α_2-Ti$_3$Al

Based on the above analyses on the interfacial microstructure of the bonded Ti$_2$AlNb/TAN joint and the microstructure of the Ti$_2$AlNb substrate adjacent to the seam, the proposed evolution of the bonded joint is as follows. Due to the concentration gradients of Al and Nb atoms for the two substrates, in the bonding process, Al atoms diffused from the TAN substrate to the Ti$_2$AlNb substrate, and Nb atoms diffused in the opposite direction. Moreover, the diffusion velocity of Al atoms was faster than that of Nb atoms, because the atomic mass of Al atoms is only a quarter of that of Nb atoms. For the Ti$_2$AlNb substrate, a part of the O phase gradually transformed into B2 phase in the processes of heating and isothermal periods, according to the diagram of Ti$_3$Al-Nb [26]. Therefore, when Al atoms diffused to the Ti$_2$AlNb substrate, the B2 phase combined with Al atoms to generate Al(Nb,Ti)$_2$ phase and Ti$_3$Al phase by eutectoid reaction of Ti$_2$AlNb(B2) + Al → Al(Nb,Ti)$_2$ + Ti$_3$Al. Hence, the (Al(Nb,Ti)$_2$ + Ti$_3$Al) mixed layer (zone II) was formed [27]. The Al(Nb,Ti)$_2$ phase possessed the same structure as the AlNb$_2$ phase, but was different from the common AlNb$_2$ phase. The reason for formation of Al(Nb,Ti)$_2$ phase was that Nb atoms in AlNb$_2$ phase were partly replaced by Ti atoms with solid solution of Ti at high temperature [24]. Therefore, zone I could be regarded as a phase transformation zone. After the mixed zone II formed, excess Al atoms passed through the mixed layer into the Ti$_2$AlNb substrate. As an Nb-rich and Al-depleted phase, B2 phase transformed into O phase when the concentration of Al exceeded its solubility in B2 phase. Afterwards, Nb atoms in B2 phase started to diffuse into other phases. Moreover, α_2 phase was an Al-rich and Nb-depleted phase, so Nb atoms easily diffused into α_2 phase and distorted its lattice, leading to the transformation of α_2 phase into O phase. Eventually, a continuous O phase near the Ti$_2$AlNb substrate in zone I was formed. In brief, the formation of zone I was mainly caused by the transformation of B2 phase and α_2 phase. Zone III, containing Ti$_3$Al phase, could also be regarded as a phase transformation zone. It was formed by the following two aspects: On one hand, a reaction of 3TiAl → Ti$_3$Al + 2Al occurred, with the Al atoms decreasing. On the other hand, partial Ti atoms diffused to the TAN alloy and took part in the reaction of TiAl + 2Ti → Ti$_3$Al [28].

3.2. Effect of Bonding Parameters on the Interfacial Microstructure of Ti$_2$AlNb/TAN Joints

It is well known that bonding parameters play important roles in the interfacial microstructure of bonded joints. Therefore, the interfacial microstructures of joints bonded at various parameters were characterized, as shown in Figures 4 and 5. The corresponding EDS results of the spots marked in Figure 5c are listed in Table 2.

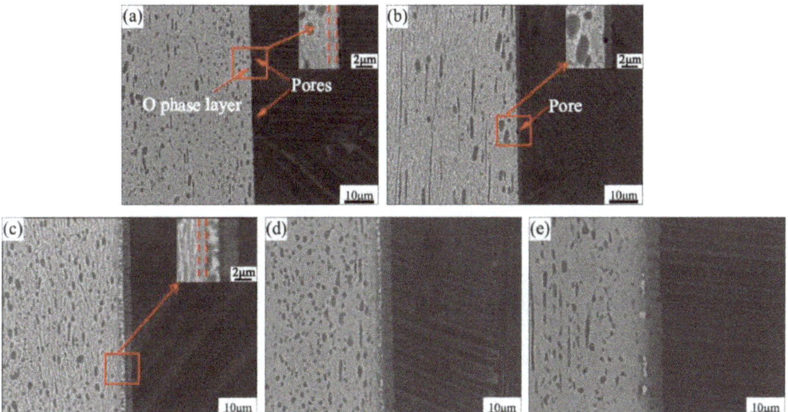

Figure 4. Interfacial microstructure of bonded joints for 60 min/5 MPa at different temperatures. (**a**) 930 °C, (**b**) 950 °C, (**c**) 970 °C, (**d**) 990 °C, and (**e**) 1010 °C.

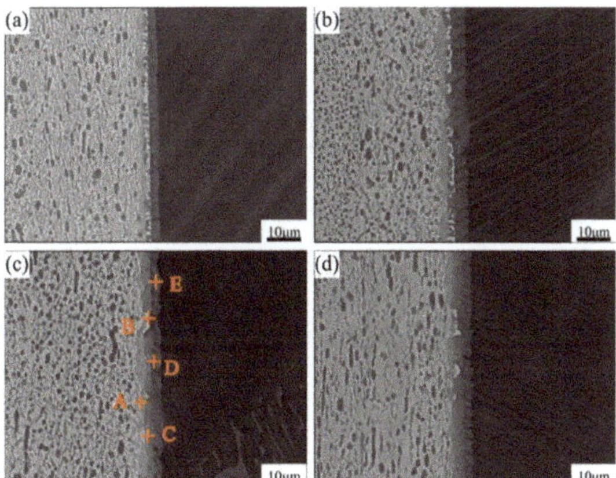

Figure 5. BSE images of interfacial microstructure of bonded joints at 970 °C/5 MPa for (**a**) 60 min, (**b**) 90 min, (**c**) 120 min, (**d**) 150 min.

Table 2. Chemical compositions and possible phases of each spot marked in Figure 5c (atom %).

Spot	Ti	Al	Nb	Possible Phase
A	52.34	27.50	20.07	O
B	43.91	34.80	21.29	Al(Nb,Ti)$_2$
C	51.33	33.45	15.23	α_2-Ti$_3$Al
D	56.78	33.78	9.43	α_2-Ti$_3$Al
E	53.22	35.89	10.89	α_2-Ti$_3$Al

Figure 4a,b shows that some pores existed at the interface. This may have been because the bonding temperature was not high enough to cause considerable plastic deformation and/or enough atomic diffusion to guarantee complete contact. In addition, only O phase and Al(Nb,Ti)$_2$ phase formed at the joints bonded at 930 °C and 950 °C, respectively, which is why the diffusion of Nb atoms was not sufficient to form Al(Nb,Ti)$_2$ phase. On the other hand, Al atoms directly diffused into the Ti$_2$AlNb substrate, causing the transformation of B2 phase and α_2 phase into O phase (Figure 4a). When the temperature was high enough, as at 950 °C, the diffusion of Nb atoms was accelerated and the Nb atoms easily accumulated in the concentrations required to form the Al(Nb,Ti)$_2$ phase [24]. In this case, Al atoms mainly participated in the formation of the Al(Nb,Ti)$_2$ phase (Figure 4b). As temperature further increased to 970 °C, in addition to taking part in the formation of Al(Nb,Ti)$_2$ phase, Al atoms also passed through the mixed layer to the Ti$_2$AlNb substrate and formed the O phase, as shown in Figure 4c. Figure 4c–e shows that the Al(Nb,Ti)$_2$ phase decreased gradually with increasing bonding temperature. High temperature enhanced the diffusion of Nb to the Ti$_2$AlNb substrate, leaving insufficient Nb at the interface to form Al(Nb,Ti)$_2$ phase [29]. As to the Ti$_3$Al layer, its thickness obviously increased with the increase of the bonding temperature.

Figure 5 shows the evolution of the interfacial microstructure of Ti$_2$AlNb/TAN joints bonded at 970 °C. Well-formed joints with no defects were obtained every time. With extended holding time, the amount of Al(Nb,Ti)$_2$ phase reduced gradually, and eventually nearly disappeared when the holding time reached 120 and 150 min. In the isothermal period, along with the diffusion of Nb atoms, the concentration of Nb in the Al(Nb,Ti)$_2$ phase decreased gradually, resulting in its decomposition. Along the TAN substrate, a Ti$_3$Al transition layer consisting of two kinds of Ti$_3$Al phases with different concentrations of Nb formed. The formation of one Ti$_3$Al phase (marked as C) was due to the decomposition of Al(Nb,Ti)$_2$ phase. The formation mechanism of the other one was the same as that of the Ti$_3$Al phase marked as G in Figure 3b. The Ti$_3$Al transition layer thickened and coarsened when holding time reached 150 min. On the other hand, the O phase layer became discontinuous, which resulted from the durative diffusion of Nb from Ti$_2$AlNb to O phase inducing the transition of O phase to B2 phase.

3.3. Bonding Properties and Fracture Morphology of Bonded Ti$_2$AlNb/TAN Joints

Figure 6 illustrates the effect of bonding temperature and holding time on the average shear strength of the bonded Ti$_2$AlNb/TAN joints. It shows that the average shear strength exhibited the same tendency, first increasing and then decreasing, in both increasing temperature and time tests.

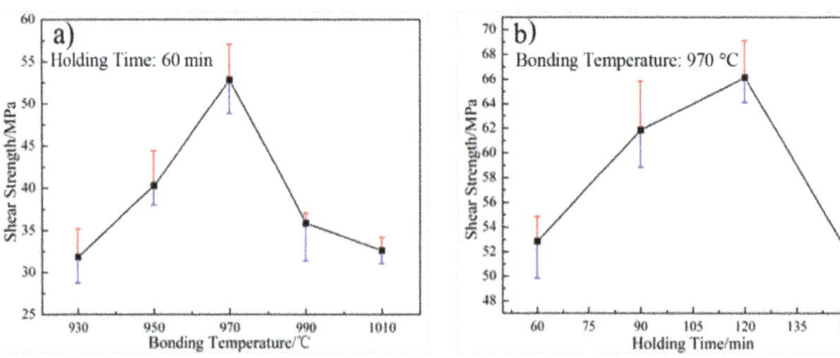

Figure 6. Effect of bonding parameters on the shear strength of joints. (**a**) Bonding temperature; (**b**) holding time.

In order to further explore the fracture mechanism of the joints, fracture analyses were carried out. Furthermore, the relationship between interfacial microstructures and joining properties was established. Figure 7 shows the cross-section BSE images and fractography of joints made at different bonding temperatures and holding times, taken after shear test. The corresponding EDS results were listed in Table 3. All the fracture presented characteristic cleavage facets. As shown in Figure 7a–b, it was observed that the cracks mainly initiated in the mixed layer, which consisted of Al(Nb,Ti)$_2$ (spot A) and Ti$_3$Al phases (spot B). The fracture, which was obtained at 970 °C for 60 min (Figure 7c–d), also included Al(Nb,Ti)$_2$ (spot C) and Ti$_3$Al phases. However, it can be clearly seen that there were two types of Ti$_3$Al phases with different contents of the Nb element. The Ti$_3$Al phase with lower content of Nb (spot D) lay in the Ti$_3$Al layer, and the one with the higher content of Nb (spot E or F) lay in the mixed layer. It can therefore be concluded that the cracks were mainly through the mixed layer and the Ti$_3$Al layer. For further analyses, XRD analysis of the fracture surface on the TAN side after shear test of the joints at 970 °C/60 min/5 MPa was carried out, for which the XRD patterns are given in Figure 8. According to the results shown in Figure 8, Al(Nb,Ti)$_2$ phase, which had the same structure as AlNb$_2$ phase [24], and Ti$_3$Al phase were detected from the fracture surfaces. The above results further confirmed the above analyses. When the holding time was extended to 120 min (Figure 7e–f), the fracture surface mainly consisted of TiAl phase (spot G) and two kinds of Ti$_3$Al phase (spots H and I). It was concluded that there was little Al(Nb,Ti)$_2$ phase either in the joints or in Ti$_3$Al transition layer. The joints mainly fractured in the Ti$_3$Al transition layer, and partly in the interface between the Ti$_3$Al transition layer and the TAN substrate, which had the maximum shear strength. With holding time extended to 150 min (Figure 7g–h), the fracture only included two types of Ti$_3$Al phase (spots J and K). Hence, the crack initiated and propagated in the transition layer.

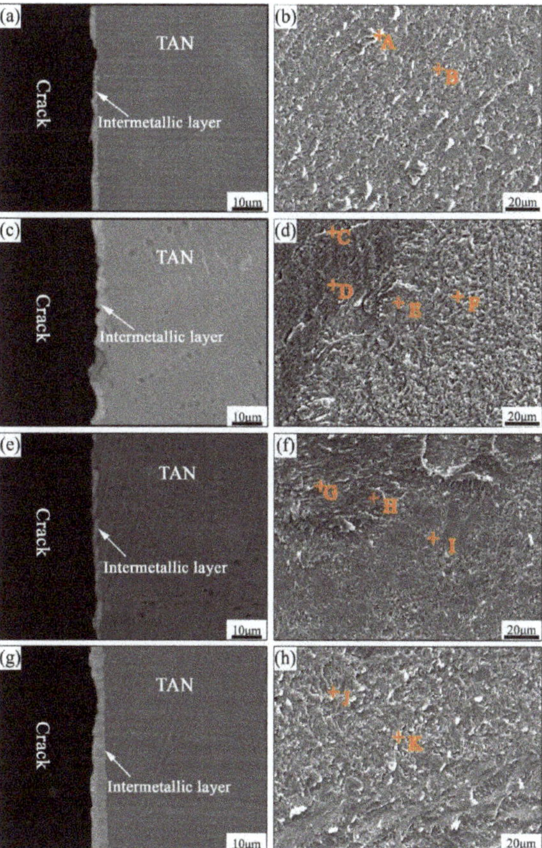

Figure 7. Fractured cross-section BSE images and fractography of the joints bonded at (**a–b**) 950 °C/60 min, (**c–d**) 970 °C/60 min, (**e–f**) 970 °C/120 min, and (**g–h**) 970 °C/150 min. (TAN side).

Table 3. Chemical compositions and possible phases of each spot marked in Figure 7 (atom %).

Spot	Ti	Al	Nb	Possible Phase	Corresponding IMC Layer
A	49.77	38.29	11.44	Al(Nb,Ti)$_2$	mixed layer
B	51.91	31.88	16.22	α_2-Ti$_3$Al	mixed layer
C	44.82	38.31	16.87	Al(Nb,Ti)$_2$	mixed layer
D	52.78	33.74	13.48	α_2-Ti$_3$Al	Ti$_3$Al phase layer
E	50.09	31.37	18.54	α_2-Ti$_3$Al	mixed layer
F	50.28	31.25	18.47	α_2-Ti$_3$Al	mixed layer
G	44.63	47.38	8.00	TiAl	TAN substrate
H	58.16	33.27	8.57	α_2-Ti$_3$Al	transition layer
H	53.09	31.47	15.44	α_2-Ti$_3$Al	transition layer
J	52.67	35.08	12.25	α_2-Ti$_3$Al	transition layer
K	51.82	30.72	17.46	α_2-Ti$_3$Al	transition layer

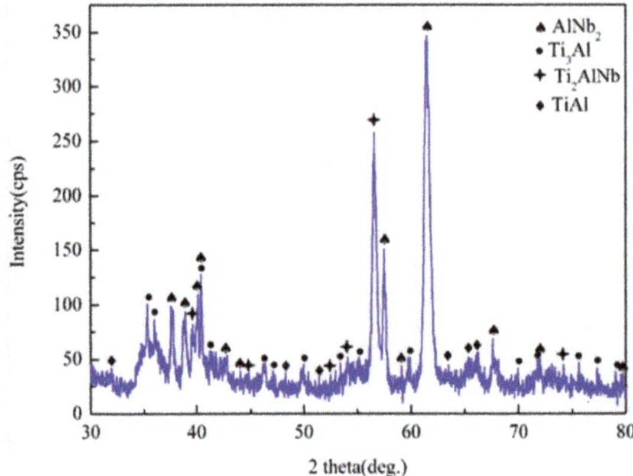

Figure 8. XRD patterns of the fracture surface after shear test of the joints bonded at 970 °C/60 min/5 MPa.

For the joints created at 930 °C for 60 min, plenty of pores existed in the joints, decreasing the real bond area. Shear strength was therefore poor. As the bonding temperature increased, a compact contact was guaranteed and sufficient reactions occurred, following the interdiffusion of atoms. Continuous intermetallic compounds (IMCs) layers, including O phase layer, (Al(Nb,Ti)$_2$ + Ti$_3$Al) mixed phase layer, and Ti$_3$Al phase layer (in Figure 4), formed at the interface. Therefore, an increasing shear strength was obtained, with the highest 52.9 MPa, achieved in joints made at 970 °C for 60 min. Whereas, with the further increase of bonding temperature, the Ti$_3$Al layer thickened and Ti$_3$Al grains coarsened, apparently decreasing the shear strength.

As the holding time increased, the Al(Nb,Ti)$_2$ phase decomposed gradually and the Ti$_3$Al transition layer formed (in Figure 5), which increased the shear strength of the joints [24]. The shear strength increased steadily and reached its highest value of 66.1 MPa at 970 °C for 120 min, as shown in Figure 6b. With the holding time further increased, despite the fact that the reduction of Al(Nb,Ti)$_2$ phase was beneficial to performance of the joints, the mechanical properties of the joint still decreased by reason of thickening of the Ti$_3$Al transition layer and coarsening of Ti$_3$Al grains.

Based on the above analysis of interfacial microstructure and fracture results, it was concluded that the Al(Nb,Ti)$_2$ phase was detrimental to the Ti$_2$AlNb/TAN bonded joints, due to its hardness and brittleness. High bonding temperature or long holding time enhanced the diffusion of atoms, which caused the decrease of Al(Nb,Ti)$_2$ phase and therefore better bonded joints. However, further increase of the bonding temperature or the holding time promoted the excessive growth of Ti$_3$Al phase, which resulted in a decrease of the shear strength of Ti$_2$AlNb/TAN bonded joints.

4. Conclusions

Bonding a Ti$_2$AlNb alloy to a TAN alloy was achieved successfully by direct diffusion bonding. The effects of bonding temperature and holding time on the interfacial microstructure and mechanical properties of Ti$_2$AlNb/TAN bonded joints were investigated in detail. Primary conclusions are summarized as follows.

(1) The typical interfacial microstructure of the Ti$_2$AlNb/TAN joints bonded at 970 °C for 15 min under a pressure of 5 MPa was Ti$_2$AlNb substrate/O phase/Al(Nb,Ti)$_2$ + Ti$_3$Al/Ti$_3$Al/TAN substrate.

(2) Bonding temperature had a great influence on the priority of the formation of O phase and Al(Nb,Ti)$_2$ phase. When bonding temperature was low, Al atoms diffused to the Ti$_2$AlNb substrate directly, which caused the formation of O phase without Al(Nb,Ti)$_2$ phase. When the bonding temperature was high enough, Nb atoms reached the desired concentration quickly. In this condition, Al atoms first reacted with B2 phase to generate Al(Nb,Ti)$_2$ phase. Then, excess Al atoms passed though the mixed layer to the Ti$_2$AlNb substrate and promoted the formation of O phase. As bonding temperature or holding time were further increased, Al(Nb,Ti)$_2$ phase gradually decomposed into Ti$_3$Al phase and a Ti$_3$Al transition layer formed. Meanwhile, the O phase layer changed from a continuous state to a discontinuous one.

(3) The Al(Nb,Ti)$_2$ phase was hard and brittle, so the initial fracture location mainly occurred in the mixed layer. With the decomposition of Al(Nb,Ti)$_2$ phase and the formation of transition layer, the shear strength was improved, and the average value reached 66.1 MPa when Ti$_2$AlNb alloy and TAN alloy were bonded at 970 °C for 120 min. The fracture location mainly occurred in the Ti$_3$Al transition layer and the fracture mode was brittle fracture.

Author Contributions: Conceptualization, H.B., Y.L. and X.S.; Methodology, X.S. and J.F.; Formal Analysis, Y.L. and S.H.; Data Curation, Y.L. and S.H.; Writing-Original Draft Preparation, H.B.; Writing-Review & Editing, W.F. and S.H.; Supervision, X.S. and J.F.; Funding Acquisition, X.S.

Funding: This project is supported by National Natural Science Foundation of China (Grant Nos. 51775138, U1537206 and U1737205) and the Key Research & Development program of Shandong Province (No. 2017GGX40103 and 2016GGA10085).

Conflicts of Interest: The authors declare no conflict of interest.

References

1. Lu, Z.-G.; Wu, J.; Guo, R.-P.; Xu, L.; Yang, R. Hot deformation mechanism and ring rolling behavior of powder metallurgy Ti2AlNb intermetallics. *Acta Metall. Sin.* **2017**, *30*, 621–629. [CrossRef]
2. Chu, Y.; Li, J.; Zhu, L.; Liu, Y.; Tang, B.; Kou, H. Microstructure evolution of a high Nb containing TiAl alloy with ($\alpha_2 + \gamma$) microstructure during elevated temperature deformation. *Metals* **2018**, *8*, 916. [CrossRef]
3. Wang, S.; Xu, W.; Zong, Y.; Zhong, X.; Shan, D. Effect of initial microstructures on hot deformation behavior and workability of Ti2AlNb-based alloy. *Metals* **2018**, *8*, 382. [CrossRef]
4. Li, T.-R.; Liu, G.-H.; Xu, M.; Fu, T.-L.; Tian, Y.; Misra, R.D.K.; Wang, Z.-D. Hot deformation behavior and microstructural characteristics of Ti–46Al–8Nb alloy. *Acta Metall. Sin.* **2018**, *31*, 933–944. [CrossRef]
5. Yong, W.; Bin, L.; Rui, Y.; Xiaodong, H.; Ping, R. Local deformation and processing maps of Ti-24Al-17Nb-0.5 Mo alloy. *Acta Metall. Sin.* **2012**, *25*, 95–101.
6. Liu, C.; Lu, X.; Yang, F.; Xu, W.; Wang, Z.; Qu, X. Metal Injection moulding of high Nb-containing TiAl alloy and its oxidation behaviour at 900 °C. *Metals* **2018**, *8*, 163. [CrossRef]
7. Xiaoguo, S.; Jian, C.; Jiakun, L.; Liyan, Z.; Jicai, F. Reaction-diffusion bonding of high Nb containing TiAl alloy. *Rare Metal Mat. Eng.* **2014**, *43*, 28–31. [CrossRef]
8. Feng, G.-J.; Li, Z.-R.; Liu, R.-H.; Feng, S.-C. Effects of joining conditions on microstructure and mechanical properties of C$_f$/Al composites and TiAl alloy combustion synthesis joints. *Acta Metall. Sin.* **2015**, *28*, 405–413. [CrossRef]

9. Dong, D.; Zhu, D.; Wang, Y.; Wang, G.; Wu, P.; He, Q. Microstructure and shear strength of brazing TiAl/Si3N4 joints with Ag-Cu binary alloy as filler metal. *Metals* **2018**, *8*, 896. [CrossRef]

10. Si, X.; Zhao, H.; Cao, J.; Song, X.; Tang, D.; Feng, J. Brazing high Nb containing TiAl alloy using Ti-28Ni eutectic brazing alloy: Interfacial microstructure and joining properties. *Mater. Sci. Eng. A* **2015**, *636*, 522–528. [CrossRef]

11. Wu, Z.; Hu, R.; Zhang, T.; Zhou, H.; Kou, H.; Li, J. Microstructure determined fracture behavior of a high Nb containing TiAl alloy. *Mater. Sci. Eng. A* **2016**, *666*, 297–304. [CrossRef]

12. Chen, G.; Zhang, B.; Liu, W.; Feng, J. Crack formation and control upon the electron beam welding of TiAl-based alloys. *Intermetallics* **2011**, *19*, 1857–1863.

13. Zhang, K.; Liu, M.; Lei, Z.; Chen, Y. Microstructure evolution and tensile properties of laser-TIG hybrid welds of Ti$_2$AlNb-based titanium aluminide. *J. Mater. Eng. Perform.* **2014**, *23*, 3778–3785. [CrossRef]

14. Chen, X.; Xie, F.; Ma, T.; Li, W.; Wu, X. Microstructure evolution and mechanical properties of linear friction welded Ti$_2$AlNb alloy. *J. Alloys Compd.* **2015**, *646*, 490–496. [CrossRef]

15. Song, X.; Ben, B.; Hu, S.; Feng, J.; Tang, D. Vacuum brazing high Nb-containing TiAl alloy to Ti60 alloy using Ti-28Ni eutectic brazing alloy. *J. Alloys Compd.* **2017**, *692*, 485–491. [CrossRef]

16. Cao, J.; Dai, X.; Liu, J.; Si, X.; Feng, J. Relationship between microstructure and mechanical properties of TiAl/Ti2AlNb joint brazed using Ti-27Co eutectic filler metal. *Mater. Des.* **2017**, *121*, 176–184. [CrossRef]

17. Tang, B.; Qi, X.S.; Kou, H.C.; Li, J.S.; Milenkovic, S. Recrystallization behavior at diffusion bonding interface of high Nb containing TiAl Alloy. *Adv. Eng. Mater.* **2016**, *18*, 657–664. [CrossRef]

18. Zou, G.-S.; Xie, E.-H.; Bai, H.-L.; Wu, A.-P.; Wang, Q.; Ren, J.-L. A study on transient liquid phase diffusion bonding of Ti-22Al-25Nb alloy. *Mater. Sci. Eng. A* **2009**, *499*, 101–105. [CrossRef]

19. Cao, J.; Feng, J.C.; Li, Z.R. Microstructure and fracture properties of reaction-assisted diffusion bonding of TiAl intermetallic with Al/Ni multilayer foils. *J. Alloys Compd.* **2008**, *466*, 363–367. [CrossRef]

20. He, P.; Zhang, J.; Zhou, R.; Li, X. Diffusion bonding technology of a titanium alloy to a stainless steel web with an Ni interlayer. *Mater. Charact.* **1999**, *43*, 287–292. [CrossRef]

21. Duarte, L.I.; Ramos, A.S.; Vieira, M.F.; Viana, F.; Vieira, M.T.; Koçak, M. Solid-state diffusion bonding of gamma-TiAl alloys using Ti/Al thin films as interlayers. *Intermetallics* **2006**, *14*, 1151–1156. [CrossRef]

22. Du, Z.; Jiang, S.; Zhang, K.; Lu, Z.; Li, B.; Zhang, D. The structural design and superplastic forming/diffusion bonding of Ti 2 AlNb based alloy for four-layer structure. *Mater. Des.* **2016**, *104*, 242–250. [CrossRef]

23. Zou, G.S.; Bai, H.L.; Xie, E.H.; Wu, S.J.; Wu, A.P.; Wang, Q.; Ren, J.L. Solid diffusion bonding of Ti-22Al-25Nb O phase alloy. *Chin. J. Nonferrous Met.* **2008**, *18*, 577–582.

24. Zou, J.; Cui, Y.; Yang, R. Diffusion bonding of dissimilar intermetallic alloys based on Ti$_2$AlNb and TiAl. *J. Mater. Sci. Technol.* **2009**, *25*, 819–824.

25. Hellwig, A.; Palm, M.; Inden, G. Phase equilibria in the Al-Nb-Ti system at high temperatures. *Intermetallics* **1998**, *6*, 79–94. [CrossRef]

26. Muraleedharan, K.; Gogia, A.K.; Nandy, T.K.; Banerjee, D.; Lele, S. Transformations in a Ti-24AI-15Nb alloy: Part I. Phase equilibria and microstructure. *Metall. Trans. A* **1992**, *23*, 401–415. [CrossRef]

27. Yang, S.J.; Nam, S.W.; Hagiwara, M. Phase identification and effect of W on the microstructure and micro-hardness of Ti2AlNb-based intermetallic alloys. *J. Alloys Compd.* **2003**, *350*, 280–287. [CrossRef]

28. Cao, S.; Xiao, S.; Chen, Y.; Xu, L.; Wang, X.; Han, J.; Jia, Y. Phase transformations of the L1$_2$-Ti$_3$Al phase in γ-TiAl alloy. *Mater. Des.* **2017**, *121*, 61–68. [CrossRef]

29. Du, X.W.; Zhu, J.; Zhang, X.; Cheng, Z.Y.; Kim, Y.W. Composition change during creep in colonies oriented for easy-slip of Ti-46.5Al-2Cr-3Nb-0.2W. *Mater. Sci. Eng. A* **2000**, *291*, 131–135. [CrossRef]

Article

A New Approach to Simulate HSLA Steel Multipass Welding through Distributed Point Heat Sources Model

Dario Magno Batista Ferreira [1], Antonio do Nascimento Silva Alves [2],
Rubelmar Maia de Azevedo Cruz Neto [2], Thiago Ferreira Martins [2] and Sérgio Duarte Brandi [2,*]

[1]   Mechanical Engineering Coordination, Instituto Federal do Espírito Santo—Ifes, Vitória 29040-780, Brazil;
      dario@ifes.edu.br
[2]   Metallurgical and Materials Engineering Department, Universidade de São Paulo—USP,
      São Paulo 05508-030, Brazil; ansa@usp.br (A.d.N.S.A.); rubelmar.neto@usp.br (R.M.d.A.C.N.);
      thiagoferreira@usp.br (T.F.M.)
*   Correspondence: sebrandi@usp.br; Tel.: +55-(11)-971105056

Received: 28 August 2018; Accepted: 18 September 2018; Published: 15 November 2018

Abstract: Mechanical properties of welded joints depend on the way heat flows through the welding passes. In multipass welding the reheating of the heat affected zone (HAZ) can form local brittle zones that need to be delimited for evaluation. The difficulty lies in the choice of a model that can simulate multipass welding. This study evaluated Rosenthal's Medium Thick Plate (MTP) and the Distributed heat Sources (DHS) of Mhyr and Gröng models. Two assumptions were considered for both models: constant and temperature-dependent physical properties. It was carried out on a multipass welding of an API 5L X80 tube, with 1016 mm (42″) external diameter, 16 mm thick and half V-groove bevel, in the 3G up position. The root pass was welded with Gas Metal Arc Welding (GMAW) process with controlled short-circuit transfer. The Flux Cored Arc Welding (FCAW) process was used in the filling and finishing passes, using filler metal E111T1-K3M-JH4. The evaluation criteria used were overlapping the simulated isotherms on the marks revealed in the macrographs and the comparison between the experimental thermal cycle and those simulated by the proposed models. The DHS model with the temperature-dependent properties presented the best results and simulated with accuracy the HAZ of root and second welding passes. In this way, it was possible to delimit the HAZ heated sub-regions.

Keywords: welding thermal cycles; medium thick plate model; distributed point heat sources model; local brittle zone; API 5L X80 steel

---

1. Introduction

In electric arc welding, the addition and the base metals are fused by the heat source and it still imposes phase changes in the solid base metal (BM), i.e., in the heat affected zone (HAZ) [1]. The retention time above certain temperatures reached along the HAZ dissolves precipitates and promotes its redistribution in the austenitic matrix together with grain growth [2–4]. Further, the energy retention times above certain temperatures, the peak temperatures and cooling rates are different along the HAZ. Hence, different microstructural regions throughout the HAZ width are created, so that its final microstructures differ significantly from that of the original BM [5,6]. Figure 1 presents the microstructures in the experimental HAZ regions of an API 5L X80 steel, in which it is possible to observe different grain sizes and microstructures. The coarse-grained HAZ (CGHAZ) region presents grains whose diameter is greater than the others. Its microstructure is composed by ferrite with a second phase of carbides, while the others are composed by ferrite with aggregate ferrite-carbide. The intercritical HAZ (ICHAZ) grains still present some orientation from the base metal grains. Therefore, the HAZ physical properties also differ from that of BM. The hardness and toughness are mechanical properties that are dependent of the welding thermal cycle and they are reference parameters for measuring the steel's weldability [7].

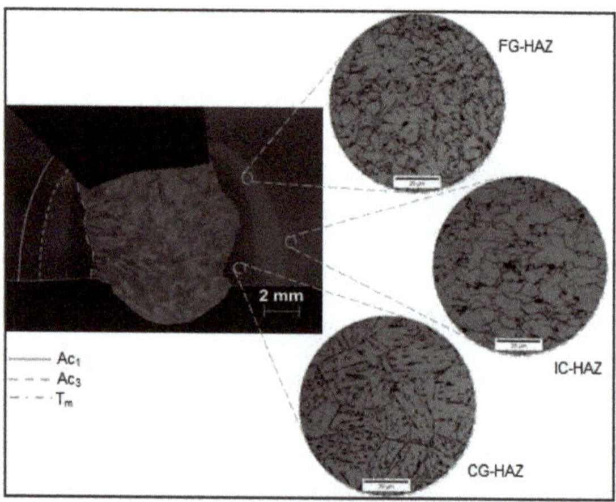

Figure 1. Experimental representation of heat affected zone (HAZ) regions, root pass welding of an API 5L X80 steel.

The study and domain of these transformations are required for the prediction and control of the resulting microstructure and welded-joint required properties, consequently [6,7]. Figure 2 shows schematically the HAZ sub-regions in a high-strength low-alloy steel (HSLA) multipass welding, HT50 [5].

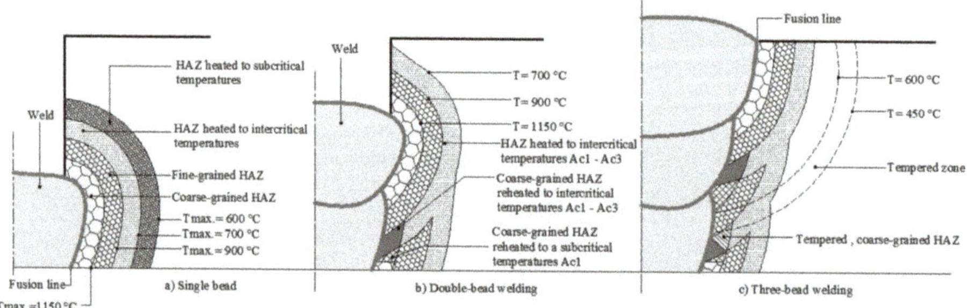

Figure 2. Schematic representation of multipass welding HAZ regions of the high-strength low-alloy steel (HSLA) HT50. In (**a**) root pass; in (**b**) root pass reheated one time and (**c**) root pass reheated twice and filling pass reheated one time, reproduced from [5], with permission from Taylor & Francis, 2000.

The HAZ root pass, represented by Figure 2a, is characterized by different microstructural regions: CGHAZ, fine-grained HAZ (FGHAZ), ICHAZ and subcritical HAZ (SCHAZ). Figure 2b presents the second welding pass that promotes the formation of a new HAZ adjacent to its corresponding pass with the same microstructural regions shown before. However, this second welding pass partially affects the previous HAZ, so that new microstructure sub-regions arise. Traveling on the root CGHAZ from the top towards to the schematic bottom face specimen, it is possible to recognize that a sub-region was reheated above the recrystallization temperature. It is named CG-CGHAZ. The next lower portion thereof CGHAZ is reheated in the temperature range in which there is grain refinement—the

RG-CGHAZ. The same CGHAZ is still reheated within the range of intercritical temperature, making arise the IR-CGHAZ. This subregion needs more attention to be known as the local brittle zone [5,8] (LBZ). The temperature range between A_{c1} and A_{c3}, is favorable to MA microconstituent formation in multipass weldments [2,5,8], as it can deteriorate the steel toughness, depending on its size and distribution in the ferritic matrix [9,10]. The following sub-region is reheated in the subcritical temperature range: it arises the SC-CGHAZ. For the last, remains the unaffected root CGHAZ. Similarly, the third welding pass, shown in Figure 2c, affects the previous HAZ and the root pass HAZ, but is less intense than the second one.

Software packages that use the Finite Element Method (FEM) have been used successfully to simulate welding, in which the material physical properties are temperature dependent [11], to calculate isotherms, thermal cycles, residual stresses [12,13], distortions [14] and uses a volumetric heat source with a Gaussian power distribution. The heat source dimensions are adjusted by weld bead macrography [15–17]. FEM has also been used to simulate multipass welding to solve these same questions, however, the previous HAZ reheating through subsequent welding pass is not evaluated [18–22]. Single pass welding simulation is a costly operation for the FEM, the more expensive it will be for multipass welding simulation [18,23]. HAZ simulation in the weld cross section through FEM presents the difficulty of the delimitation of the isotherms maximum widths that are in different planes for different values of thickness. HAZ reheating simulation is another difficulty presented for FEM software packages because the user does not have flexibility to save the isotherms coordinates of each weld pass.

Due to the presented difficulties, analytical solutions for the heat flux in the welding will be presented.

The analytical solutions developed by Rosenthal [24] to describe the welding heat flow considered the quasi-stationary heat flow regime in an autogenous welding. The analytical solutions or Rosenthal's model is based on the Fourier differential equation, described by Equation (1). In this equation, λ is thermal conductivity and ρc_p is the volumetric thermal capacity and Q_0 is a component of energy generation.

$$\frac{\partial}{\partial x}\left(\lambda\frac{\partial T}{\partial x}\right) + \frac{\partial}{\partial y}\left(\lambda\frac{\partial T}{\partial y}\right) + \frac{\partial}{\partial z}\left(\lambda\frac{\partial T}{\partial z}\right) + Q_0 = \rho c_p\frac{\partial T}{\partial t} \tag{1}$$

The heat transfer models developed by Rosenthal [24–26], called thin plate (2D) model and thick plate (3D) model are most widely used for their simplicity, in order to analyze heat flow within a welded joint. Later, Jhaveri et al. [27] proposed a dimensionless parameter, called relative thickness (τ), which defines the relative thickness interval to use on 2D or 3D heat transfer model. Therefore, the thin plate heat transfer model should be used when the value of τ is less than 0.6. On the other hand, when the value of τ is more than 0.9 the thick plate solution is employed. There is a gap among 0.6–0.9 relative thickness, in which both models (2D and 3D) are not applied. The medium thick plate (MTP) is another model developed by Rosenthal [24,26] that can be used in this τ interval, in which the other two models do not present good results [2].

Among the mentioned heat flow models, the MTP model is more robust as it can predict temperature isotherms for thick plate (3D) and for thin plate (2D) heat flows [2]. To find the solution of medium thick plate model, some hypotheses were assumed [24]:

1. Physical properties constant at room temperature (λ and α), independent of temperature;
2. Point heat source moves in a straight line with constant speed v;
3. No energy generated or consumed within the plate, i.e., $Q_0 = 0$;
4. All the heat flow is transmitted by conduction. Radiation and convection through surfaces are neglected;
5. The plate is semi-infinite, with a thickness d;

6.   The initial condition is defined by Equation (2).

$$T(0) = T_0 \tag{2}$$

7.   The boundary conditions are defined by Equation (3);

$$\begin{cases} \frac{\partial T}{\partial z} = 0, \ z = 0 \\ \frac{\partial T}{\partial z} = 0, \ z = d \end{cases} \tag{3}$$

The simplification imposed by Equation (3) implies that there is no heat loss by radiation and convection through the upper and lower plate surfaces, i.e., both surfaces are adiabatic. This condition is achieved by the method of images, which considers specular reflections of the actual source by imaginary ones ($\ldots$, $2q_{-2}$, $2q_{-1}$, $2q_1$, $2q_2$, $\ldots$), in relation to the planes $z = 0$ and $z = d$, as shown in Figure 3.

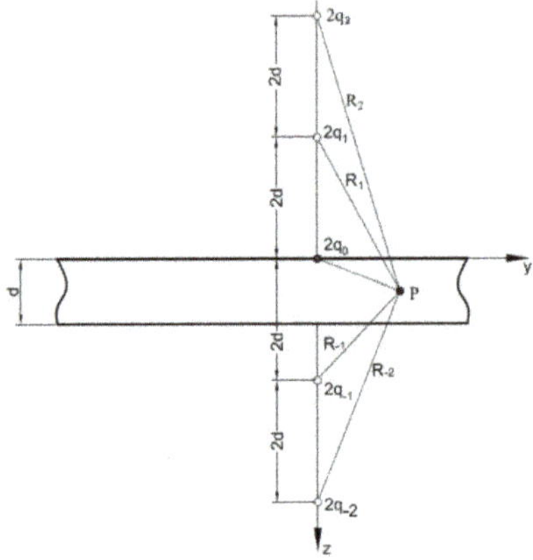

Figure 3. Disposition of actual and imaginary heat sources on MTP model [2].

The coordinate system used to solve Equation (1) defines that the x-axis is related to the welding direction and the weld pool length. The y-axis is related to the width of HAZ and weld pool, while the z-axis is related to the plate thickness and the weld pool depth. The contribution from these heat sources allows to obtain the temperature distribution solution, in the form of a convergent series, according to Equation (4a).

$$T - T_0 = \frac{q_0}{2\pi\lambda} exp\left(-\frac{vx}{2\alpha}\right)\left[\sum_{i=-\infty}^{i=+\infty} exp\left(-\frac{v}{2\alpha}R_i\right)\right] \tag{4a}$$

where

$$R_i = \sqrt{x^2 + y^2 + (z - 2id)^2} \tag{4b}$$

Equation (4a,b) define the solution of Equation (1) with its initial and boundary conditions in quasi stationary regime, where R_i is the radius vector which measures the distance between the imaginary source q_i and the point P where it is intended to calculate its peak temperature. The real heat source is presented by q_0 at the origin of the coordinate system, α is the thermal diffusivity, (x, y, z) are the coordinates of the point P and v is the weld speed.

Figure 4 shows the temperature isosurfaces of 1520 °C, 1000 °C, 700 °C and 500 °C simulating the welding of a low alloy steel with the MTP model, heat input of 1.2 kJ/mm and 14.5 mm of thickness.

The shape of the isotherm of 1520 °C indicates a 3D heat flow, while the shape of the isotherm of 500 °C indicates a heat flow between 3D and 2D, because it touches both plate surfaces.

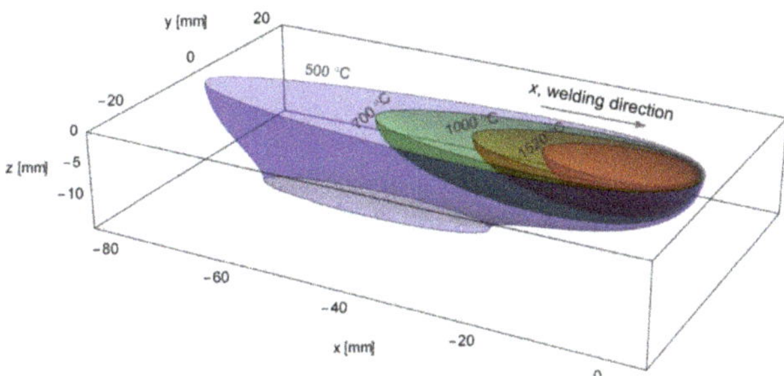

Figure 4. Isosurfaces of a medium thick steel plate.

However, a new proposal for the study of welding heat flux was based on the analytical model called discretely distributed point heat sources model (DHS) developed by Myhr and Grong [2], initially intended to predict the convection effects within the weld pool. These effects cause changes in the bead format, called finger, as shown in Figure 5.

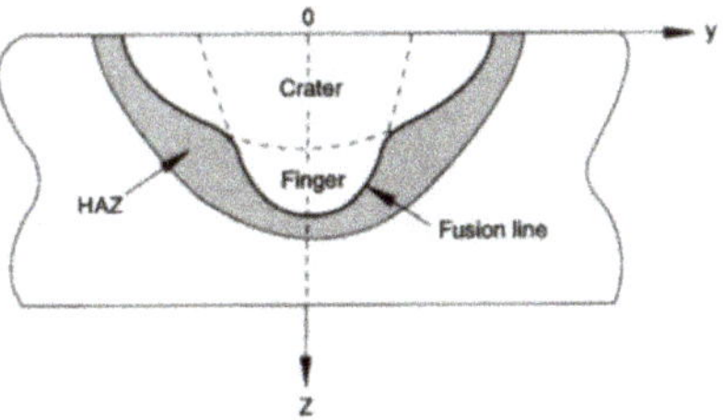

Figure 5. Convection effects in weld bead shape [2].

The DHS model is based on the point heat sources distribution of Rosenthal's MTP on a fusion zone cross section, so that the increase in temperature at point P located within the plate is calculated based on the method of images [2], as defined schematically in Figure 6.

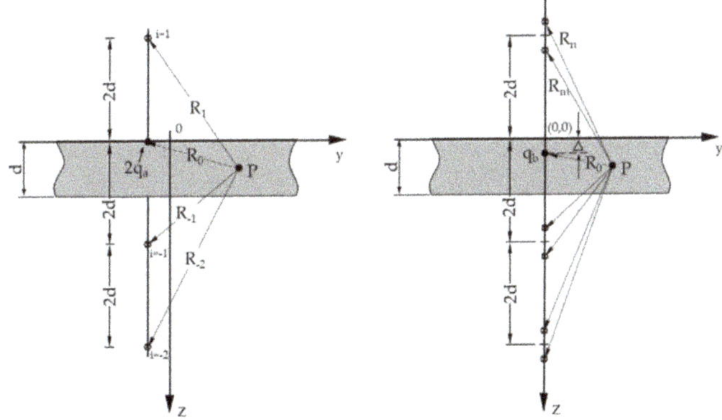

Figure 6. Contribution of sources q_a and q_b in increasing the temperature in the point P [2].

The plate surfaces are adiabatic like they are in the MTP model [2,24]. The point heat sources, in this case, are subdivided into ones that are located on the plate (q_a) along the y-axis and those immersed in the molten pool (q_b) along the z-axis, as shown in Figure 7.

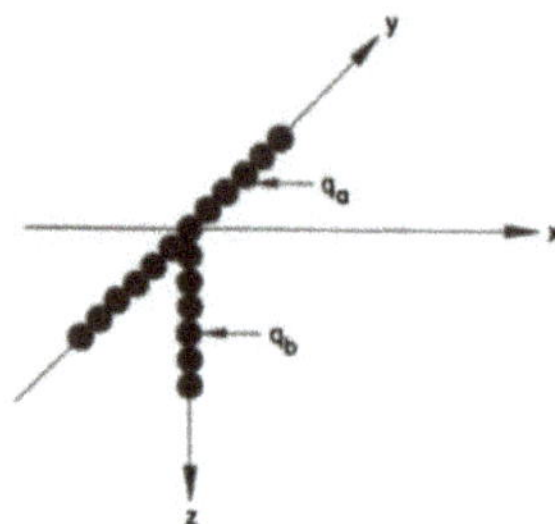

Figure 7. Point heat sources distribution in y-z plane to simulate finger effect in weld bead shape [2].

Finally, the temperature at a point is the sum of the power contributions of all point heat sources, as presented in Equation (5).

$$T - T_0 = \sum_i \left[T\left(q_a^i\right) + T\left(q_b^i\right) \right]$$

(5)

The heat source total power (q_0) is the sum of the contributions of all point heat sources, considering the welding arc efficiency (η), arc voltage (V) and welding current (I), as presented by Equation (6).

$$q_0 = \sum_i \left[q_a^i + q_b^i \right] = \eta V I$$

(6)

This model, however, was modified by Ramirez and Brandi [28], so that sources q_a^i located on the y-axis could be displaced from a value Δ_z into the weld pool, in the z direction. Thus, the point sources could be located on the y-axis, or below it, located anywhere in the y-z plane, since confined inside the weld pool [28]. This model was developed due to the need to simulate multipass welding [28]. The authors considered physical properties independent of temperature and they still used Gleeble® system equipment (Dynamic Systems Inc., Poestenkill, NY, USA) to simulate physically the HAZ. According to the authors, the results were quite satisfactory when compared with experimental and physically simulated data.

Models presented evolution as the boundary conditions imposed on the differential equation and the size of the heat source are close to actual welding situations. The difficulty of applying these models lies in the definition of the values of the physical properties, since they are temperature-dependent. In this way, errors can be made when trying to delimit the HAZ regions, or to simulate thermal cycles with different maximum temperatures.

Thus, this work proposes to evaluate the effectiveness of MTP and DHS models by overlapping the simulated isotherms on the marks revealed in the macrographs and by means of experimental thermal cycle. For both models, the hypothesis of constant and temperature-dependent physical properties will be considered. The best option will be used to simulate the HAZ of multipass welding.

2. Materials and Methods

A girth welding was carried out between two API 5L X80 steel pipe with 1016 mm (42″) diameter and 16 mm wall thickness, whose chemical composition is shown in Table 1.

Table 1. Chemical composition of API 5L X80 steel.

Alloy	C	Mn	Si	Cr	Ni	Mo	Al	Cu	Ti	V	Nb	B
wt%	0.06	1.597	0.216	0.192	0.198	0.002	0.049	0.012	0.015	0.027	0.0649	0.0003

The joint was manually welded using only half perimeter of the pipes. The root pass was welded by Gas Metal Arc Welding (GMAW) using Surface Tension Transfer Mode (STT®, The Lincoln Electric Company, Cleveland, OH, USA) throughout the mentioned pipe extension. Subsequent passes were welded by the Flux Cored Arc Welding (FCAW) process. The beginning of each weld pass has been displaced to maintain an extension of the previous one free of microstructural changes due to the subsequent beads, as shown in Figure 8.

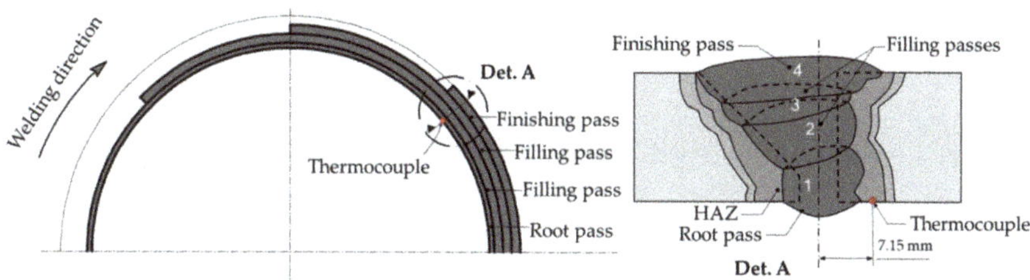

Figure 8. Schematic view of welding passes.

According to the schematic view of the welded joint, depicted in Figure 8 (Det. A), thermocouple K-type 0.25 mm diameter was welded on the inner pipe surface and in the beginning of welding finishing pass section, at 7.15 mm from the weld centerline. Thus, it was possible to record the thermal cycle during all welding passes from a data acquisition system. K-type thermocouple has a measuring range of -220–$1260\ ^\circ$C with an error of $\pm 4.4\ ^\circ$C [29]. Welding parameters were also recorded, as voltage (V) and current (I), in function of time, in the range of 0.0002 s. The average source power was calculated from the instantaneous power values.

Extended Myhr and Grong's Model

The DHS model was evaluated in this work with some additional considerations to better represent the fusion zone contour, the HAZ isotherms and the thermal cycle, starting with the model proposed by Ramirez and Brandi [28]. That is, since the sources q_a^i have been displaced from z axis into the weld pool, they will have the same nature as the sources q_b^j. Therefore, both kind of heat sources will be treated just as q_i, and their displacements will assume values of their own coordinates related to system origin, (y_{qi}, z_{qi}). The quasi-stationary heat flow regime, in this case, is represented by Equation (7).

$$T(q_i) = \frac{q_i}{4\pi\lambda} e^{-\frac{vx}{2\alpha}} \left[\sum_{m=-\infty}^{m=\infty} \frac{-R_m \frac{v}{2\alpha}}{R_m} + \sum_{n=-\infty}^{n=\infty} \frac{-R_n \frac{v}{2\alpha}}{R_n} \right] \tag{7}$$

where,

$$R_m = \sqrt{x^2 + \left(y_{q_i} - y_p\right)^2 + \left(z_{q_i} - 2md - z_p\right)^2} \tag{7a}$$

$$R_n = \sqrt{x^2 + \left(y_{q_i} - y_p\right)^2 + \left(z_{q_i} - 2nd + z_p\right)^2} \tag{7b}$$

Thus, the point heat sources positions and their respective power values became more flexible and η is assumed as a constant value for all of them, which is different from the previous works [2,16,17,28]. However, there is still a question regarding the values to be attributed to the physical properties, as they are dependent on the temperature. Rosenthal, although he formulated the MTP model considering such properties independent of temperature, suggests that they are adjusted step by step [24]. This approach was partially adopted by Azar et al. [16], when using the DHS to simulate the fusion zone isotherm and other two HAZ isotherms in a hyperbaric welding. In their simulation, the physical properties changed by temperature ranges, whose base metal was an API 5L X70 steel.

Starting from the principle that the HAZ has different microstructures, which had undergone different thermal histories along HAZ width, then, it is possible to obtain some inaccuracies when using physical properties by temperature ranges. For this reason, this work proposes to adopt the values of physical properties varying with each simulated isotherm temperature, as well as with simulated thermal cycles. This work still proposes to utilize point heat sources positions and their respective power values to adjust the melting temperature isotherm through the fusion zone contour on the macrograph. Once the heat sources distributions are adjusted, this will be kept simulating other HAZ isotherms and thermal cycles.

The thermal cycle at the *P*-point within the HAZ was simulated with discretized temperatures by variation of 1 °C from the preheat temperature T0. That is, during the heating stage, the evolution of the temperature versus time was defined by an increase of 1 °C, in relation to the previous temperature (T_{i-1}) until it reaches a maximum value. The instant t_i at which the temperature at the *P*-point is increased by 1 °C is calculated by Equation (8).

$$T_1 = T0 \tag{8a}$$

$$T_i = T_{i-1} + 1, \; i > 1 \tag{8b}$$

$$t_1 = t_{(T_1)} = 0 \tag{8c}$$

$$t_i = t_{i-1} + \frac{\Delta t_1 + \Delta t_2}{2}, \; i > 1 \tag{8d}$$

The values of the time intervals Δt_1 and Δt_2 were defined as the time required to vary from 1 °C between the temperatures T_{i-1} and T_i. For the calculation of Δt_1, the physical properties were considered constant and calculated at the temperature T_{i-1}. An analytical thermal cycle was then calculated from these physical property values and other variables defined in Equation (7). The calculation of Δt_1 was determined from reading in the thermal cycle, when the temperatures ranged from T_{i-1}–T_i in the *P*-point.

The value of the time interval Δt_2 was calculated using the same methodology, but with the physical properties calculated at temperature T_i. This calculation scheme is shown in Figure 9, which depicts part of the analytical thermal cycles during heating at temperatures (T) and ($T + 1$), using the physical properties depending on the respective temperatures. The Δt_{mean} was calculated by the arithmetic mean of Δt_1 and Δt_2.

For each value of T_i, the values of the thermal properties, the respective analytical thermal cycle and its maximum temperature—$T_{\max(\lambda(T_i))}$—are calculated. The maximum temperature of the discretized heat cycle—$T_{\max(T_i)}$—is calculated according to Equation (9).

$$T_{\max(T_i)} = T_{\max(\lambda(T_i))} \; if \; \left| T_i - T_{\max(\lambda(T_i))} \right| \leq 2\,°C \tag{9}$$

The difference of 2 °C between T_i and $T_{\max(\lambda(T_i))}$ was chosen to calculate $T_{\max(T_i)}$ because the temperature T_i is increased by 1 °C in the discretized thermal cycle. This value was enough to ensure the convergence criterion.

The cooling stage of the thermal cycle started from the last temperature reached in the heating step and the calculations of Δt_1 and Δt_2, following the same criterion. However, the next temperature value will decrease by 1 °C with respect to the latest one. Finally, the cooling time interval from 800 °C–500 °C (Δt_{8-5}) was obtained from the final numerically simulated thermal cycle at the interest point.

For the simulation purpose of the HAZ isotherms and thermal cycles, as suggested, the thermal conductivity and diffusivity properties of AISI 1020 steel, as a function of temperature, were used, according to Figure 10.

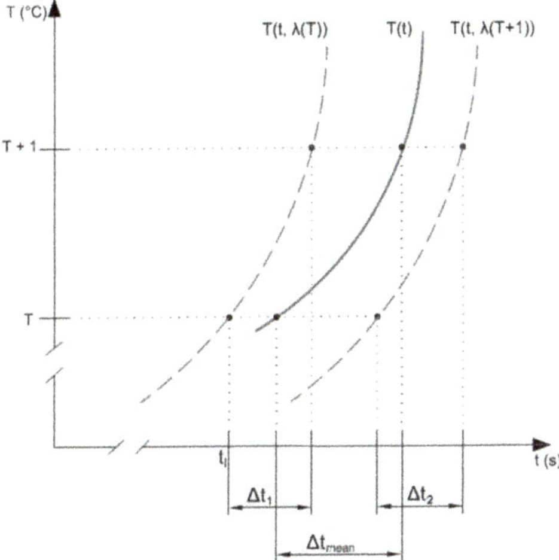

Figure 9. Thermal cycle calculation scheme using physical properties temperature dependent at temperatures T and $T + 1$.

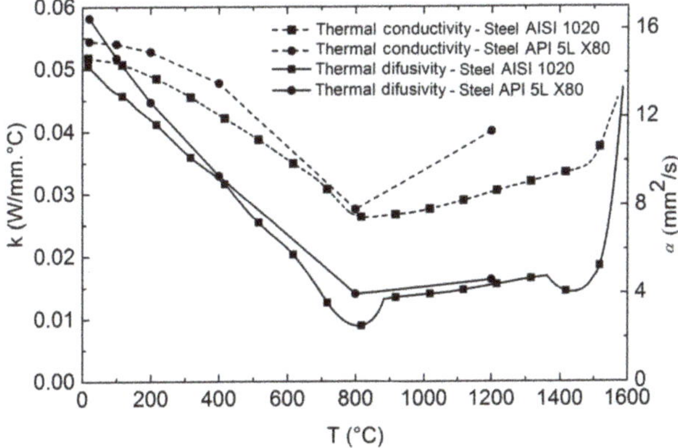

Figure 10. Thermal conductivity and diffusivity of API 5L X80 and AISI 1020 steels as a function of temperature, adapted from [11,13], with permission from Taylor & Francis, 2002.

In this picture, one can notice that the values of AISI 1020 thermal conductivity and thermal diffusivity are close enough to the API 5L X80 steel. The AISI 1020 steel physical properties were used to simulate the welding heat flow for the API 5L X80 steel pipe joint, as these can be estimated from the liquidus temperature to the room temperature.

3. Results

3.1. Comparison between Different Analytical Models

Conventionally, the analytical solutions for welding simulations consider the physical properties as constant, i.e., independent of temperature, just as they were developed [2,24,26,30]. However, in engineering materials, the physical properties are temperature-dependent, such as those shown in Figure 10. To evaluate the DHS proposed model, a comparative study will be carried out with MTP model, initially considering the temperature-independent physical properties and later with the temperature-dependent ones, through the methodology suggested in this work. The welding parameters were the same for all simulations, i.e., the heat source power was equal to 2376 W and the welding speed equal to 1.05 mm/s, that resulted in a heat input of 2.26 kJ/mm. Considering the case in which the physical properties were assumed constant, they were selected at the temperature of 25 °C, in which values were obtained from Figure 10: λ was equal to 0.052 W/mm °C and α equal to 14,208 mm^2/s.

The simulation codes were developed through Wolfram Mathematica® software (Version 10.2.0.0 Student Edition, Wolfram Research, Champaign, IL, USA) to determine the isotherms profiles and thermal cycles, using the resource of finding roots, like Newton-Raphson and bisection methods, because the temperature field equations presented before are transcendental functions.

Figures 11 and 12 present the results of simulation in which the physical properties were constants and based on MTP and DHS models, respectively.

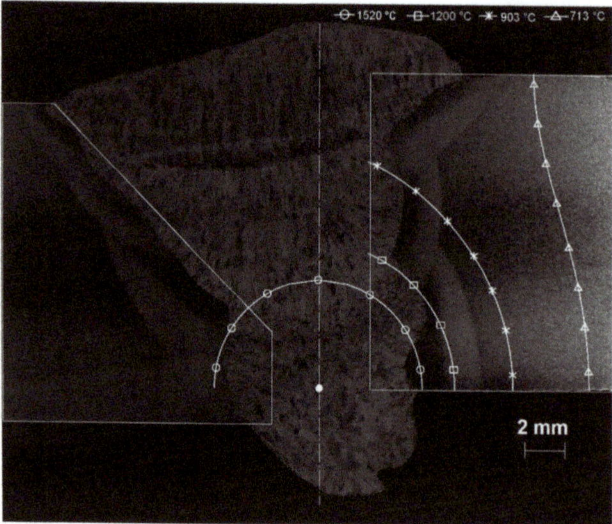

Figure 11. Macrograph and front view of the root pass welding simulation through the MTP model with properties independent of temperature.

In Figure 11, the contours of the original bevel and the weld centerline are highlighted, the point heat source is presented through the white point on the specimen's bottom surface. The simulated HAZ isotherms are superimposed on the multipass welded joint macrograph to assess the fit between simulated and experimental results. The isotherms of 1520 °C, 1200 °C, 903 °C and 713 °C represent respectively the base metal solidus, recrystallization, A_{c3} and A_{c1} temperatures. The first one was obtained through Thermo-Calc® software (Version 2.2.1.1, Foundation for Computational Thermodynamics, Stockholm, Sweden), the second value was based on Pickering [31] and the last two of them were determined using dilatometry analysis.

Figure 12 presents the DHS simulation using 19-point heat sources distributed within the melt zone, each one using 8 imaginary sources. Typically, the simulations by DHS use less point heat sources [2,16,17,28]. The greater the amount of point heat sources used in a simulation, the easier to adjust their position and respective power source values. The innermost heat sources are those with the highest power values. The final adjustment of the simulated fusion isotherm to the weld pool cross section is promoted by the most external heat sources, which have the lowest power values. This power distribution of heat sources suggests a Gaussian distribution, as proposed in numerical methods [32,33].

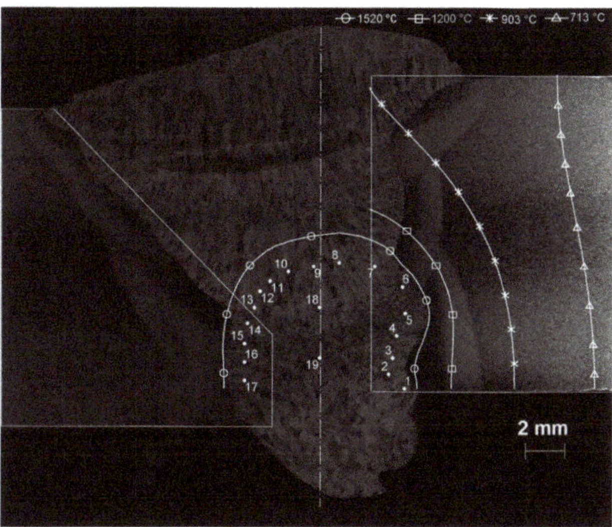

Figure 12. Macrograph and front view of the root pass welding simulation through the DHS model with properties independent of temperature.

A better fit can be observed between the melt temperature isotherm with the molten zone contour. However, for the other isotherms the fit was worse, i.e., their positions did not coincide with the contours revealed in the macrograph. Table 2 shows the relative error between the positions of 1520 °C and 903 °C simulated isotherms with respect to three reference points located at 1.5 mm, 3.0 mm and 4.5 mm, respectively, above the specimen lower surface.

Table 2. Relative error between MTP and DHS isotherms in relation to reference positions within the HAZ with temperature-independent physical properties.

Position	MTP		DHS	
	1520 °C	903 °C	1520 °C	903 °C
1	12.4%	19.1%	9.2%	20.7%
2	−18.3%	18.7%	2.1%	22.3%
3	−48.8%	19.8%	1.5%	27.0%

The following isotherm simulations were carried out with the temperature-dependent physical properties, in which each property value was selected based on the current isotherm temperature value. Figures 13 and 14 present the results considering MTP and DHS models, respectively. It can be observed in Figure 13 that the isotherm's positions provided a better adjustment in relation to the macrograph revealed contours when compared to the results presented in Figure 11.

Figure 14—which represents the proposed model in this work—shows the DHS simulation using 19-point heat sources distributed within the melt zone, whose power and position distributions were adjusted to simulate molten zone contour. Once the necessary adjustment was done, this configuration remained the same to simulate the other isotherms. The power value of 2376 W used to simulate the 1520 °C isotherm was within the calculated average source power of 1877 ± 920 W. This high standard deviation value was related to the welding process used, which, in this case, was a curt circuit one, in which there is a great current variation. This behavior of the power values is also reported by Azar [16] that calculates the average power from its instantaneous values.

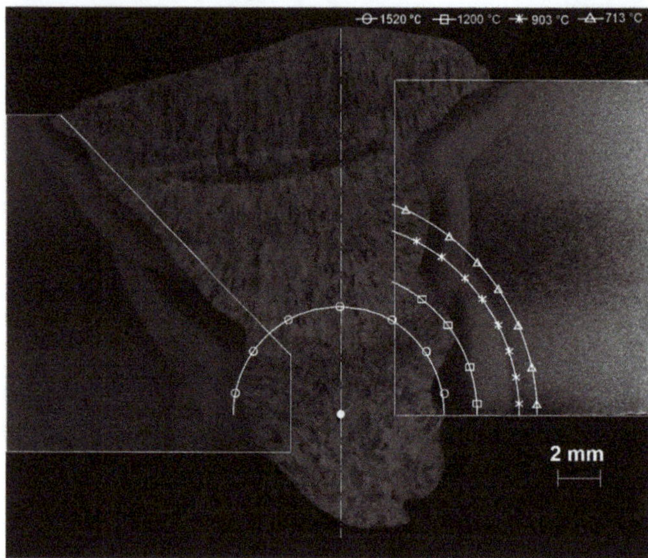

Figure 13. Macrograph and front view of the root pass welding simulation through MTP model with temperature-dependent properties.

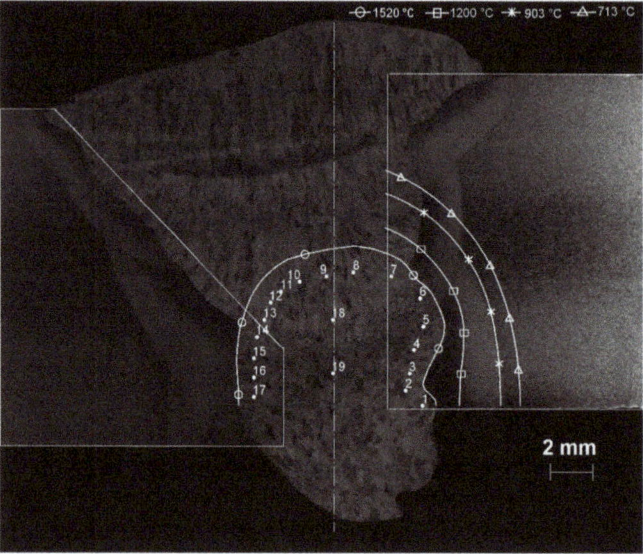

Figure 14. Macrograph and front view of the root pass welding simulation through DHS model with temperature-dependent properties.

As can be seen, the adjustment of the isotherms on the contour revealed in the macrographs is more accurate in Figure 14 than in Figure 13. Table 3 presents the relative error for both simulations, in the same positions of the reference used before. Negative error values indicate the isotherm position was on the left side of the revealed macrograph contours.

Table 3. Relative error between MTP and DHS isotherms in relation to reference positions within the HAZ with temperature-dependent physical properties.

Position	MTP		DHS	
	1520 °C	903 °C	1520 °C	903 °C
1	6.9%	3.4%	−1.1%	−1.9%
2	−24.9%	1.4%	−2.3%	−0.9%
3	−60.0%	−0.8%	−2.3%	1.6%

The performance of models was also compared with respect to the simulated thermal cycles. For the cases in which the properties were considered temperature-dependent, the methodology proposed in topic 3 was adopted for both MTP and DHS models. The results of the simulations are summarized in Figure 15 that presents thermal cycles simulated through the same criteria used before in the isotherm profiles determination.

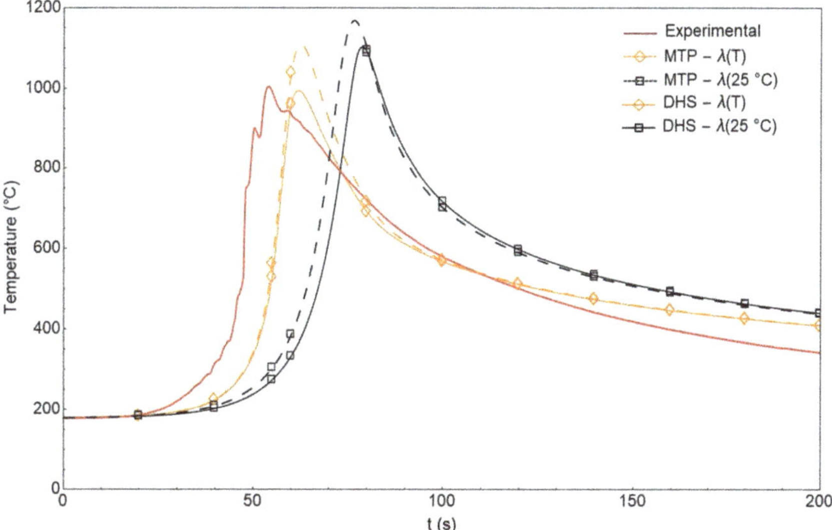

Figure 15. Experimental and simulated thermo cycles considering all the criteria.

Table 4 presents a thermal cycle performance study based on the maximum temperature (T_{max}) reached by a thermocouple and its Δt_{8-5}. The thermal cycles whose physical properties are temperature independent are presented by $\lambda(25\,°C)$ and those whose physical properties are temperature dependent are presented by $\lambda(T)$. The position used to simulate them was the same where the thermocouple was installed, i.e., 7.15 mm right of the weld center line or 2.35 mm from the molten zone, in which $T_{max} = 1004.34\,°C$ and $\Delta t_{8-5} = 47.98$ s.

Table 4. Thermal cycle evaluation based on T_{max} and Δt_{8-5} criteria.

Physical Property	MTP Simulation				DHS Simulation			
	T_{max} (°C)	T_{max} Error (%)	Δt_{8-5} (s)	Δt_{8-5} Error (%)	T_{max} (°C)	T_{max} Error (%)	Δt_{8-5} (s)	Δt_{8-5} Error (%)
$\lambda(T)$	1108.58	10.38	50.82	5.92	994.62	−0.97	52.04	8.47
$\lambda(25\,°C)$	1168.18	16.31	62.43	30.12	1103.53	9.88	63.18	31.68

It can be seen that the DHS proposed model presented the best answer in relation to $T_{\max}$, but not in relation to Δt_{8-5}, whose calculated value was higher than the experimental one. It means that the effect of neglecting heat losses has a small but consistent effect on simulation of the maximum temperature, and probably also has a consistent effect on the time lag difference shown in Figure 15 for thermal cycling.

Another criterion for the thermal cycle evaluation through the proposed DHS model was the Δt_{8-5} variation depending on the maximum temperature reached on the HAZ width. In this case, the z coordinate was set at 1.2 mm above the specimen's bottom face on the right side of the groove and the results are shown in Figure 16.

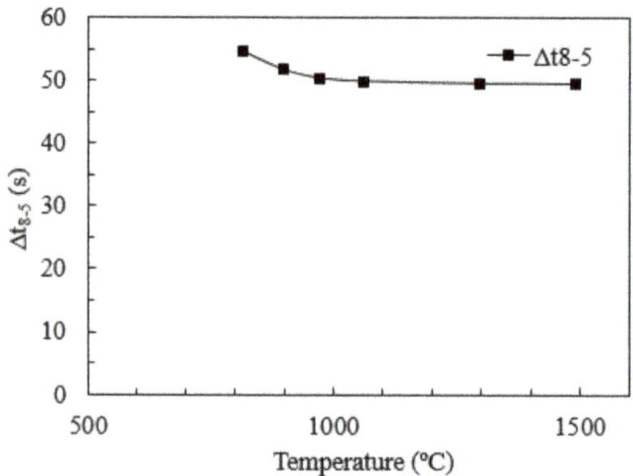

Figure 16. Δt_{8-5} variation depending on the maximum temperature reached on the HAZ width.

In the neighborhood of the molten zone, in the 1489 °C maximum temperature position, the Δt_{8-5} value was 49.5 s. At the lower temperature HAZ end, in the 814 °C maximum temperature position, the Δt_{8-5} value was 54.6 s in the simulated thermal cycle. That is, there was a variation of only 10.3% compared to the minimum Δt_{8-5} value for a variation of 83% in temperature values throughout the HAZ width. This behavior is typical of values found in analytical solutions [2,4,24,27]. In other words, in the weld centerline on the plate upper surface and behind the weld pool, the value of Δt_{8-5} is constant because the coordinate values y and z are null [4,24,34]. However, when these coordinates are inside the HAZ, the Δt_{8-5} variation is very small [4,27,35]. Therefore, a Δt_{8-5} value measured by a thermocouple may be used in whole HAZ extent, whose error is not greater than 10%.

3.2. Multipass Welding Simulation through DHS Model

Figure 14 shows how DHS as proposed was able to simulate the fusion zone contour of the root pass. The melting temperature isotherm profile generated by the MTP model cannot adjust to that of the fusion zone contour because its profile is semicircular in the y-z plane [24,30,34]. Therefore, this model is even less suitable to simulate multipass welding. As in the previous studies, the DHS point heat sources were moved into the fused zone, to simulate a multipass welding. It is enough that they are redistributed within the fused zone of the subsequent welding pass. In this case, to simulate the welding of the first filling pass by the DHS model, 26-point heat sources were used, each one with 8 imaginary sources, whose results are shown in Figure 17. In this figure, the simulated first filling pass isotherms are overlapped by those of root pass presented in Figure 14.

These overlaps of HAZ isotherms take a main role in the welded joint evaluation, since mechanical properties, such as hardness and toughness, as well as microstructures formation are dependent on the welding thermal history.

Thus, the intersection between them presents the previous HAZ sub-regions that were affected by heating. Although the macrograph already provides an idea of this phenomenon, the isotherm simulation of the filling pass indicates in which temperature ranges it occurred for each root pass HAZ sub-region. The isotherm intersections allowed delimitating nine microstructural sub-regions in the root pass, which arose from the heating by subsequent welding pass, as presented in Figure 18.

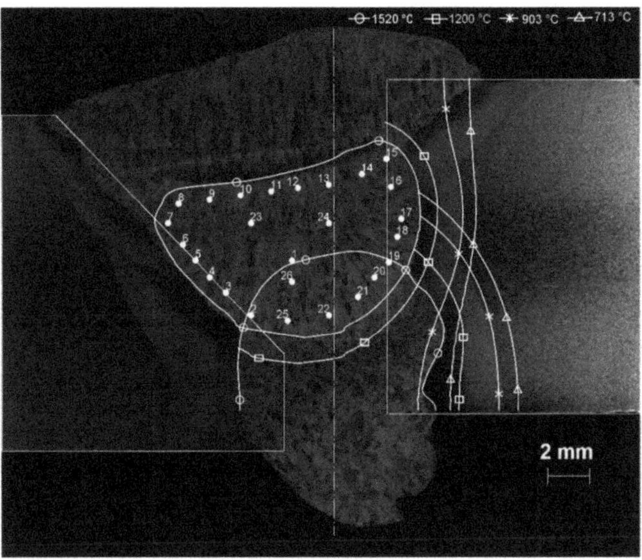

Figure 17. First filling pass simulated isotherms by DHS model overlapped those of root pass.

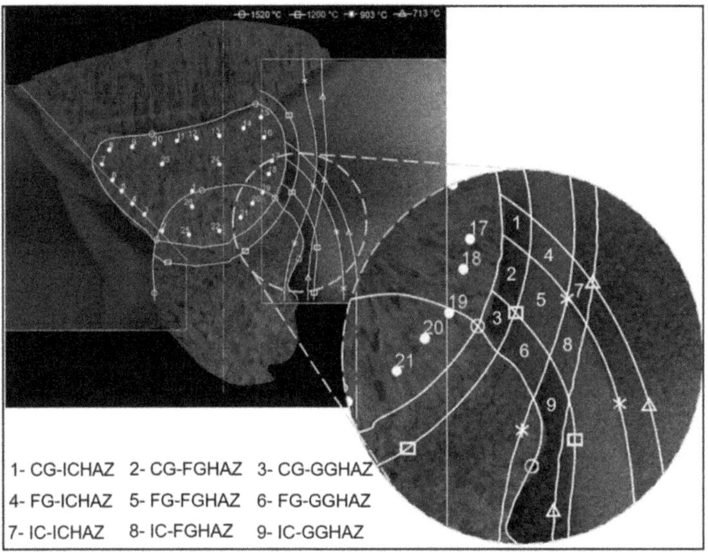

Figure 18. Reheated root pass sub-regions.

Regions 1–3 (CG-ICHAZ, CG-FGHAZ and CG-CGHAZ) were reheated beyond recrystallization temperature, while regions 4–6 were reheated in the fine-grained region temperature range. Region 9 is the coarse-grained HAZ intercritically reheated, IC-CGHAZ, the local brittle zone previously cited. This sub-region is very narrow, whose base is measured 1.3 mm and its height is equal to 3.7 mm.

4. Conclusions

The flexibility of defining the heat sources positions and their respective power values, based on the DHS model, allowed to fit more easily the weld pool contour on the mark revealed in the macrography of the API 5L X80 steel, when considering the physical properties to be temperature dependent. The other HAZ isotherms also fitted to marks revealed on the macrograph. Consequently, it is possible to simulate thermal cycles at any position within the HAZ, without the inconvenience of thermocouples installed inside it, in which this kind of installation changes the heat flow conditions in the thermocouple vicinity.

By using the DHS model with the physical properties temperature dependent and appropriate to the material, it was possible to obtain a good fit of the simulated isotherms to the HAZ revealed contours, only by adjusting the melting temperature isotherm to the shape of the fusion zone without the need of the weld pool length.

The advantage of DHS as proposed in relation to the analytical classic models is the possibility of simulating a multipass welding by displacing the point heat sources into a current pass fused zone.

As a result, it is possible to obtain not only the adjacent HAZ of the current pass, as it is possible to identify HAZ sub-regions of previous passes that were reheated. It is possible to identify regions of local embrittlement because of multipass welding and evaluate them with tests like microhardness measurements, impact Charpy, CTOD and others.

The IC-CGHAZ would be difficult to represent by any FEM software package because the cross-section that can be performed on the heat source constitutes a single plane that does not represent the locations where the maximum isotherm widths pass through the thickness of the specimen.

Author Contributions: Conceptualization, D.M.B.F., A.d.N.S.A., R.M.d.A.C.N., T.F.M. and S.D.B.; Methodology, D.M.B.F., A.d.N.S.A., T.F.M. and R.M.d.A.C.N.; Validation, D.M.B.F. and S.D.B.; Formal Analysis, D.M.B.F. and S.D.B.; Investigation, D.M.B.F. and S.D.B.; Writing-Original Draft Preparation, D.M.B.F. and A.d.N.S.A.; Writing-Review & Editing, D.M.B.F. and A.d.N.S.A.; Visualization, D.M.B.F. and S.D.B.; Supervision, S.D.B.; Funding Acquisition, D.M.B.F. and S.D.B.

Funding: This research was funded by FUSP—Fundação de Apoio à Universidade de São Paulo and IFES—Instituto Federal de Educação, Ciência e Tecnologia do Espírito Santo (Prodif 03/2018).

Acknowledgments: The work was supported by Coordenação de Aperfeiçoamento de Pessoal de Nível Superior (CAPES) through Dinter Program, Instituto Federal de Educação, Ciência e Tecnologia do Espírito Santo (IFES) and Fundação de Ampara à Pesquisa do Estado do Amazonas (FAPEAM), for funding the scholarship. The authors also appreciate CBMM for providing the pipes, Lincoln Electric Company and Voith Hydro Ltd. the support given to the welding of pipes and the supply of consumables.

Conflicts of Interest: The authors declare no conflict of interest.

References

1. Easterling, K.E. *Introduction to the Physical Metallurgy of Welding*, 2nd ed.; Butterworth-Heinemann: London, UK, 1992.
2. Grong, O. *Metallurgical Modelling of Welding*, 2nd ed.; The Institute of Materials: London, UK, 1997.
3. ASM HANDBOOK. *Properties and Selection: Irons Steels and High Perfomance Alloys*, 3rd ed.; ASM International: Cleveland, OH, USA, 1993.
4. Adams, C.M., Jr. Cooling rates and peak temperatures in fusion welding. *Weld. J.* **1958**, *37*, 210s–215s.
5. Lomozik, M. Effect of the welding thermal cycles on the structural changes in the heat affected zone and on its properfies in joints welded in low-alloy steels. *Weld. Int.* **2000**, *14*, 845–850. [CrossRef]
6. Bang, K.S.; Kim, W.Y. Estimation and Prediction of HAZ Softening in Thermomechanically Controlled-Rolled and Accelerated-Cooled Steel. *Weld. J.* **2002**, *8*, 174–179.
7. Yurioka, N.; Suzuki, H.; Okumura, M.; Ohshita, S.; Saito, S. Carbon Equivalents to Assess Cold Cracking Sensitivity and hardness of steel welds. *Nippon Steel Tech. Rep.* **1982**, *20*, 61–73.
8. Fairchild, D.P.; Bangaru, N.V.; Koo, J.Y.; Harrison, P.L. A Study Concerning Intercritical HAZ Microstructure and Toughness in HSLA Steels. *Weld. J.* **1991**, *70*, 321s–330s.

9. Di, X.J.; Cai, L.; Xing, X.-X.; Chen, C.-X.; Xue, Z.-K. Microstructure and mechanical properties of intercritical heat-affected zone of X80 pipeline steel in simulated in-service welding. *Acta Met. Sin.* **2015**, *28*, 883–891. [CrossRef]

10. Bhadeshia, H.K.D.H. About calculating the characteristics of the martensite-austenite constituent. In Proceedings of the International Seminar on Welding of High Strength Pipeline Steels, Araxa, Brazil, 28–30 November 2013; pp. 99–106.

11. Chunyan, Y.; Cuiying, L.; Bo, Y. 3D modeling of the hydrogen distribution in X80 pipeline steel welded joints. *Comput. Mater. Sci.* **2014**, *83*, 158–163.

12. Nóbrega, J.A.; Diniz, D.D.S.; Silva, A.A.; Maciel, T.M.; Albuquerque, V.H.C.; Tavares, J.M.R.S. Numerical evaluation of temperature field and residual stresses in an API 5L X80 steel welded joint using the finite element method. *Metals* **2016**, *6*, 28. [CrossRef]

13. Cho, S.H.; Kim, J.W. Analysis of residual stress in carbon steel weldment incorporating phase transformations. *Sci. Technol. Weld. Join.* **2002**, *7*, 212–216. [CrossRef]

14. Pasternak, H.; Launert, B.; Krausche, T. Welding of girders with thick plates—Fabrication, measurement and simulation. *J. Constr. Steel Res.* **2015**, *115*, 407–416. [CrossRef]

15. Bate, S.K.; Charles, R.; Warren, A. Finite element analysis of a single bead-on-plate specimen using SYSWELD. *Int. J. Press. Vessel. Pip.* **2009**, *86*, 73–78. [CrossRef]

16. Azar, A.S.; Ås, S.K.; Akselsen, O.M. Analytical modeling of weld bead shape in dry hyperbaric GMAW using Ar-He chamber gas mixtures. *J. Mater. Eng. Perform.* **2013**, *22*, 673–680. [CrossRef]

17. Azar, A.S.; Ås, S.K.; Akselsen, O.M. Determination of welding heat source parameters from actual bead shape. *Comput. Mater. Sci.* **2012**, *54*, 176–182. [CrossRef]

18. Lindgren, L.; Runnemalm, H.; Näsström, M.O. Simulation of multipass welding of a thick plate. *Int. J. Numer. Methods Eng.* **1999**, *44*, 1301–1316. [CrossRef]

19. Börjesson, L.; Lindgren, L.E. Simulation of Multipass Welding with Simultaneous Computation of Material Properties. *J. Eng. Mater. Technol.* **2001**, *123*, 106. [CrossRef]

20. Deng, D.; Kiyoshima, S. Numerical Investigation on Welding Residual Stress in 2.25Cr-1Mo Steel Pipes. *Trans. JWRI* **2007**, *36*, 73–90.

21. Deng, D.; Murakawa, H. Numerical simulation of temperature field and residual stress in multi-pass welds in stainless steel pipe and comparison with experimental measurements. *Comput. Mater. Sci.* **2006**, *37*, 269–277. [CrossRef]

22. Maekawa, A.; Kawahara, A.; Serizawa, H.; Murakawa, H. Fast three-dimensional multipass welding simulation using an iterative substructure method. *J. Mater. Process. Technol.* **2015**, *215*, 30–41. [CrossRef]

23. Duranton, P.; Devaux, J.; Robin, V.; Gilles, P.; Bergheau, J.M. 3D modelling of multipass welding of a 316L stainless steel pipe. In Proceedings of the International Conference on Advances in Materials and Processing Technologies—AMPT2003, Dublin, Ireland, 8–11 July 2003; pp. 974–977.

24. Rosenthal, D. Mathematical Theory of Heat Distribution during Welding and Cutting. *Weld. J.* **1941**, *20*, 220s–234s.

25. Rosenthal, D.; Schmerber, R. Thermal study of arc welding: experimental verification of theoretical formulas. *Weld. J. Res. Suppl.* **1938**, *17*, 2s–8s.

26. Rosenthal, D. The theory of moving sources of heat and its application to metal treatments. *Trans. ASME* **1946**, *43*, 849–866.

27. Jhaveri, P.; Moffatt, W.G.; Adams, C.M., Jr. The effect of plate thickness and radiation on the heat flow in welding and cutting. *Weld. J.* **1962**, *41*, 12s–16s.

28. Ramirez, A.J.L.; Brandi, S.D. Application of discrete distribution point heat source model to simulate multipass weld thermal cycles in medium thick plates. *Sci. Technol. Weld. Join.* **2004**, *9*, 72–82. [CrossRef]

29. Astm Committee E20 On Temperature Measur. *Manual on the Use of Thermocouples in Temperature Measurement*, 4th ed.; ASTM International: Philadelphia, PA, USA, 1993.

30. Eagar, T.W.; Tsai, N.S. Temperature fields produced by traveling distributed heat sources. *Weld. J. Res. Suppl.* **1983**, *62*, 346s–355s.

31. Pickering, F.B. High-Strength, Low-Alloy Steels—A Decade of Progress. In *Microalloying 75, Proceedings of an International Symposium on High-Strength, Low-Alloy Steels*; Union Carbide Corp.: New York, NY, USA, 1977.

32. Goldak, J.; Chakravarti, A.; Bibby, M. A new Finite Element Model for welding heat sources. *Metall. Trans. B* **1984**, *15B*, 299–305. [CrossRef]

33. Pavelic, V.; Tanbakuchi, R.; Uyehara, O.A.; Myers, P.S. Experimental and Computed Temperature Histories in Gas Tungsten-Arc Welding of Thin Plate. *Weld. J. Res. Suppl.* **1969**, *48*, 295s–305s.
34. Myhr, O.R.; Grong, Ø. Dimensionless maps for heat flow analyses in fusion welding. *Acta Metall. Mater.* **1990**, *38*, 449–460. [CrossRef]
35. Poorhaydari, K.; Patchett, B.M.; Ivey, D.G. Estimation of cooling rate in the welding of plates with intermediate thickness. *Weld. J.* **2005**, *84*, 149s–155s.

Article

Study on the Effect of Energy-Input on the Joint Mechanical Properties of Rotary Friction-Welding

Guilong Wang [1,2], Jinglong Li [1], Weilong Wang [1], Jiangtao Xiong [1] and Fusheng Zhang [1,*

[1] Shaanxi Key Laboratory of Friction Welding Technologies, Northwestern Polytechnical University, Xi'an 710072, China; wangguilongnos@126.com (G.W.); lifwnpu@163.com (J.L.); dargonwang@126.com (W.W.); dearrivertao@sina.com (J.X.)

[2] State Key Laboratory of Solidification Processing, Northwestern Polytechnical University, Xi'an 710072, China

* Correspondence: zhangfusheng@nwpu.edu.cn; Tel.: +86-029-8849-1426

Received: 17 October 2018; Accepted: 3 November 2018; Published: 6 November 2018

Abstract: The objective of the present study is to investigate the effect of energy-input on the mechanical properties of a 304 stainless-steel joint welded by continuous-drive rotary friction-welding (RFW). RFW experiments were conducted over a wide range of welding parameters (welding pressure: 25–200 MPa, rotation speed: 500–2300 rpm, welding time: 4–20 s, and forging pressure: 100–200 MPa). The results show that the energy-input has a significant effect on the tensile strength of RFW joints. With the increase of energy-input, the tensile strength rapidly increases until reaching the maximum value and then slightly decreases. An empirical model for energy-input was established based on RFW experiments that cover a wide range of welding parameters. The accuracy of the model was verified by extra RFW experiments. In addition, the model for optimal energy-input of different forging pressures was obtained. To verify the accuracy of the model, the optimal energy-input of a 170 MPa forging pressure was calculated. Three RFW experiments in which energy-input was equal to the calculated value were made. The joints' tensile strength coefficients were 90%, 93%, and 96% respectively, which proved that the model is accurate.

Keywords: rotary friction-welding; energy-input; joint mechanical properties

---

1. Introduction

Rotary friction-welding (RFW) is a method of manufacturing that has been used extensively in recent times due to its advantages such as low heat input, production time, ease of manufacture, and environment friendliness [1,2]. The main parameters of rotary friction-welding are welding pressure, welding rotational speed, welding time, and forging pressure [3,4]. The influence of welding parameters on joint performance has always been a research focus of rotary friction-welding, whether in terms of numerical models or experiments. According to previous literature, it has been found that the RFW joints' tensile strength increases with the increase of friction time until it reaches a maximum point, after which the tensile strength decreases [5]. The welding time determines the amount of heat flux in the joint and the width of the heat-affect zone (HAZ), which increases as the welding time increases [6]. Regarding welding pressure, it was found that friction pressure has a significant effect on tensile strength and the width of HAZ, both of which increased along with the increase in friction pressure until reaching a certain value [7]. In addition, friction pressure also affects the temperature gradient, the required weld power, and the burn-off rate during the RFW process [2]. For the rotational speed, numerical analyses found that an increase in rotational speed causes the weld interface to reach a quasi-stable temperature quicker, which starts material extrusion earlier and increases the upsetting rate remarkably [8]. The rotational speed also affects the heat input of the joint; thus, it has a significant effect on the notch tensile strength and impact toughness [9]. Moreover, the overall width of the weld zone reduces with the increase of rotational speed [10].

As discussed above, the parameters of welding time, welding pressure, and rotational speed all have a significant effect on joint performance when considered separately. However, in practical applications of RFW, it is often necessary to adjust several welding parameters simultaneously according to the actual condition, such as the welding material characteristics and the welding machine's condition. Therefore, it is necessary to evaluate the significance of each parameter on joint performance when considering all parameters synthetically. Such an understanding will help to improve the weld quality, to eliminate weld defects and transfer knowledge of welding procedures to new conditions (e.g., for new geometry or new materials). It is generally known that welding pressure and rotation speed determine the characteristics of the power input during the RFW process, and the integral of power for welding time is the energy-input. For the RFW method, heat flux is generated by the conversion of mechanical energy into thermal energy at the interface of the work pieces [11]. In addition, the amount of energy-input dictates a successful welding process, the quality of joint, the shape of flash, the micro-structure, as well as the residual stress [12,13]. Therefore, it is reasonable to use the energy-input as an index to characterize the joint performance comprehensively. In this work, the effect of the energy-input on the joint's mechanical property was investigated; an energy-input model as a function of welding pressure, rotational speed, and welding time was obtained, and the accuracy of the model was verified by experimental data. The model that describes the variation of optimal energy-input with forging pressure also was established, and an excellent performance joint was obtained based on the calculated optimal energy-input value.

2. Experimental Procedures

The specimen used in the present investigation was 304 stainless steel (304SS) rods with diameters of 25 mm. The specimens were welded by the C320 (Friction Welding Eng. & Tech. Co. Ltd., Hanzhong, Shaanxi, China) continuous-drive rotary friction-welding machine (maximum of 45 kW and 320 kN forging load). According to our preliminary experiments and the literature on 304SS RFW [5,6,9,14], the welding experiments were conducted using welding parameters in which the welding pressure varied from 40 MPa to 200 MPa, rotational speed varied from 500 rpm to 2300 rpm, and welding time varied from 4 s to 20 s, all while under forging pressures of 100 MPa, 120 MPa, 140 MPa, 160 MPa, 180 MPa, and 200 MPa, respectively. During the RFW process, the welding power was recorded by a computer through an A/D converter with a sampling time of 0.01 s, and the surface temperature of the joint was measured by an infrared thermal imaging instrument (Tec VarioCAM@hr, head-HS Infra, Dresden, Germany) with a sampling time of 0.02 s. The data collected by the infrared thermography instrument had been calibrated using the data that thermocouple collected, and the emissivity of 304SS was determined to 0.8. Figure 1 shows the location of the experimental equipment and the diagram of the temperature-detected area.

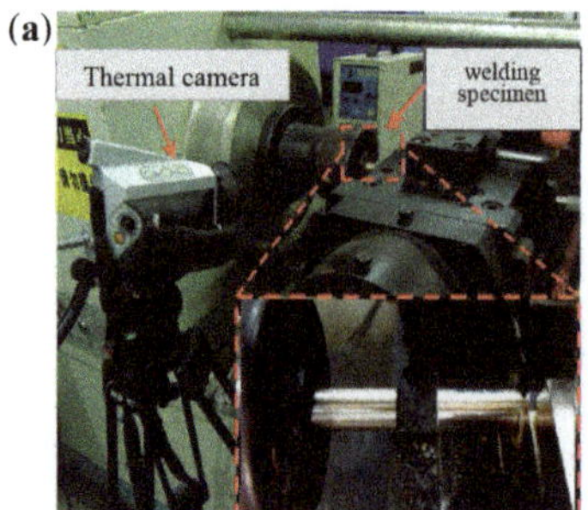

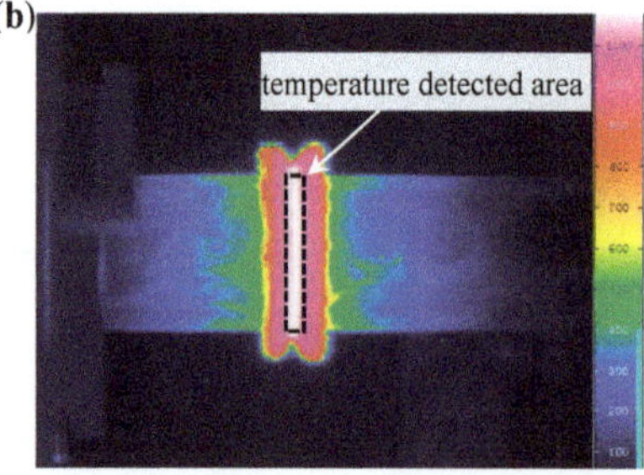

Figure 1. The (**a**) image of experimental equipment and (**b**) diagram of temperature detected area.

After welding, the axial cross-section of the joint was obtained for microstructural examination. The test specimens for Optical Microscope examination were mounted for polishing, then etched by a reagent of 5 g of $FeCl_3$ dissolved in dilute hydrochloric acid (15 mL HCl + 60 mL H_2O). The microstructure of the specimens was observed using a polarizing microscope (Olympus PMG3, Missouri, TX, USA), and the microhardness was measured on the cross-section perpendicular to the friction interface using a microhardness tester (Shimadzu HMV-2T, Dallas, TX, USA) under a load of 2 N. The microhardness test area includes all of the welding zone. Tensile test specimens were prepared according to GB/T228.1-2010. Figure 2 shows the geometry of the tensile test specimen. Three sets of tensile testing were conducted on a tensile-testing machine (Instron 3382, loading range 0~100 kN) using a crosshead speed of 1mm/min at room temperature, and its mean value was taken. Figure 3 displays a typical engineering stress-strain curve of the 304SS RFW joint. After tensile testing, the fracture specimens were etched by the same reagent as described above, and the macro-morphology of fracture surface were observed by SEM (JSM–6390A, JEOL, Tokyo, Japan).

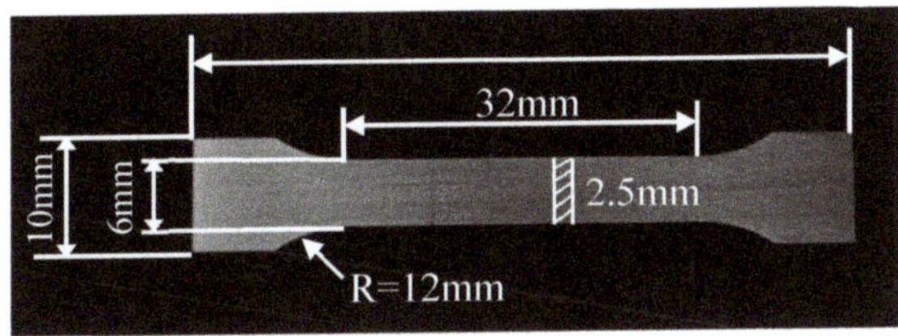

Figure 2. The schematic diagram of tensile specimen.

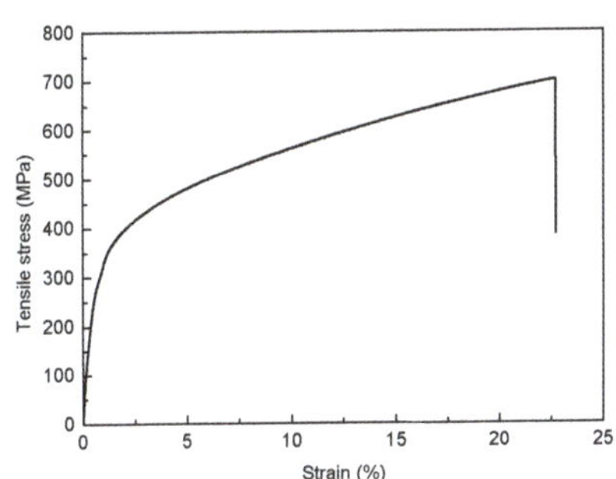

Figure 3. Typical engineering stress-strain curve of 304 stainless steel (SS) rotary friction-welding (RFW) joint (welding pressure is 40 MPa, rotation speed is 500 rpm, welding time is 9 s, and forging pressure is 200 MPa).

3. Results and Discussion

3.1. The Relationship between Energy-Input and Tensile Strength

Figure 4 shows the variation trend of the joint's tensile strength with energy-input under different forging pressures. It can be seen that the tensile strength has the same variation trend with the change of the energy-input, i.e., with an energy-input increase, the tensile strength increases until it reaches a maximum and then decreases. In addition, an optimum energy-input exists for different forging pressures that makes the joint have the highest tensile strength. (The variation characteristics of optimum energy-input corresponding to forging pressure will be discussed in Section 4). Four group experiments under a forging pressure of 200 MPa were selected to study the influence mechanism of energy-input on the joint strength. They are numbered from No.1 to No.4, as indicated in Figure 4. Table 1 lists the welding parameters, energy-input value, and tensile strength of these four experiments.

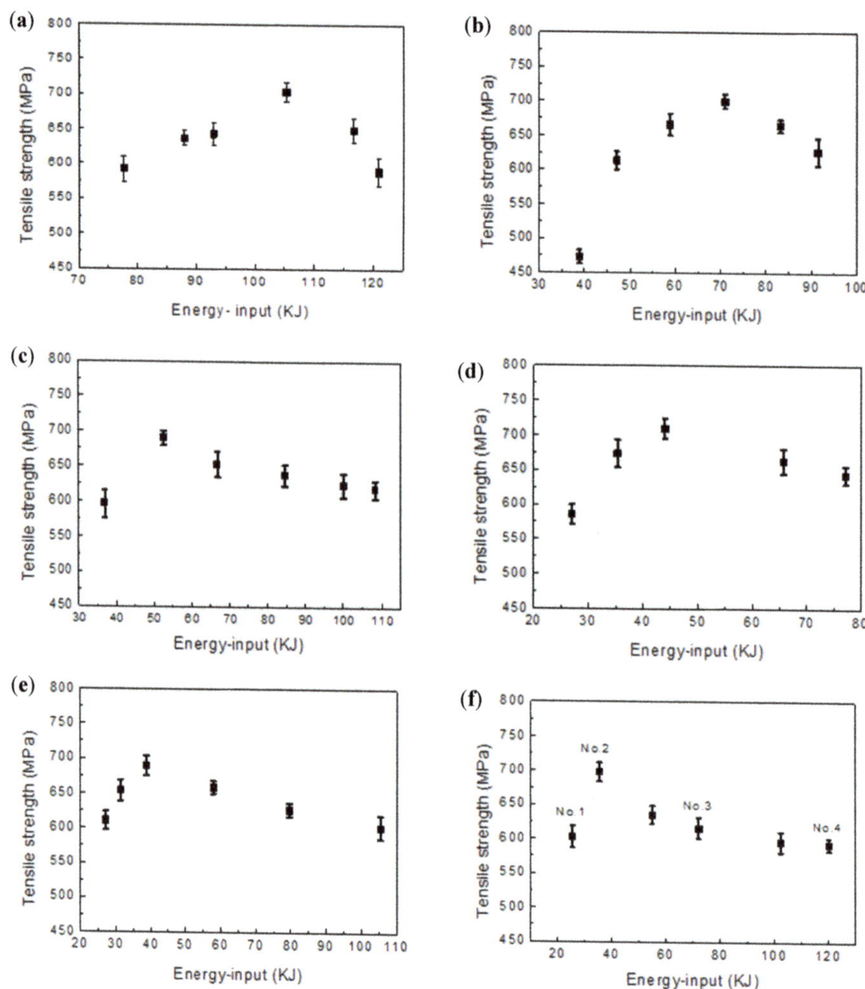

Figure 4. The variation trend of the 304SS RFW joint's tensile strength with energy-input under forging pressures of (**a**) 100 MPa, (**b**) 120 MPa, (**c**)140 MPa, (**d**) 160 MPa, (**e**) 180 MPa, and (**f**) 200 MPa. The bars associated with the tensile strength (the mean value of three tensile strength measurements) correspond to standard deviations.

Table 1. Experiment parameters and the joint's tensile strength under a forging pressure of 200 MPa.

No.	Friction Pressure (MPa)	Rotation Speed (rpm)	Friction Time (s)	Energy-Input (kJ)	Average Tensile Strength (MPa)
No.1	40	1200	4	25	603
No.2	40	500	9	35	698
No.3	120	800	7	55	635
No.4	160	1400	19	120	592

3.2. The Thermal Cycle of Different Joints

Figure 5 shows the evolution curves of the joint's temperature of No.1 to No.4 specimens. As shown in Figure 5a, the highest temperature of No.1 to No.4 specimens during the RFW process are 1100 °C, 939 °C, 1046 °C, and 1017 °C, respectively. On the other hand, the energy-input of No.1 to

No.4 is 25 kJ, 35 kJ, 55 kJ, and 120 kJ, respectively. We can see that the maximum temperature has no obvious variation regularity with the change of energy input, i.e., with the increase of energy-input, the maximum temperature has no obvious increase or decrease trend. In fact, the joint's temperature is mainly influenced by the rotational speed and welding pressure [11,15]. However, the energy-input determines the thermal cycle time of joint, i.e., the time from the weld beginning to the joint cooling to room temperature. As seen in Figure 5b, when the energy-input increases from 25 kJ to 120 kJ, the thermal cycle time increases from 478 s to 1157 s. The reason for this is that the more the energy-input, the wider the area of joint that is heated during the welding process, which results in a longer cooling time after welding.

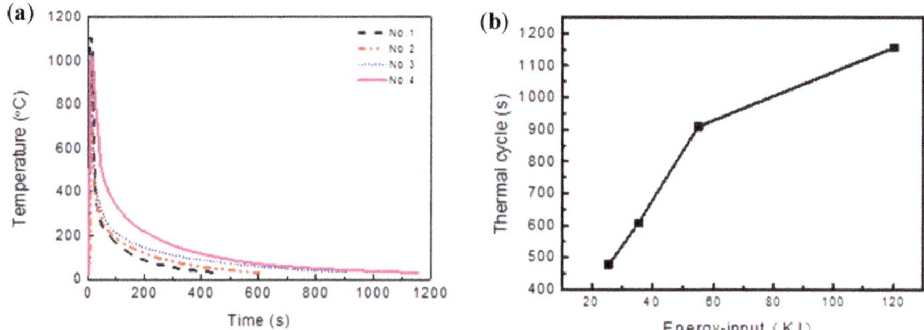

Figure 5. The evolution of (a) temperature during RFW process and (b) cycle time with energy-input.

3.3. Variation of Microstructure with Energy-Input

Figure 6 shows the microstructure of base metal (BM). The BM employed in this study is 304SS rods, of which the supply state is annealing treatment after rolling; thus, the initial microstructure consists of uniform equiaxed grains with an average size of 26 μm.

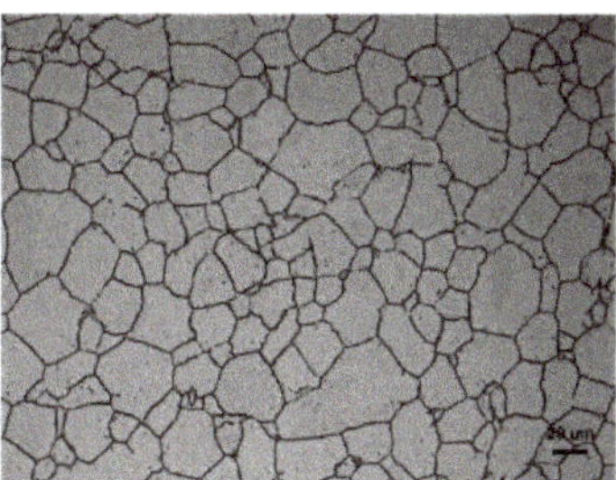

Figure 6. Microstructure of the 304SS base metal.

Figure 7 shows the typical microstructure morphology of a welded joint. Three zones consisting of weld zone (WZ), thermo-mechanically affected zone (TMAZ), and base metal (BZ) can be observed.

The WZ consists of equiaxed grains. The reason is that the WZ undergoes the most severe plastic deformation and the highest temperature thermal cycle compared to other zones, which results in sufficient dynamic recrystallization (DRX) occurring in the WZ. On other hand, partial DRX or no DRX occurs in the TMAZ, and a large degree of bending deformation occurs in the TMAZ that results in a clear radial flow direction existing in the microstructure.

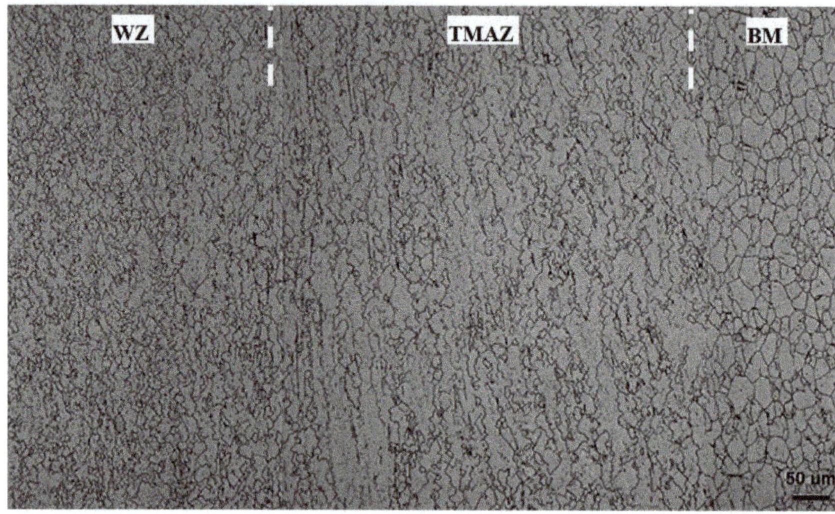

Figure 7. Typical microstructure cataloguing of 304SS joint welded by RFW.

Figure 8 shows the TMAZ microstructure of specimens No.1–No.4. The TMAZ grain sizes of specimens No.1–No.4 are 50 μm, 30 μm, 32 μm and 40 μm, respectively. It should be noted that the TMAZ grain size of specimen No.1 shows obvious coarsening (nearly two times that of the BM). This is due to the energy-input of No.1, which is insufficient to induce partial DRX in the TMAZ. As Figure 5 shows, specimen No.1 experiences the shortest thermal cycle. Thus, the TMAZ grains of specimen No.1 mainly undergo torsion and elongation, which cause grain coarsening, as Figure 8a shows. However, as Figure 8b shows, with the energy-input increasing, the average TMAZ grain size of specimen No.2 decreases to 30 μm. The main reason for the grain size reduction is that partial DRX occurs in the TMAZ under the effect of welding thermal and plastic deformation. As the energy-input further increases, the thermal cycle grows longer, and the grains of the TMAZ grow under the thermal effect.

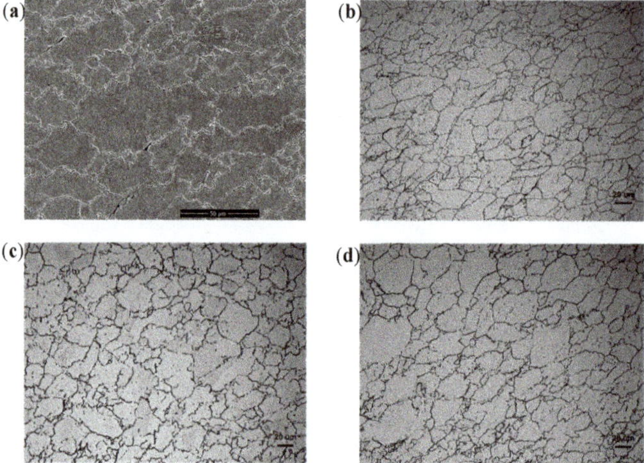

Figure 8. The thermo-mechanically affected zone (TMAZ) microstructure of specimens (**a**) No.1, (**b**) No.2, (**c**) No.3, and (**d**) No.4.

Figure 9 presents the WZ microstructure of different specimens. The average WZ grain size of specimens No.1–No.4 is 19 μm, 17 μm, 25 μm, and 40 μm, respectively. The WZ undergoes the most severe plastic deformation and the highest temperature thermal cycle compared to other zones, which results in adequate DRX occurring in this zone [16]. For specimens No.1 and No.2, the DRX results in the fine grain in the WZ. However, when the energy-input increases, the size of WZ grains increase due to the longer thermal cycle, which eventually results in the larger grains being generated in the WZ of specimen No.4.

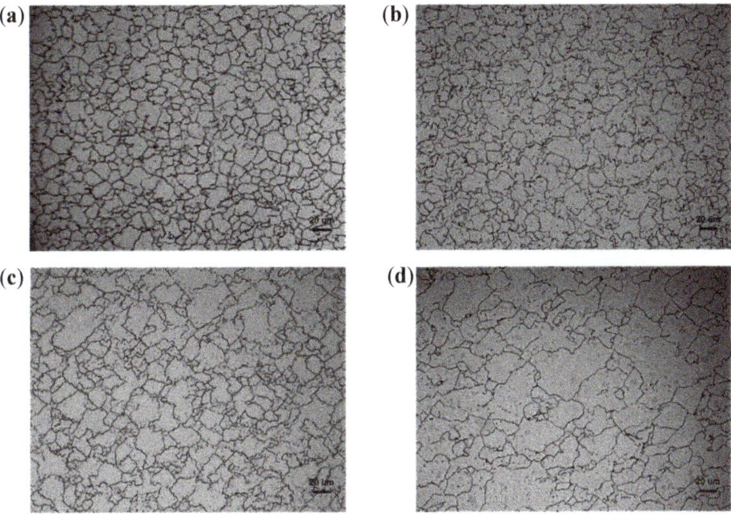

Figure 9. The weld zone (WZ) microstructure of specimens (**a**) No.1, (**b**) No.2, (**c**) No.3, and (**d**) No.4.

3.4. Hardness Distribution

The microhardness distribution throughout the welding surface from weld zone to base material was measured. Figure 10a illustrates the direction of microhardness measurement, the hardness of the WZ, TMAZ, HAZ, and BM, which were measured, respectively, for each joint. Figure 7b–e shows the microhardness distribution of different joints. The results show that the average hardness value of the BM is 213 Hv, and the average WZ hardness value of specimens No.1–No.4 were 252, 242, 230, and 218 Hv, respectively. It can be observed that the WZ hardness of all four specimens was higher than that of the BM. This is due to the strain hardening effect during the welding process [17]. In addition, the WZ of sample No.1 is the area with the highest hardness of the whole joint; the same result also exists in sample No.2. However, with the energy-input increasing, the highest hardness value of specimens No.3 and No.4 occurs in the TMAZ. This is due to the grain size of the WZ becoming large as the energy-input increases, as Figure 9 shows; moreover, the coarsening grains results in the decrease in hardness [18,19]. When the energy-input increases to 120 kJ, i.e., specimen No.4, the hardening of the WZ caused by strain is almost offset by the softening caused by grain growth [20]. Moreover, the plastic deformation in the TMAZ turns severely as the energy-input increases, which results in a more severe deformable hardening. It should be noted that an obvious softening phenomenon appeared in the TMAZ of specimen No.1. The lowest TMAZ hardness of specimen No.1 is 198 Hv, which decreases 7% compared with that of the BM. As discussed above, the average TMAZ grain size of specimen No.1 is nearly two times of that of the BM, which leads to a decrease in hardness. The low TMAZ hardness will make this zone a weak area of the joint, which we will discuss in next section. In addition, the TMAZ hardness of specimen No.2 is slightly higher than that of the BM. It also can be observed from Figure 10c that the hardness distribution of specimen No.2 is smooth and has no catastrophe point, which corresponds to the uniformity of the microstructure, as discussed in Section 3.2.

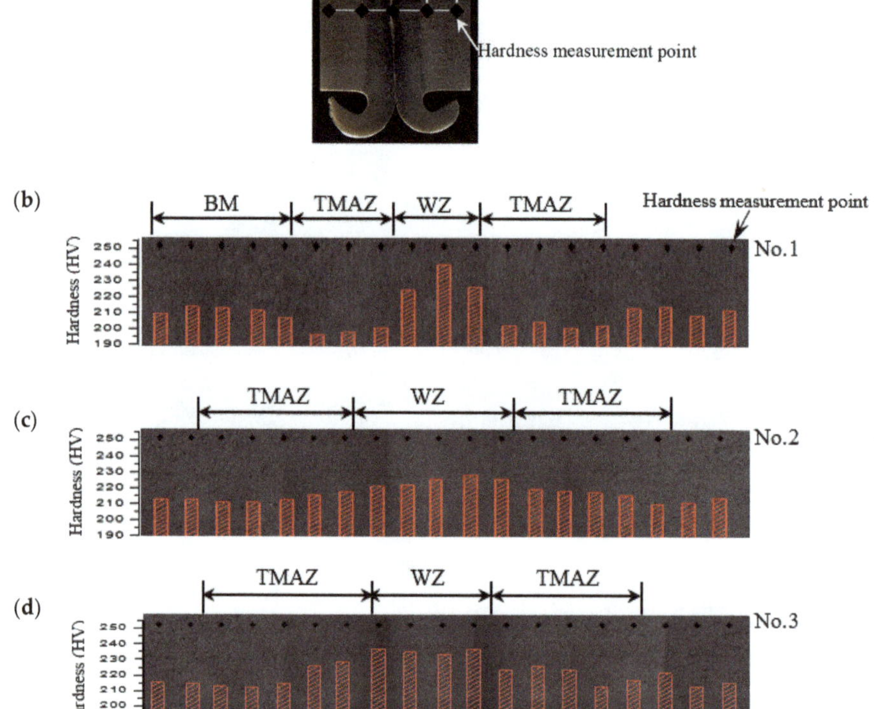

Figure 10. A diagram of the (**a**) microhardness test point and (**b**–**e**) microhardness distribution of specimens No.1–No.4.

3.5. The Relationship of Tensile Specimen Fracture Zone and Energy-Input

Figure 11 represents the morphology of the fracture tensile specimen and the microstructure near the fracture zone. From Figure 11a, it can be observed that the microstructure near the fracture zone of specimen No.1 is the elongated coarse grain. By comparing this microstructure to the TMAZ microstructure of specimen No.1, shown by Figure 8a, it is clear that the No.1 tensile specimen fractured at the TMAZ. As discussed above, the TMAZ average grain size of specimen No.1 is 50 μm, and the hardness of the TMAZ is 29% lower than that of BM. The coarse grain and low hardness lead to the TMAZ becoming the weak zone of joint No.1. When the energy-input increases, no obvious grain coarsening area and softening zone occurs in joint No.2, which gives the joint its excellent mechanical properties. Though the tensile specimen of No.2 also fractured at the TMAZ, the fracture surface of No.2 is a curvilinear form, and the tensile strength of joint No.2 is 698 MPa, which is 94% of that of the BM. As Figure 9c,d shows, the WZ grain size of joints No.3 and No.4 is obviously coarser than that of the BM, which results in the No.3 and No.4 tensile specimens fracturing at the WZ, as Figure 11c,d shows.

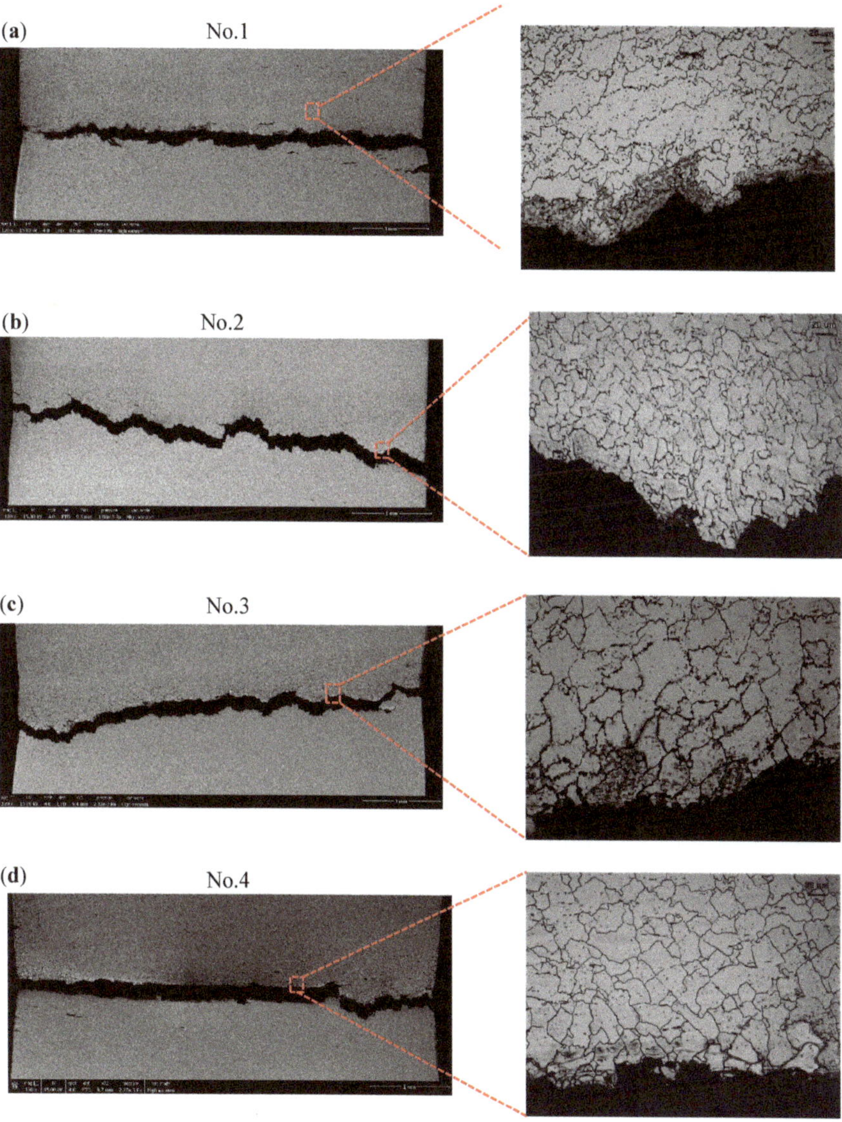

Figure 11. Photograph of macroscopic and microstructure of fracture tensile specimen.

4. Modeling of the Energy-Input

To establish a model for energy-input with welding parameters as variables, we must first fit the energy-input curve as a quadratic function, i.e.,

$$Q = \sum_{i=0}^{2} C_i t^i = C_0 + C_1 t + C_1 t^2 \tag{1}$$

in which Q is energy-input, t is welding time, and C_0, C_1, and C_2 are fitting coefficients.

Figure 12 shows the comparison of the calculated curves and experimentally measured curves of energy-input. The fitting equation described by Equation (1) can be seen to fit very well with the experimental data of energy-input. Table 2 lists the values of fitting coefficients C_0, C_1, and C_2 for different welding parameters.

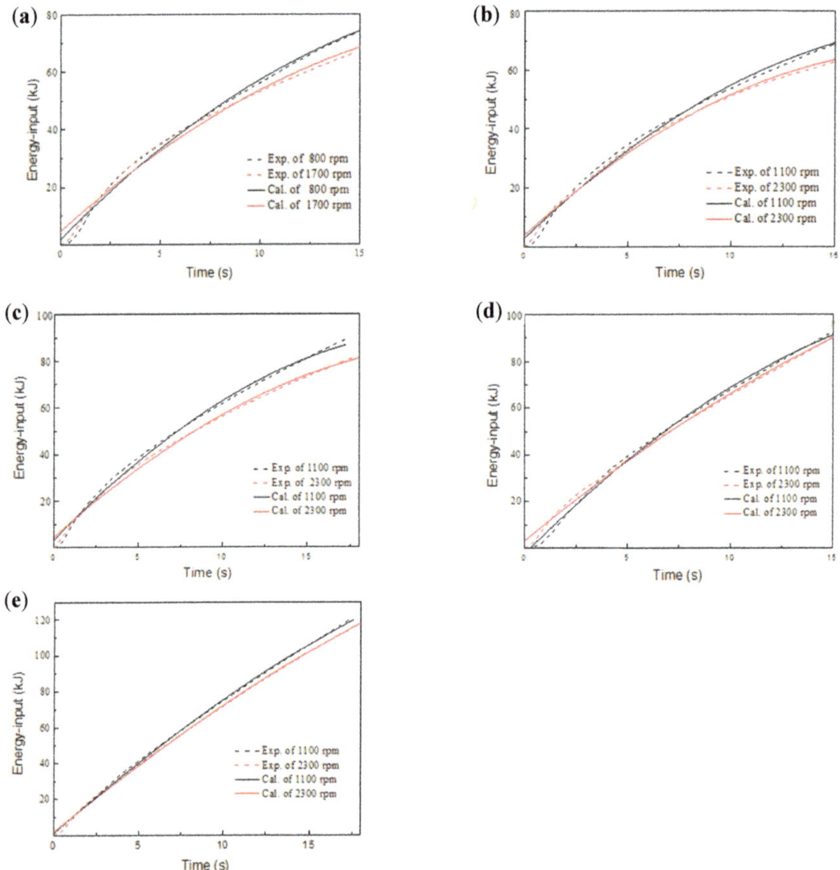

Figure 12. Comparison between calculated and experimental values of energy-input under different rotational speeds when welding pressure was constant at (**a**) 25 MPa, (**b**) 40 MPa, (**c**) 80 MPa, (**d**) 120 MPa, and (**e**) 160 MPa.

Table 2. The coefficient of different welding experiments.

P (MPa)	N (rpm)	C_0	C_1	C_2
25	500	−3969.91	7720.94	−160.97
	800	1792.98	6995.59	−144.81
	1100	1594.45	6956.77	−160.58
	1400	4350.03	6335.06	−137.91
	1700	4423.71	6233.98	−131.23
	2000	4319.92	5931.05	−121.72
40	500	−4249.60	7053.29	−128.14
	1100	2637.38	6802.22	−157.76
	1700	4130.76	6353.79	−149.41
	2300	3590.85	6437.81	−163.03
80	500	−3599.73	8111.83	−121.67
	800	2395.92	7738.01	−158.53
	1100	3134.80	7545.50	−156.31
	1400	3832.45	6973.78	−158.59
	1700	4268.07	6899.27	−131.95
	2300	4324.83	6509.14	−124.84
120	500	−4163.32	9006.37	−131.23
	800	−2473.33	8864.62	−174.24
	1100	1636.11	8242.70	−147.53
	1400	1745.32	7831.08	−133.24
	1700	3049.07	7313.38	−100.93
	2300	3238.76	3238.76	−73.20
160	1100	825.70	8365.38	−90.80
	1400	897.32	8165.46	−80.14
	1700	1452.71	7813.79	−74.77
	2300	1436.52	6701.22	−15.79
200	1100	359.98	9759.89	−94.88
	1400	140.55	7950.32	−16.90

The equation of C_0, C_1, and C_2, in which the variables are welding pressure and rotation speed, can be obtained by regression analysis of the data in Table 2, i.e.,

$$C_0 = -2305 - 13.20P + 3.69N \tag{2}$$

$$C_1 = 8237 + 12.31P - 1.58N \tag{3}$$

$$C_2 = -219 + 0.49P + 0.04N \tag{4}$$

Taking Equations (2)–(4) into account in Equation (1), the energy-input can be expressed as a formula of welding pressure, welding rotational speed, and welding time, i.e.,

$$Q = (-2305 - 13.20P + 3.69N) + (8237 + 12.31P - 1.58N)t + (-219 + 0.49P + 0.04N)t^2 \tag{5}$$

To verify the accuracy of the model, extra welds were made with new parameters. Figure 13 shows the comparison of the calculated and experimental curves. As the results show, the model closely predicts the actual energy-input data. As the weld experiments of Figure 13 were made with parameters that are outside of those used in creating the model, they demonstrate that the model accurately predicts the energy-input beyond the initially investigated parameters.

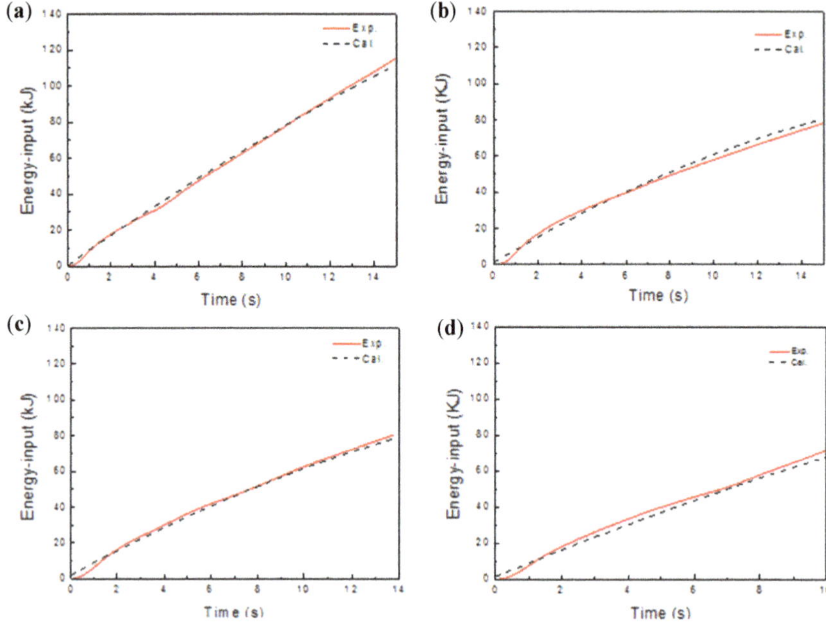

Figure 13. Comparison between calculated and experimental values of energy-input under welding parameters of (**a**) p = 200 MPa, N = 1400 rpm; (**b**) p = 80 MPa, N = 1200 rpm; (**c**) p = 100 MPa, N = 1500 rpm; and (**d**) p = 140 MPa, N = 1500 rpm.

As discussed in Section 3.1, there is an optimal energy-input value that causes the joints to have the maximum tensile strength under different forging pressures. Figure 14 shows the variation trend of optimal energy-input for different forging pressures. The results indicate that the larger the forging pressure, the less the optimal energy-input. The optimal energy-input decreases from 105 kJ to 35 kJ when the forging pressure rises from 100 MPa to 200 MPa. The deformation is a necessary factor for forming a RFW joint [21]. When the forging pressure is large, the joint is easily deformed, which means a small amount of energy-input could obtain a stable joint [14,22]. As seen in Figure 14, a three-order function fits the experimental data very well, which is in the following form:

$$Q = 690545 - 10642.84\,p_f + 58.96\,p_f{}^2 - 1106 \times 10^{-4}\,p_f{}^3 \tag{6}$$

in which p_f is the forging pressure.

Extra experiments under a forging pressure of 170 MPa, which are not included in those used in establishing the model, were made to verify the accuracy of Equation (6). Firstly, the optimal energy-input for a forging pressure of 170 MPa was calculated based on Equation (6), which was 41 kJ. Subsequently, two welding parameters (welding pressure, and speed and time) were randomly set, and the other parameter can be calculated according to Equation (5). In this study, three group experiments were made to verify the calculated optimal energy-input accuracy or otherwise disprove it. The welding pressure and rotational speed of first group was 100 MPa and 1300 rpm, respectively, and the calculated welding time was 6.1 s. The welding pressure and welding time of the second group was 40 MPa and 8 s, respectively, and the calculated rotational speed was 2000 rpm. The rotational speed and welding time of third group was 800 rpm and 5 s, respectively, and the calculated welding pressure was 160 MPa. Table 3 shows the welding parameters and tensile strength of these three verified experiments. As shown in Table 3, the strength of all the three joints was over 90% that of the BM. It is proven that the optimal energy-input model (Equation (6)) is accurate.

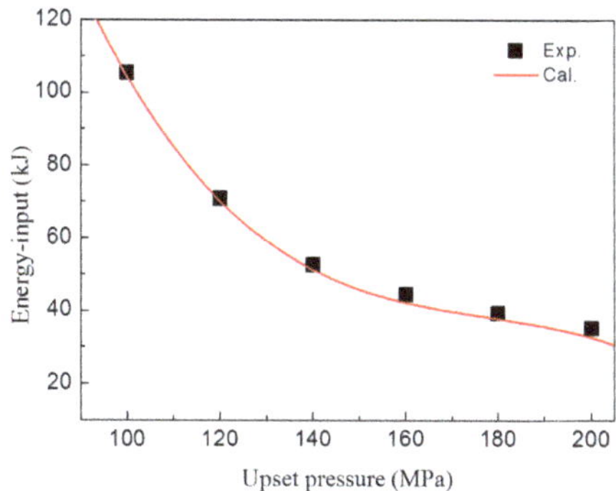

Figure 14. Variation trend of optimum energy-input with forging pressure.

Table 3. Welding parameters and tensile strength of verification experiments.

No.	Friction Pressure (MPa)	Rotation Speed (rpm)	Friction Time (s)	Tensile Strength (MPa)
No.5	100	1300	6	93%
No.6	40	2000	8	90%
No.7	160	800	5	96%

5. Conclusions

In this study, the effect of energy-input on the mechanical properties of RFW 304SS joints was investigated. RFW experiments were conducted over a wide range of welding parameters. Based on the results and discussion above, the conclusions are as follows:

(1) When forging pressure is kept constant, the tensile strength increases with the increase of energy-input until reaching a maximum and then decreases.

(2) Though the maximum temperature has no obvious variation regularity with the change of energy-input, the thermal cycle has a positive correlation with the increase of energy-input, i.e., as energy-input increases, the duration time from the commencement of welding to joint cooling to room temperature increases.

(3) An empirical model was established that describes energy-input as a function of the welding parameters. The accuracy of the model was verified by extra RFW experiments.

(4) An empirical model for the optimal energy-input of different forging pressures was obtained. Then, the optimal energy-input for 170 MPa forging pressure was calculated. Three group experiments were made based on the calculated energy-input value. The joints' tensile strength coefficients of these three experiments were 93%, 90%, and 96%, respectively. The results proved the accuracy of the model for optimal energy-input.

Author Contributions: J.L., F.Z. and G.W. conceived and designed the experiments. G.W. and W.W. carried out the experiments. J.X. and G.W. analyzed the experimental data. G.W. wrote the manuscript. J.L., F.Z. and J.X. reviewed the manuscript.

Funding: This work is supported by the National Natural Science Foundation of China (Grand No: 51475376 and 51575451).

Conflicts of Interest: The authors declare no conflict of interest.

References

1. Bouarroudj, E.; Chikh, S.; Abdi, S.; Miroud, D. Thermal analysis during a rotational friction welding. *Appl. Therm. Eng.* **2017**, *110*, 1543–1553. [CrossRef]

2. Li, W.Y.; Vairis, A.; Preuss, M.; Ma, T.J. Linear and rotary friction welding review. *Int. Mater. Rev.* **2016**, *61*, 71–100. [CrossRef]

3. Maalekian, M. Friction welding–critical assessment of literature. *Sci. Technol. Weld Join.* **2007**, *12*, 738–759. [CrossRef]

4. Li, P.; Li, J.L.; Li, X.; Xiong, J.T.; Zhang, F.S.; Liang, L. A study of the mechanisms involved in initial friction process of continuous drive friction welding. *J. Adhes. Sci. Technol.* **2015**, *29*, 1246–1257. [CrossRef]

5. Sahin, M. Evaluation of the joint-interface properties of austenitic-stainless steels (AISI 304) joined by friction welding. *Mater. Des.* **2007**, *28*, 2244–2250. [CrossRef]

6. Sathiya, P.; Aravindan, S.; Haq, A.N. Some experimental investigations on friction welded stainless steel joints. *Mater. Des.* **2008**, *29*, 1099–1109. [CrossRef]

7. Ates, H.; Turker, M.; Kurt, A. Effect of friction pressure on the properties of friction welded MA956 iron-based superalloy. *Mater. Des.* **2007**, *28*, 948–953. [CrossRef]

8. Li, W.Y.; Wang, F.F. Modeling of continuous drive friction welding of mild steel. *Sci. Eng. A-Struct.* **2011**, *528*, 5921–5926. [CrossRef]

9. Chander, G.S.; Reddy, G.M.; Rao, A.V. Influence of Rotational Speed on Microstructure and Mechanical Properties of Dissimilar Metal AISI 304-AISI 4140 Continuous Drive Friction Welds. *J. Iron Steel Res. Int.* **2012**, *19*, 64–73. [CrossRef]

10. Palanivel, R.; Laubscher, R.F.; Dinaharan, I. An investigation into the effect of friction welding parameters on tensile strength of titanium tubes by utilizing an empirical relationship. *Measurement* **2017**, *98*, 77–91. [CrossRef]

11. Maalekian, M.; Kozeschnik, E.; Brantner, H.P.; Ceriak, H. Comparative analysis of heat generation in friction welding of steel bars. *Acta Mater.* **2008**, *56*, 2843–2855. [CrossRef]

12. Guo, W.; You, G.Q.; Yuan, G.Y.; Zhang, X.L. Microstructure and mechanical properties of dissimilar inertia friction welding of 7A04 aluminum alloy to AZ31 magnesium alloy. *J. Alloy Compd.* **2017**, *695*, 3267–3277. [CrossRef]

13. Xiong, J.T.; Zhou, W.; Li, J.L.; Zhang, F.S.; Huang, W.D. The thermodynamic analytical models for steady-state of linear friction welding based on the maximum entropy production principle. *Mater. Des.* **2017**, *129*, 53–62. [CrossRef]

14. Sahin, M. Characterization of properties in plastically deformed austenitic-stainless steels joined by friction welding. *Mater. Des.* **2009**, *30*, 135–144. [CrossRef]

15. Xiong, J.T.; Li, J.L.; Wei, Y.N.; Zhang, F.S.; Huang, W.D. An analytical model of steady-state continuous drive friction welding. *Acta Mater.* **2013**, *61*, 1662–1675. [CrossRef]

16. Hazra, M.; Rao, K.S.; Reddy, G.M. Friction welding of a nickel free high nitrogen steel: Influence of forge force on microstructure, mechanical properties and pitting corrosion resistance. *J. Mater. Res. Technol.* **2014**, *3*, 90–100. [CrossRef]

17. Li, P.; Li, J.L.; Salman, M.; Liang, L.; Xiong, J.T.; Zhang, F.S. Effect of friction time on mechanical and metallurgical properties of continuous drive friction welded Ti6Al4V/SUS321 joints. *Mater. Des.* **2014**, *56*, 649–656. [CrossRef]

18. Rajasekhara, S.; Ferreira, P.J.; Karjalainen, L.P.; Kyrolkainen, A. Hall–Petch behavior in ultra-fine-grained AISI 301LN stainless steel. *Metall. Mater. Trans. A* **2007**, *38*, 1202–1210. [CrossRef]

19. Di Schino, A.; Kenny, J.M. Grain refinement strengthening of a micro-crystalline high nitrogen austenitic stainless steel. *Mate. Lett.* **2003**, *57*, 1830–1834. [CrossRef]

20. Fu, H.H.; Benson, D.J.; Meyers, M.A. Analytical and computational description of effect of grain size on yield stress of metals. *Acta Mater.* **2001**, *49*, 2567–2582. [CrossRef]

21. Mori, K.I.; Bay, N.; Fratini, L.; Micari, F.; Tekkaya, A.E. Joining by plastic deformation. *CIRP Ann-Manuf. Technol.* **2013**, *62*, 673–694. [CrossRef]

22. Uday, M.B.; Ahmad Fauzi, M.N.; Zuhailawati, H.; Ismail, A.B. Advances in friction welding process: A review. *Sci. Technol. Weld. Join.* **2010**, *15*, 534–558. [CrossRef]

Article

Dissimilar Materials Joining of Carbon Fiber Polymer to Dual Phase 980 by Friction Bit Joining, Adhesive Bonding, and Weldbonding

Yong Chae Lim [1], Hoonmo Park [2], Junho Jang [2], Jake W. McMurray [1], Bradly S. Lokitz [3], Jong Kahk Keum [3,4], Zhenggang Wu [1] and Zhili Feng [1,*]

[1] Materials Science and Technology Division, Oak Ridge National Laboratory, Oak Ridge, TN 37830, USA; limy@ornl.gov (Y.C.L.); mcmurrayjw1@ornl.gov (J.W.M.); wuz2@ornl.gov (Z.W.)

[2] Central Advanced Research and Engineering Institute, Hyundai Motor Company, Uiwang, Gyeonggi 16802, Korea; hmpark@hyundai.com (H.P.); junhojang@hyundai.com (J.J.)

[3] Center for Nanophase Materials Science, Oak Ridge National Laboratory, Oak Ridge, TN 37830, USA; lokitzbs@ornl.gov (B.S.L.); keumjk@ornl.gov (J.K.K.)

[4] Neutron Scattering Division, Oak Ridge National Laboratory, Oak Ridge, TN 37830, USA

* Correspondence: fengz@ornl.gov; Tel.: +1-865-576-3797

Received: 11 September 2018; Accepted: 19 October 2018; Published: 24 October 2018

Abstract: In the present work, joining of a carbon fiber-reinforced polymer and dual phase 980 steel was studied using the friction bit joining, adhesive bonding, and weldbonding processes. The friction bit joining process was optimized for the maximum joint strength by varying the process parameters. Then, the adhesive bonding and weld bonding (friction bit joining plus adhesive bonding) processes were further developed. Lap shear tensile and cross-tension testing were used to assess the joint integrity of each process. Fractured specimens were compared for the individual processes. The microstructures in the joining bit ranged from tempered martensite to fully martensite in the cross-section view of friction bit-joined specimens. Additionally, the thermal decomposition temperature of the as-received carbon fiber composite was studied by thermogravimetric analysis. Fourier-transform infrared–attenuated total reflectance spectroscopy and X-ray diffraction measurements showed minimal variations in the absorption peak and diffraction peak patterns, indicating insignificant thermal degradation of the carbon fiber matrix due to friction bit joining.

Keywords: dissimilar material joining; carbon fiber-reinforced polymer; dual-phase steel; friction bit joining; adhesive bonding; weld bonding; mechanical strength

---

1. Introduction

Achieving lightweight, multi-material auto body structures is a critical goal for the automotive industry to comply with government regulations (i.e., to improve fuel efficiency and reduce greenhouse gas emissions) [1,2]. Currently, four types of lightweight materials—high-strength aluminum alloys, magnesium alloys, ultra-high-strength/advanced high-strength steels (AHSSs), and polymer composites (i.e., carbon fiber-reinforced polymers (CFRPs) and glass fiber-reinforced polymers)—have been identified as substitutes for current steel and/or cast iron auto body structures. The selection of lightweight materials should be carefully explored to produce multi-material structures while satisfying the structural stability and safety performance requirements of vehicles.

Although various combinations of the four candidate materials are possible for potential lightweight vehicles, one of the most interesting material combinations is CFRPs and AHSSs. This is because CFRPs have high mechanical strength with a high strength-to-weight ratio [3,4]. Additionally, a high-strength dual-phase (DP) steel with good mechanical properties [5,6] could be used as a thinner-gauge substitute in existing vehicle structures, leading to substantial weight reductions. However, because of the physical and chemical dissimilarities of individual materials, the development of suitable joining technologies is essential for enabling multi-material auto bodies. Recently, extensive research has been conducted for joining carbon fiber composites to various metals (e.g., aluminum or magnesium alloys), using laser welding [7,8], friction stir blind riveting [9], friction spot joining [10–12], ultrasonic welding [13], adhesive bonding [14], friction lap welding [15,16], friction-based injection clinching joining [17], self-piercing riveting [18,19], and hybrid bonding (adhesive + mechanical fastening) [20–22]. In addition, very limited work has been focused on joining carbon fiber composites to steel [8,14,23,24]. However, the hole clinching and self-piercing rivet processes have been shown to be difficult to apply to AHSSs (σ_{TS} > 780 MPa) [25–27]. For this reason, no previous study has attempted to join carbon fiber composites to DP980, potentially because of the limitations of the joining mechanism for each individual process. Although adhesive bonding can be used for CFRPs and different grades of AHSSs, weak peel strength and environmental degradation are key concerns [28]. Therefore, none of the joining technologies mentioned is an easy or the sole way to join a carbon fiber composite and DP980.

In the present work, the authors applied the friction bit joining (FBJ) process to spot-weld carbon fiber composites to DP980. Initially, various welding parameters were used to achieve the highest lap joint strength for FBJ specimens. Next, adhesive bonding and weldbonding (FBJ + adhesive) processes were further developed for the selected materials. The joint integrity for each process was assessed by lap shear tensile and cross-tension testing. Fractured samples were observed and compared for the individual processes. Vickers microhardness was measured on the cross-sectioned FBJ samples to characterize hardness profile changes due to the evolution of microstructures during the joining process. Optical and scanning electron microscopy were used to analyze the interfaces between the joining bit and the CFRP. Additionally, thermogravimetric analysis (TGA) was conducted to study the thermal decomposition temperature of the as-received CFRP. In a cross-sectioned FBJ specimen, Fourier transform infrared–attenuated total reflectance (FTIR-ATR) spectroscopy and X-ray diffraction (XRD) were used to study the thermal degradation of the CFRP close to the joining bit.

2. Materials and Methods

For the study, the authors purchased a commercially available carbon fiber composite with 2.0 mm-thick epoxy resin (TB Carbon PN.008, T_g = 120 °C) laminate reinforced with 34–36 wt% of unidirectional carbon fibers (CP150NS, TB Carbon, Yangsan-si, Korea), which was used as a top sheet material. The physical properties of carbon fiber are summarized in Table 1. The stacking sequence of the carbon fiber layers followed the configuration [0°/45°/90°] with four plies. The tensile strength of carbon fiber composites was measured at 655 MPa. A 1.2 mm-thick DP980 was used as a bottom sheet material. Joining bits were fabricated using the American Iron and Steel Institute (AISI, Washington, DC, USA) 4140 alloy steel. The joining bit head was a hexagonal shape with 9.525 mm in width. The shank diameter and length of joining bit were 6.6 and 4.57 mm, respectively.

Nominal chemical compositions and mechanical properties of each material are provided in Tables 2 and 3. Each material was cut and prepared with coupon dimensions of 100 mm in length × 25 mm in width for lap joint specimens. The overlap area for the lap joint configuration was 25 × 25 mm². A cross-tension specimen composed of carbon fiber composite and DP980 had coupon dimensions of 150 mm in length × 50 mm in width [25]. The hole size in the cross-tension specimen was 20 mm. The overlap area for the cross-tension specimen was 50 by 50 mm². Before the materials were joined, acetone and isopropyl alcohol were used to remove any dirt and grease on both material surfaces.

Table 1. Physical properties of carbon fiber.

Grade	Diameter (μm)	Density (kg/m^3)	Carbon Fiber Weight (g/m^2)	Tensile Strength (MPa)	Tensile Modulus (GPa)
24 ton, 12 K	6–8	1740–1900	150	2250	127

Table 2. Nominal chemical compositions of DP980 and American Iron and Steel Institute (AISI) 4140 (wt%).

Element	C	Mn	Si	Cr	S	P	Ni	B	Mo	V	N	Cu	Cb	Nb	Sn	Ti	Al
DP980	0.113	2.147	0.965	0.380	0.003	0.014	0.01	0.0004	0.05	–	0.004	0.01	–	0.002	0.001	0.003	0.038
AISI 4140	0.42	0.84	0.25	0.99	0.02	0.017	0.16	–	0.16	0.003	–	0.003	0.003	–	–	–	0.021

Table 3. Mechanical properties of DP980 and AISI 4140.

Material	Yield Strength (MPa)	Tensile Strength (MPa)	Elongation (%)
DP980	703	1009	16
AISI 4140	906	1010	18

An STA 449 F1 Jupiter simultaneous thermal analyzer (Netzsch Gerätebau-GmbH, Selb, Germany) was used to investigate the thermochemistry of the as-received carbon fiber composite. Differential scanning calorimetry (DSC) was coupled with thermogravimetry with the aim of determining incipient decomposition temperatures. Disc-shaped specimens taken from the as-received carbon fiber composite, weighing 61.98 mg with a diameter of 6.4 mm and a thickness of around 2 mm, were prepared to fit snugly and flush to the bottom of platinum crucibles supplied by Netzsch (Selb, Germany). During FBJ, the peak temperature in the joining bit can be expected to exceed the austenite transition temperature (A_3) of steels [29]. For this reason, the maximum temperature was set at 1000 °C for the thermogravimetry measurements. The samples were heated at a constant ramp rate of 5 °C/min to 1000 °C in flowing ultra-high-purity argon gas at 100 cc/min. The change in weight and a microvolt signal, due to the temperature difference between the sample and the reference crucible, were recorded as functions of temperature.

FBJ is a two-stage joining process [30,31]. In the first stage (Figure 1b), the joining bit cuts and plunges into the top material using a spindle speed of 2000 rev·min^{-1} and a plunge rate in the z-direction of 171.5 mm·min^{-1}. In the second stage (Figure 1c), the joining bit is bonded to the bottom steel sheet by frictional heat and z-direction axial force. Various welding conditions were tested to achieve the highest lap shear strength. After preliminary tests, the total plunge depth by the joining bit and joining feed rate were fixed at 4.32 mm and 171.5 mm·min^{-1}, respectively. Then, different joining speeds in the range from 1500 to 2500 rev·min^{-1} were used to achieve the maximum lap joint strength. Figure 2 is a schematic view of the lap joint-configured sample holder, tool holder, joining bit, and clamping parts used for the present work.

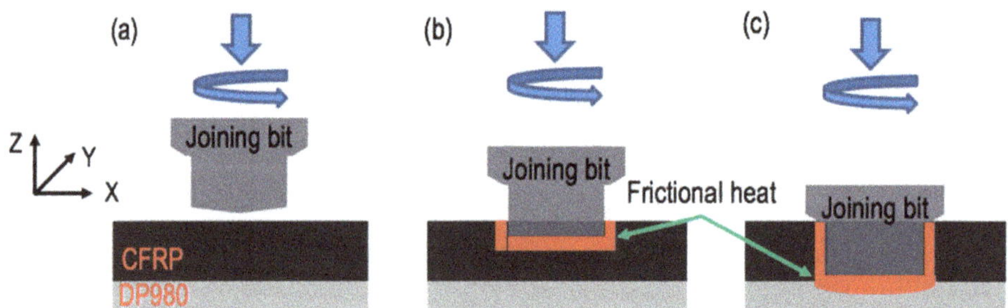

Figure 1. A schematic of the friction bit joining (FBJ) process: (**a**) initial stage, (**b**) cutting and plunge stage, and (**c**) joining stage.

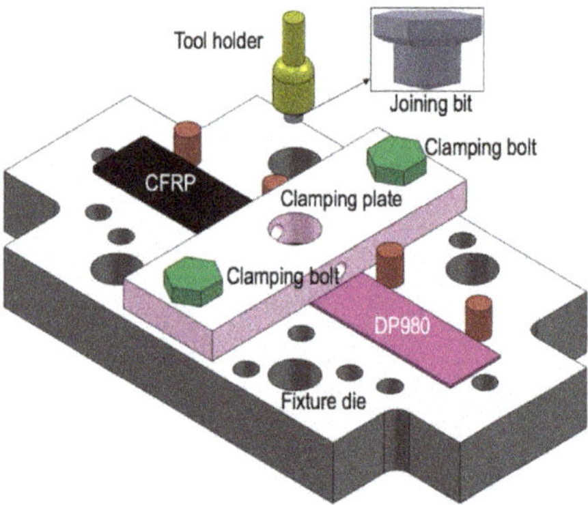

Figure 2. Schematic view of the lap shear joint-configured sample holder and the clamped parts used in this study.

A commercially available adhesive, 3M Scotch-weld Epoxy DP460, was used for the adhesive bonding and the weld bonding (FBJ + adhesive) processes. An applicator with a mixing nozzle that supplies a 2:1 mixing ratio of resin to a hardener was used. A 250 µm glass bead was applied in the joint area to maintain a constant adhesive bond line thickness. Then the adhesive was cured in an oven for 1 h at 60 °C, as recommended by the manufacturer. For weldbonding, FBJ was conducted on a cured specimen with the same welding process parameters.

Lap shear tensile tests were conducted using an MTS hydraulic test frame at a cross head rate of 10 mm·min^{-1} and at room temperature. Spacers (25×25 mm^2) were used to grip the samples to prevent bending of the specimen during the lap shear test. The gripping area on the both ends was 25×25 mm^2. Similarly, cross-tension tests were conducted with the same cross head rate as used for the lap shear tensile tests and at room temperature. Each mechanical test was triplicated for each joining process.

For cross-sectional analysis, FBJ samples were cross-sectioned and then mounted in epoxy resin. Then the samples were finely polished using a common metallographic procedure. A 5% Nital solution was used to etch the sample surface to reveal the microstructure. An optical microscope (Nikon Epiphot, Nikon, Tokyo, Japan) and a high-resolution scanning electron microscope (SEM) (JELO 6500 FEG-SEM, Hitachi High-Technologies, Krefeld, Germany) were used to characterize the cross sections of FBJ specimens. A Leco LM100 AT (Leco, Saint Joseph, MI, USA) was used to measure the Vickers microhardness of FBJ specimens with the following conditions: a 500 g load, 13 s of dwell time, and 250 µm spacing.

XRD patterns of a thermally affected carbon fiber composite directly mated with DP980 and the as-received CFRP were recorded using a PANalytical X'Pert Pro-MPD powder diffractometer (Malvern Panalytical, Amelo, The Netherlands) equipped with a Si-based position-sensitive one-dimensional detector and a nickel-filtered copper Kα radiation source. For the measurements, X-ray was generated at 45 kV/40 mA at a beam wavelength of $\lambda = 1.5416 \times 10^{-10}$ m (copper Kα radiation). The XRD measurements of the epoxy/carbon fiber multilayer were carried out in the reflection mode; hence, the XRD probed the surface morphology of the outer epoxy layer.

FTIR-ATR spectroscopy (Bruker Optik GmbH, Ettlingen, Germany) with a Bruker Lumos FTIR microscope in the ATR mode was used to probe the surface of the epoxy composite in close proximity to the joining bit. The spectra represented an average of 128 scans with a spectral resolution of 4 cm^{-1}. Spectra from five different spots with a 300 µm spacing over a 1000 µm distance were obtained using a line scan that originated at the interface between the joint bit and the carbon fiber composite. A spectrum for the as-received carbon fiber composite was also collected as a reference.

3. Results and Discussion

3.1. Thermogravimetry and Differential Scanning Calorimetry Measurements for the As-Received Carbon Fiber Composites

The results of the thermogravimetry-coupled DSC analysis of the as-received CFRP sample are plotted in Figure 3. The STA 449 F1 instrument and software use the fundamentals of heat-flux DSC to determine both reaction temperatures and the associated enthalpy (ΔH_{rxn}). Figure 3 indicates the as-received CFRP tested is thermally stable up to 290.5 °C. The mass of the sample precipitously dropped immediately after the beginning of a sharp exothermic DSC signal. The sharp exothermic peak could be due to interactions among leftover elements that might occur during thermal degradation of the epoxy.

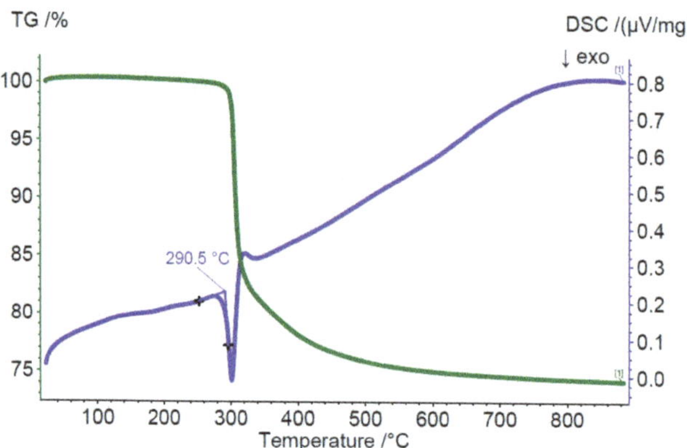

Figure 3. The Thermogravimetry (TG) (green) and Differential Scanning Calorimetry (DSC) (blue) signals from thermal analysis of the as-received carbon fiber reinforced polymer (CFRP) sample showing an exothermic reaction accompanied by mass loss at 290.5 °C.

3.2. Lap Shear Tensile Testing and Fractorgraphy

The assembly for lap shear testing of the FBJ specimen is presented in Figure 4a. Lap shear tensile testing was conducted to determine the static mechanical strength of the FBJ joints under optimized welding conditions, as already summarized in Table 4.

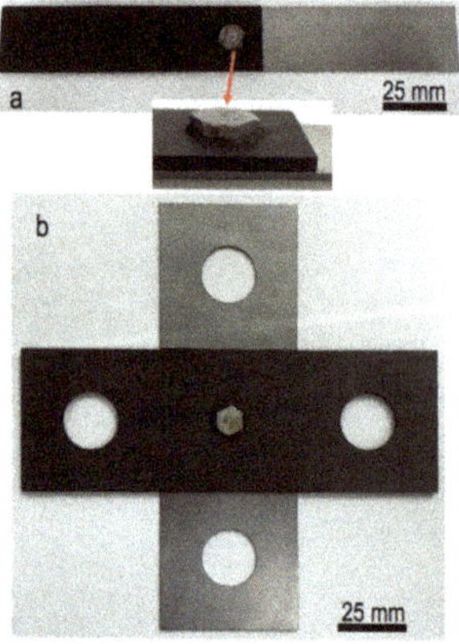

Figure 4. Assembled FBJ specimen: (**a**) lap shear specimen and (**b**) cross-tension specimen.

Table 4. FBJ welding process parameters for joining of CFRP to DP980.

Plunge Speed (rev·min^{-1})	Plunge Feed Rate (mm·min^{-1})	Joining Speed (rev·min^{-1})	Joining Feed Rate (mm·min^{-1})
2000	171.5	2100	171.5

A summary of the average lap joint failure loads and lap shear strengths for each process is plotted in Figure 5. A maximum lap joint failure load of 6.4 kN was achieved for FBJ specimens. The average lap shear failure load was 6.0 kN when three samples were repeatedly tested. The average lap shear failure loads for adhesive-bonded and weld-bonded specimens were 14.8 and 13.3 kN, respectively. For weld-bonded coupons, a small reduction in the lap joint strength (approximately 10%), compared with the adhesive-bonded coupons, was attributed to thermal degradation of the cured adhesive on the periphery of the joining bit due to frictional heating. However, the lap joint failure load of the weld-bonded specimen was almost twice higher than the failure load for the FBJ-only coupon, indicating the mechanical joint strength was improved because the load distribution was shared between the adhesive bonding and FBJ.

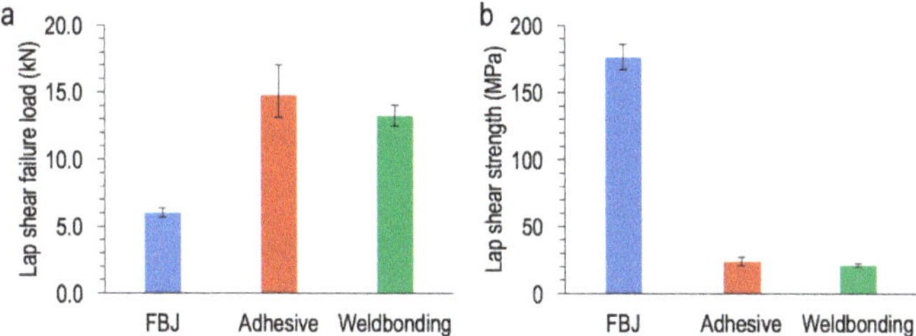

Figure 5. (**a**) Average lap shear failure load for FBJ, adhesive, and weld-bonded specimens. (**b**) Average lap shear strength of FBJ, adhesive-, and weld-bonded specimens.

Similarly, Squires et al. [29] observed an improvement in the mechanical joint strength for a high-strength aluminum alloy joined to steel by weld bonding. Di Franco et al. [32] also found increased joint strength when self-pierce riveting was combined with adhesive bonding to join a carbon fiber composite and aluminum sheet AA2024-T6. In addition, the presence of an adhesive layer in the joint area can effectively act as an insulator or a barrier against corrosion media, resulting in improvement in the corrosion resistance [33–35]. For this reason, weld bonding of carbon fiber composites to lightweight metals may have the highest success rates for protecting against corrosion under environmental conditions and for improving mechanical joint performance.

Because of the different joining processes, joint dimensions, and material combinations, a direct comparison of FBJ, adhesive-bonded, and weld-bonded joint strength with data from the available literature is not possible. For this reason, the maximum lap shear strength for the FBJ coupon was calculated based on the shank diameter (6.6 mm) of the joining bit, yielding a value of 186 MPa. Maximum lap shear strengths of 27.3 and 22.5 MPa were calculated for the adhesive bonding and weld bonding cases, respectively, by using the overlap area (25×25 mm^2) in the lap joint. This normalized calculation has been used as a common practice [11,36]. Table 5 allows readers to compare the results from this study with values for current state-of-the-art techniques for joining polymers to metals. Again, direct comparison of different processes would be not applicable, as previously explained, but general information of various joining techniques can be beneficial to readers. Overall, the results from this study show higher or comparable lap shear strengths, compared with the strengths demonstrated by other joining technologies.

Table 5. Summary of state of the art for joining polymers to metals.

Joining Process	Polymer/Metal (Thickness, mm)	Maximum Lap Shear Failure Load (kN)	Maximum Lap Shear Strength (MPa)	Joining Cycle Time (s)	Reference
Friction bit joining	CFRP (2.0)/DP980 (1.2)	6.4	186	1.51	Present
Adhesive bonding	CFRP (2.0)/DP980 (1.2)	17	27.3	Nonapplicable	Present
Weld bonding (FBJ + adhesive)	CFRP (2.0)/DP980 (1.2)	14	22.5	Nonapplicable	Present
Laser welding	CFRP-PA6 (3.0)/Galvanized steel (0.7)	3.3	Nonapplicable	1.54	[8]
Friction stir blind riveting	CFRP-PA6/6T-CF30 (3.0)/AA6111 (0.9)	3.4	Nonapplicable	3.9	[9]
Friction stir refill welding	CFRP-PPS (2.1)/Mg AZ31B-O (2.0)	2.13	21.8	8	[10]
Friction stir refill welding	CFRP-PPS (2.17)/AA2024-T3 (2.0)	1.28 ± 0.18	27 ± 2.8	4.8	[11]
Friction stir refill welding	CFRP-PPS (2.17)/AA6181-T4 (1.5)	3.52 ± 0.53 (double lap shear)	Nonapplicable	6	[12]
Ultrasonic welding	CFRP-PA66 (2.0)/AA5754 (1.0)	2.46	25	Nonapplicable	[13]
Adhesive bonding	CFRP (8.0 mm)/Marine grade steel (8.0 mm)	8.5	14.1	Nonapplicable	[14]
Friction lap welding	PE (2.0)/Mg alloy (2.0)	Nonapplicable	4.67 (surface treatment)	11.25	[15]
Friction lap welding	CFRP-PA6 (3.0)/AA5052 (2.0)	2.9 (surface treatment)	12.8 (surface treatment)	5.63	[16]
Friction-based injection clinching joining	PEI (6.35)/AA6082 (2.0)	1.42 ± 0.43	17.4	7.5	[17]
Self-pierce riveting	CFRP-PA6 (3.0)/AA5754 (2.0)	2.5	Nonapplicable	Nonapplicable	[18]
Self-pierce riveting	CFRP: Angle ply (1.5)/AA2024-T6 (2.7)	3.8	Nonapplicable	2	[32]
Adhesive bonding	CFRP: Angle ply (1.5)/AA2024-T6 (2.7)	4.99 (heat treat) 3.84 (untreated)	Nonapplicable	Nonapplicable	[32]
Hybrid (SPR + adhesive)	CFRP: Angle ply (1.5)/AA2024-T6 (2.7)	5.85 (heat treat) 5.0 (untreated)	Nonapplicable	Nonapplicable	[32]
Hole clinching	CFRP (1.2 mm)/SPRC440 (1.6 mm)	3.36	Nonapplicable	Nonapplicable	[23]

PA: polyamide, PPS: polyphenylene sulfide, PE: polyethylene, PEI: polyetherimide, AA: aluminum alloy.

Figure 6 shows fractography results for different specimens from lap shear tensile testing. For FBJ, the joining bit and DP980 were still intact, indicating good consolidated joining between the joining bit and DP980. The failure mode was found to be shear out—one of the common fracture modes (i.e., net tension, shear out, cleavage, and bearing) for composite joints—in cases, where the distance between the hole edge and the edge of the laminate was small [37]. For adhesive bonding, delamination of the carbon fiber composite matrix was observed, indicating good adhesion between the carbon fiber composite surface and the adhesive. Mixed failure modes (i.e., shear out and delamination of carbon fiber composites) were observed for weld-bonded samples, as a result of combining two joining processes (i.e., FBJ and adhesive bonding).

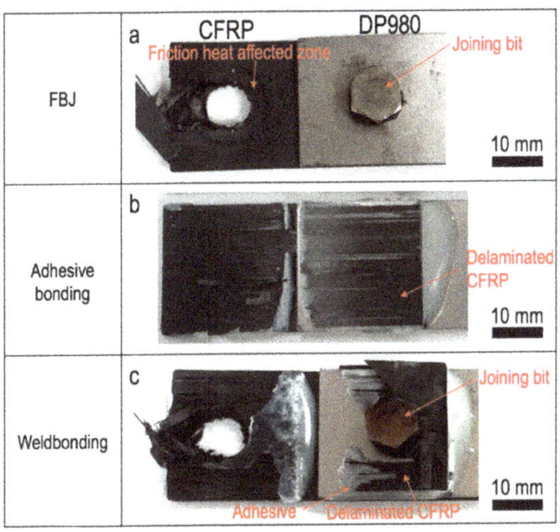

Figure 6. Comparison of fractography results for lap shear-tested (**a**) FBJ, (**b**) adhesive-bonded, and (**c**) weld-bonded specimens.

3.3. Cross Tension Testing and Fractorgraphy

Examples of final assembled cross-tension specimens joined by FBJ are shown in Figure 4b. Representative load and displacement curves from cross-tension testing of the individual processes are plotted in Figure 7. For FBJ, the peak failure load was found to be 2 kN; a maximum failure load of 4.82 kN was achieved for the adhesive-bonded specimen. Two peak failure loads (4.51 and 2.59 kN) were found for the weld-bonded specimen because of the nature of the two combined joining processes. Although the peak failure loads for adhesive-bonded (4.82 kN) and weld-bonded (4.51 kN) specimens were similar, different failure modes were seen from the load and displacement curves. The adhesive-bonded sample immediately failed at a displacement distance of 3.35 mm, whereas the weld-bonded specimen failed after reaching the second peak load. This is because the friction bit joint remained after the first failure of the adhesive-bonded coupon. It is worth mentioning that the second peak failure load for the weld-bonded sample was close to the maximum failure load of the FBJ-only sample.

Absorption energy is the ability to absorb energy under the mechanical testing condition, which is important for crash or dynamic performance for autobody structures. Absorption energy for each joining process was calculated by integration of load and displacement curved in Figure 7. Summary of peak failure load, absorption energy, displacement at failure for each joining process is presented in Table 6. Absorption energy of weld-bonded specimen is nearly twice higher than the absorption energy of adhesive bonded case. Therefore, the weld bonding process can provide improved structural safety performance compared with adhesive bonding and FBJ only.

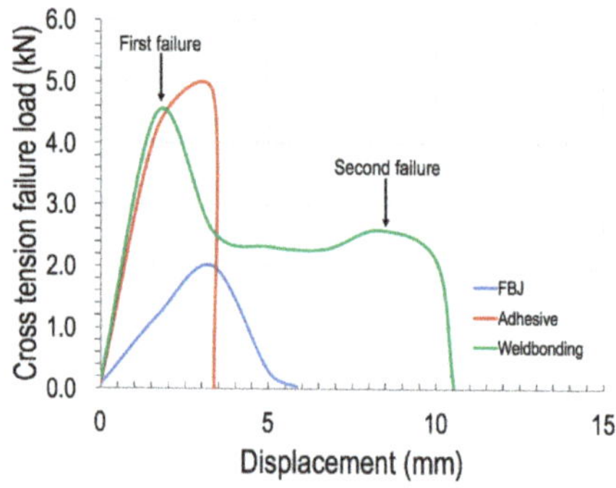

Figure 7. Representative load and displacement curves for cross-tension-tested FBJ, adhesive-bonded, and weld-bonded specimens.

Table 6. Summary of cross-tension testing for each joining process.

Joining Process	Peak Failure Load (kN)	Absorption Energy (J)	Displacement at Failure (mm)
FBJ	2	5.76	5.84
Adhesive bonding	4.82	11.19	3.35
Weld bonding	4.51 (1st peak) 2.59 (2nd peak)	26.19	10.52

Fractography results for cross-tension-tested individual specimens are provided in Figure 8. For the FBJ specimen, the joining bit and DP980 remained in an intact condition, indicating good metallurgical bonding between them, whereas the carbon fiber composite was pullout. For adhesive-bonded specimens, mixed failure modes were observed, such as adhesive failure (i.e., fracture at the interface between the adhesive and the carbon fiber composite) and some delamination of the carbon fiber composite layer, as shown in Figure 8b. Finally, the weld-bonded specimens showed complex failure modes, including carbon fiber composite pullout, some adhesive failures, and some delamination of the composite matrix, as presented in Figure 8c. These mixed failure modes for weld-bonded samples are thought to have resulted from the combined FBJ and adhesive bonding processes.

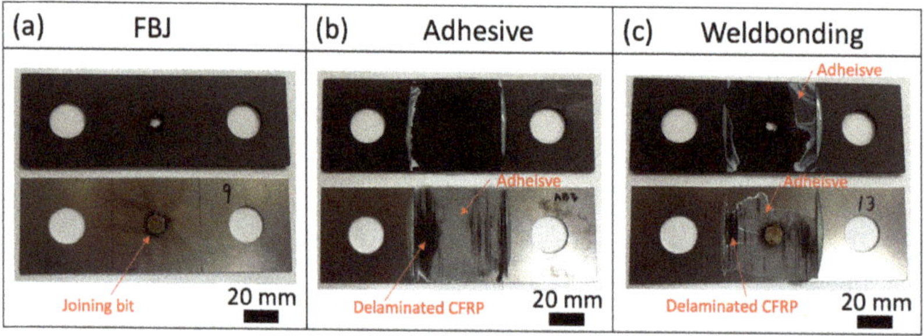

Figure 8. Comparison of fractography for cross-tension-tested specimens: (**a**) FBJ, (**b**) adhesive-bonded, and (**c**) weld-bonded.

3.4. Cross Sectional Analysis of Firction Bit-Joined Specimen

The Vickers microhardness distribution in the joining bit and DP980 (Figure 9a) is mapped in Figure 9b, and microhardness profiles along the x direction with three dashed lines are shown in Figure 9c. The average Vickers microhardness values for the joining bit and DP980 were measured at 322 and 320 HV, respectively. In the joining bit, the Vickers microhardness increased up to 760 HV because of the evolution of the microstructure from tempered martensite to fully martensite, as a result of rapid heating and cooling cycle during the joining stage. This result indicates that the peak temperature during the heating cycle was above the A_3 temperature of typical steels [29,30].

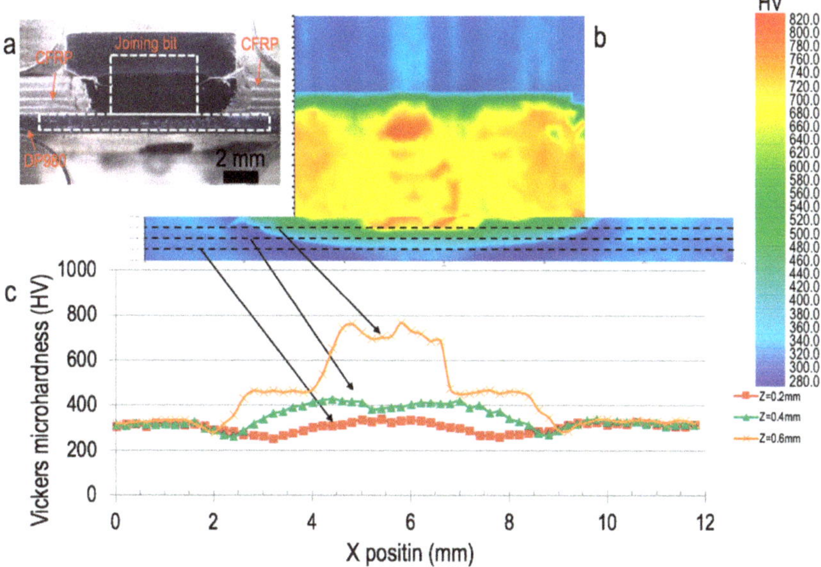

Figure 9. (**a**) Cross-sectional view of FBJ specimen; (**b**) Vickers microhardness mapping at cross section of FBJ specimen; (**c**) Vickers microhardness profile along the x direction.

Figure 10 shows optical and magnified SEM images from the cross-sectioned FBJ specimen. In Figure 10a, consolidated bonding between the joining bit and DP980 is observed without any cracks. Figure 10b,c shows SEM images near the joining bit and carbon fiber composite. At the distance ranging from 0.5 to 1.0 mm away from the edge of the joining bit, the gap between the CFRP and the joining bit was filled by the CFRP matrix. In this region, carbon fibers near the joining bit were redirected by the rotational motion of the joining bit during the joining step, as seen in Figure 10d,e. This redirection of the carbon fibers was also seen in other studies, in which rotation of a rivet occurred [38]. The remaining area was the base CFRP material, in which the carbon fibers were aligned as manufactured.

Based on the joint configuration and bonding mechanism, the CFRP matrix closer to the polymer and metal (i.e., joining bit and DP980) interface will be affected by frictional heating generated during the plunge and joining stages, as illustrated in Figure 1b,c. In the plunge stage, the joining bit cuts and plunges into the CFRP, resulting in increasing temperature in the CFRP. Because no outgassing was observed in the plunge stage during the FBJ process, the peak temperature generated in the carbon fiber composite was expected to be lower than the decomposition temperature of 290.5 °C for the composite.

During the joining step, great frictional heat is generated when the joining bit engages the steel substrate with a higher spindle rotational speed and a higher axial plunge load. Two scenarios of frictional heat conduction can be anticipated. First, frictional heat will radially diffuse into the steel substrate from the joining bit. Then, frictional heat will be conducted from the bottom steel sheet to carbon fiber composites when they directly contact the steel substrate, as a result of being clamped to it, as shown in Figure 2. However, the peak temperature of the steel substrate will decrease because of the large thermal mass from the fixture substrate. Figure 6a shows a thermally affected radial area of the carbon fiber composite, which whitened slightly. XRD was used to find any noticeable peak changes in the thermally affected radial areas of the CFRP and as-received carbon fiber composites. Figure 11 exhibits the typical amorphous phase of the epoxy [39]. Note that the XRD patterns of the two zones are essentially identical, implying that the thermal degradation is ignorable.

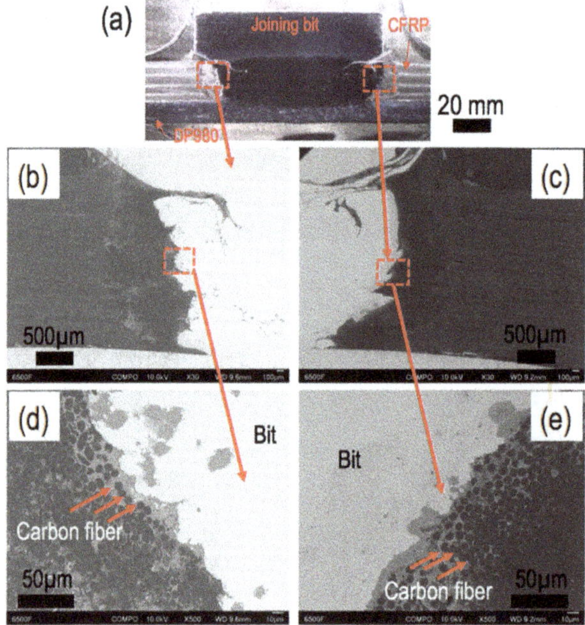

Figure 10. (**a**) Macrograph of FBJ cross-section view. (**b,d**) SEM images of the left side of FBJ cross section (**c,e**). SEM images of the right side of FBJ cross section.

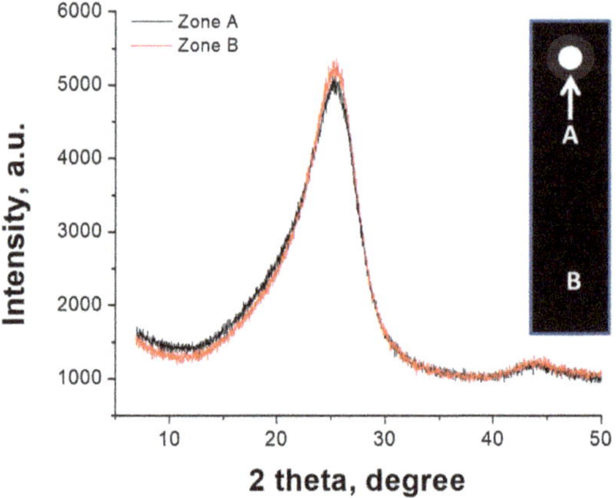

Figure 11. X-ray diffraction patterns measured the indicated area of lap joint. Zone A: thermally affected area shown in Figure 6a. Zone B: as-received CFRP.

Next, heat conduction from the joint bit to the surrounding carbon fiber composite was considered. As mentioned earlier, the peak temperature of the joining bit during the joining step can exceed the A_3

temperature, as evidenced by the Vickers microhardness measurements. The high temperature can lead to localized degassing of the polymer matrix adjacent to the joining bit due to the relatively short weld time (~0.8 s) and low thermal diffusivity (1.2–2.0 $m^2 \cdot s^{-1}$) of the carbon fiber composite [40,41]. For this reason, the remaining polymer composite was not expected to suffer from frictional heat. A cross-sectioned FBJ sample was subjected to FTIR-ATR spectroscopy to examine the surface of the epoxy composite in close proximity to the joining bit to determine if frictional heat from the joining process led to significant polymer degradation. Figure 12 shows characteristic absorption peaks for epoxy resin systems at five different locations and for a reference layer [42,43]. Although there are minor variations in the data, they are qualitatively similar; it can be concluded that if any thermal degradation occurs during the joining process, it is insignificant.

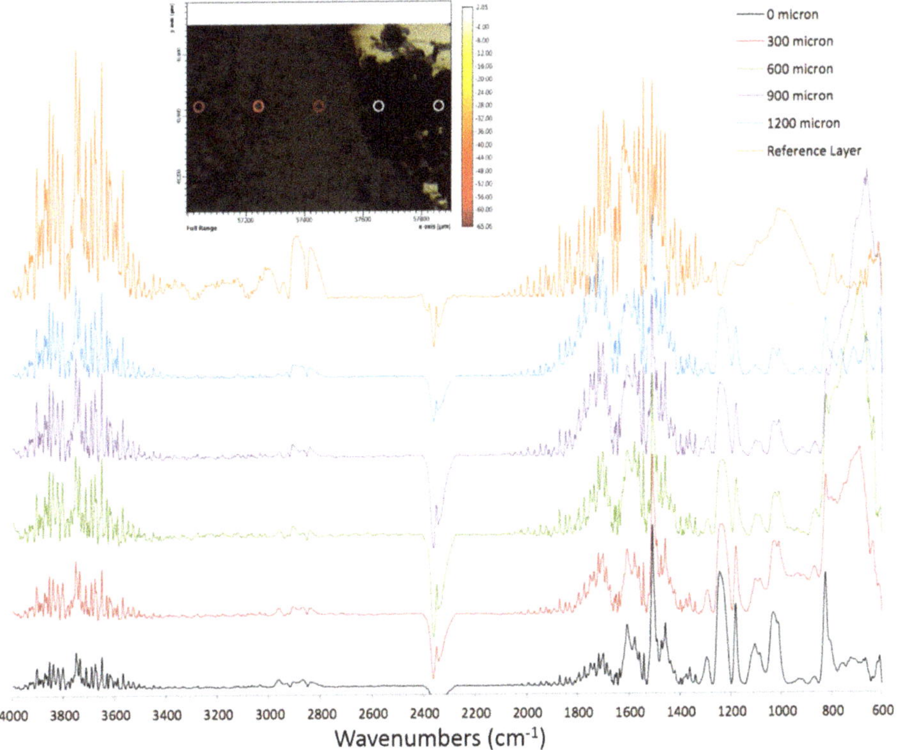

Figure 12. FTIR spectra with different locations (colored dots in the image) measured at the interface between the joining bit and CFRP.

4. Conclusions

Joining of carbon fiber composites and DP980 was successfully demonstrated by FBJ, adhesive bonding, and weld bonding. Average lap shear failure peak loads were found to be 6.0, 14.8, and 13.3 kN for FBJ, adhesive-bonded, and weld-bonded specimens, respectively. The obtained lap shear failure load for each joining process was higher than or comparable to the joint strengths found in the open literature. Cross-tension failure loads were 2.0 and 4.82 kN for FBJ and adhesive-bonded coupons, respectively; and the weld-bonded specimens showed two distinctive peak failure loads of 4.51 and 2.59 kN due to the combination of FBJ and adhesive joining. Thermogravimetry measurements indicated that the as-received CFRP will decompose starting at 290.5 °C. FTIR and XRD results showed limited variations in the peak patterns, including insignificant thermal degradation of the carbon fiber matrix as a result of FBJ.

Author Contributions: Conceptualization, Y.C.L., H.P. and Z.F.; methodology, Y.C.L., H.P., J.J.; investigation, J.W.M., B.S.L., J.K.K. and Z.W.; writing of the original draft preparation, Y.C.L.; writing of review and editing, Y.C.L. and Z.F.; supervision, Z.F.

Funding: This research was funded by Hyundai Motor Company in Korea under the contract number NFE-15-05748.

Acknowledgments: FTIR and XRD measurements for CFRP were conducted at the Center for Nanophase Materials Sciences, which is a DOE Office of Science User Facility. Oak Ridge National Laboratory is managed by UT-Battelle, LLC, for the US Department of Energy under contract DE-AC05-00OR22725. This manuscript has been authored by UT-Battelle, LLC, under contract DE-AC05-00OR22725 with the US Department of Energy (DOE). The US government retains and the publisher, by accepting the article for publication, acknowledges that the US government retains a nonexclusive, paid-up, irrevocable, worldwide license to publish or reproduce the published form of this manuscript, or allow others to do so, for US government purposes. DOE will provide public access to these results of federally sponsored research in accordance with the DOE Public Access Plan (http://energy.gov/downloads/doe-public-access-plan).

Conflicts of Interest: The authors declare no conflicts of interest.

References

1. Dilthey, U.; Stein, L. Multimaterial car body design: Challenge for welding and joining. *Sci. Technol. Weld. Join.* **2006**, *11*, 135–141. [CrossRef]
2. Meschut, G.; Janzen, V.; Olfermann, T. Innovative and highly productive joining technologies for multi-material lightweight car body structures. *J. Mater. Eng. Perf.* **2014**, *23*, 1515–1523. [CrossRef]
3. Adam, H. Carbon fibre in automotive applications. *Mater. Des.* **1997**, *18*, 349–355. [CrossRef]
4. Edwards, K.L. An overview of the technology of fibre-reinforced plastics for design purpose. *Mater. Des.* **1998**, *19*, 1–10. [CrossRef]
5. Bhagavathi, L.R.; Chaudhari, G.P.; Nath, S.K. Mechanical and corrosion behavior of plain low carbon dual-phase steels. *Mater. Des.* **2011**, *32*, 433–440. [CrossRef]
6. Mintz, B. Hot dip galvanising of transformation induced plasticity and other intercritically annealed steels. *Int. Mater. Rev.* **2001**, *46*, 169–197. [CrossRef]
7. Jung, K.W.; Kawahito, Y.; Katayama, S. Laser direct joining of carbon fibre reinforced plastic to stainless steel. *Sci. Technol. Weld. Join.* **2011**, *16*, 676–680. [CrossRef]
8. Jung, K.W.; Kawahito, Y.; Takahashi, M.; Katayama, S. Laser direct joining of carbon fiber reinforced plastic to zinc-coated steel. *Mater. Des.* **2013**, *47*, 179–188. [CrossRef]
9. Min, J.; Li, Y.; Li, J.; Carlson, B.E.; Lin, J. Friction stir blind riveting of carbon fiber-reinforced polymer composite and aluminum alloy sheets. *Int. J. Adv. Manuf. Technol.* **2015**, *76*, 1403–1410. [CrossRef]
10. Amancio-Filho, S.T.; Bueno, C.; dos Santos, J.F.; Huber, N.; Hage, E., Jr. On the feasibility of friction spot joining in magnesium/fiber-reinforced polymer composite hybrid structures. *Mater. Sci. Eng. A* **2011**, *528*, 3841–3848. [CrossRef]
11. Goushegir, S.M.; dos Santos, J.F.; Amancio-Filho, S.T. Friction spot joining of aluminum AA2024/carbon-fiber reinforced poly(phenylene sulfide) composite single lap joints: Microstructure and mechanical performance. *Mater. Des.* **2014**, *54*, 196–206. [CrossRef]
12. Esteves, J.V.; Goushegir, S.M.; dos Santos, J.F.; Canto, L.B.; Hage, E., Jr.; Amancio-Filho, S.T. Friction spot joining of aluminum AA6181-T4 and carbon fiber-reinforced poly(phenylene sulfide): Effects of process parameters on the microstructure and mechanical strength. *Mater. Des.* **2015**, *66*, 437–445. [CrossRef]
13. Balle, F.; Wagner, G.; Eifler, D. Ultrasonic metal welding of aluminium sheets to carbon fibre reinforced thermoplastic composites. *Adv. Eng. Mater.* **2009**, *11*, 35–39. [CrossRef]
14. Anyfantis, K.N.; Tsouvalis, N. Loading and fracture response of CFRP-to-steel adhesively bonded joints with thick adherents—Part I: Experiment. *Compos. Struct.* **2013**, *96*, 850–857. [CrossRef]
15. Liu, F.C.; Liao, J.; Gao, Y.; Nakata, K. Effect of plasma electrolytic oxidation coating on joining metal to plastic. *Sci. Technol. Weld. Join.* **2015**, *20*, 291–296. [CrossRef]
16. Nagatsuka, K.; Yoshida, S.; Tsuchiya, A.; Nakata, K. Direct joining of carbon-fiber-reinforced plastic to an aluminum alloy using friction lap joining. *Compos. B* **2015**, *73*, 82–88. [CrossRef]

17. Abibe, A.B.; Sonego, M.; dos Santos, J.F.; Canto, L.B.; Amancio-Filho, S.T. On the feasibility of a friction-based staking joining method for polymer-metal hybrid structures. *Mater. Des.* **2016**, *92*, 632–642. [CrossRef]

18. Zhang, J.; Yang, S. Self-piercing riveting of aluminum alloy and thermoplastic composites. *J. Compos. Mater.* **2014**, *49*, 1493–1502. [CrossRef]

19. Haque, R. Quality of self-piercing riveting (SPR) joints from cross-sectional perspective: A review. *Arch. Civ. Mech. Eng.* **2018**, *18*, 83–93. [CrossRef]

20. Kelly, G. Load transfer in hybrid (bonded/bolted) composite single-lap joints. *Compos. Struct.* **2005**, *69*, 35–43. [CrossRef]

21. Kolesnikov, B.; Herbeck, L.; Fink, A. CFRP/titanium hybrid material for improving composite bolted joints. *Compos. Struct.* **2008**, *83*, 368–380. [CrossRef]

22. Chowdhury, N.; Chiu, W.K.; Wang, J.; Chang, P. Static and fatigue testing thin riveted, bonded and hybrid carbon fiber double lap joints used in aircraft structures. *Compos. Struct.* **2015**, *121*, 315–323. [CrossRef]

23. Lee, S.H.; Lee, C.J.; Lee, K.-H.; Lee, J.-M.; Kim, B.-M.; Ko, D.C. Influence of tool shape on hole clinching for carbon fiber-reinforced plastics and SPRC440. *Adv. Mech. Eng.* **2014**, *6*, 810864. [CrossRef]

24. Wagner, J.; Wilhelm, M.; Baier, H.; Fussel, U.; Richter, T. Experimental analysis of damage propagation in riveted CFRP-steel structures by thermal loads. *Int. J. Adv. Manuf. Technol.* **2014**, *75*, 1103–1113. [CrossRef]

25. Sun, X.; Khaleel, M.A. Strength estimation of self-piercing rivets using lower bound limit load analysis. *Sci. Technol. Weld. Join.* **2005**, *10*, 624–635. [CrossRef]

26. Abe, Y.; Kato, T.; Mori, K. Self-pierce riveting of three high strength steel and aluminum alloy sheets. *J. Mater. Forum* **2008**, *1*, 1271–1274. [CrossRef]

27. Lee, C.-J.; Kim, J.-Y.; Lee, S.-K.; Ko, D.-C.; Kim, B.-M. Parametric study on mechanical clinching process for joining aluminum alloy and high-strength steel sheets. *J. Mech. Sci. Technol.* **2010**, *24*, 123–126. [CrossRef]

28. Xiao, G.Z.; Shanahan, M.E.R. Water absorption and desorption in an epoxy resin with degradation. *J. Polym. Sci. B* **1997**, *35*, 2659. [CrossRef]

29. Squires, L.; Lim, Y.C.; Miles, M.; Feng, Z. Mechanical properties of dissimilar metal joints composed of DP 980 steel and AA 7075-T6. *Sci. Technol. Weld. Join.* **2015**, *20*, 242–248. [CrossRef]

30. Miles, M.P.; Kohkonen, K.; Packer, S.; Steel, R.; Siemssen, B.; Sato, Y.S. Solid state spot joining of sheet materials using consumable bit. *Sci. Technol. Weld. Join.* **2009**, *14*, 72–77. [CrossRef]

31. Huang, T.; Sato, Y.S.; Kokawa, H.; Miles, M.P.; Kohkonen, K.; Siemssen, B.; Steel, R.J.; Packer, S. Microstructural evolution of DP980 steel during friction bit joining. *Metall. Mater. Trans. A* **2009**, *40A*, 2994–3000. [CrossRef]

32. Di Franco, G.; Fratini, L.; Pasta, A. Analysis of the mechanical performance of hybrid (SPR/bonded) single-lap joints between CFRP panels and aluminum blanks. *Int. J. Adhes. Adhes.* **2013**, *41*, 24–32. [CrossRef]

33. LeBozec, N.; LeGac, A.; Thierry, D. Corrosion performance and mechanical properties of joined automotive materials. *Mater. Corros.* **2012**, *63*, 408–415. [CrossRef]

34. Lim, Y.C.; Squires, L.; Pan, T.-Y.; Miles, M.; Song, G.-L.; Wang, Y.; Feng, Z. Study of mechanical joint strength of aluminum alloy 7075-T6 and dual phase steel 980 welded by friction bit joining and weld-bonding under corrosion medium. *Mater. Des.* **2015**, *69*, 37–43. [CrossRef]

35. Lim, Y.C.; Squires, L.; Pan, T.-Y.; Miles, M.; Kuem, J.K.; Song, G.-L.; Wang, Y.; Feng, Z. Corrosion behaviour of friction-bit-joined and weld-bonded AA7075-T6/galvannealed DP980. *Sci. Technol. Weld. Join.* **2017**, *22*, 455–464. [CrossRef]

36. Balle, F.; Eifler, D. Statistical test planning for ultrasonic welding of dissimilar materials using the example of aluminum–carbon fiber reinforced polymers (CFRP) joints. *Materialwiss. Werkstofftech.* **2012**, *43*, 286–292. [CrossRef]

37. Mallick, P.K. *Fiber-Reinforcd Composites*; Marcel Dekker Inc.: New York, NY, USA, 1993.

38. Altmeyer, J.; Suhuddin, U.F.H.; dos Santos, J.F.; Amancio-Filho, S.T. Microstructure and mechanical performance of metal-composite hybrid joints produced by FricRiveting. *Compos. Part B* **2005**, *81*, 130–140. [CrossRef]

39. Wang, Z.; Zhao, G.-L. Microwave absorption properties of carbon nanotubes-epoxy composites in a frequency range of 2-20 GHz. *Open J. Compos. Mater.* **2013**, *3*, 17–23. [CrossRef]

40. Mathew, J.; Goswami, G.L.; Ramakrishnan, N.; Naik, N.K. Parametric studies on pulsed Nd:YAG laser cutting of carbon fibre reinforced plastic composites. *J. Mater. Process. Technol.* **1999**, *89–90*, 198–203. [CrossRef]

41. Wrobel, G.; Rdzawski, Z.; Muzia, G.; Pawlak, S. Determination of thermal diffusivity of carbon/epoxy composites with different fiber content using transient thermography. *J. Achiev. Mater. Manuf. Eng.* **2009**, *37*, 518–525.
42. Liu, F.; Yin, M.; Xiong, B.; Zheng, F.; Mao, W.; Chen, Z.; He, C.; Zhao, X.; Fang, P. Evolution of microstructure of epoxy coating during UV degradation progress studied by slow positron annihilation spectroscopy and electrochemical impedance spectroscopy. *Electrochim. Acta* **2014**, *133*, 283–293. [CrossRef]
43. Cecen, V.; Seki, Y.; Sarikanat, M.; Tavman, I.H. FTIR and SEM analysis of polyester- and epoxy-based composites manufactured by VARTM process. *J. Appl. Polym. Sci.* **2008**, *108*, 2163–2170. [CrossRef]

Article

Ultrasonic Inspection of a 9% Ni Steel Joint Welded with Ni-based Superalloy 625: Simulation and Experimentation

João da Cruz Payão Filho [1], Elisa Kimus Dias Passos [1,*], Rodrigo Stohler Gonzaga [1], Ramon Fonseca Ferreira [2], Daniel Drumond Santos [2] and Diego Russo Juliano [3]

[1] Programa de Engenharia Metalúrgica e de Materiais, Instituto Alberto Luiz Coimbra de Pós-Graduação e Pesquisa de Engenharia, Universidade Federal do Rio de Janeiro (PEMM/COPPE/UFRJ), Cidade Universitária, Ilha do Fundão, Caixa Postal 68505, CEP 21941-972 Rio de Janeiro-RJ, Brazil; jpayao@metalmat.ufrj.br (J.d.C.P.F.); rodrigovr@metalmat.ufrj.br (R.S.G.)

[2] Serviço Nacional de Aprendizagem Industrial do Rio de Janeiro, Instituto Senai de Tecnologia Solda (IST Solda), R.São Francisco Xavier, no 601, Maracanã, CEP 20550-011 Rio de Janeiro-RJ, Brazil; rfferreira@firjan.org.br (R.F.F.); dadsantos@firjan.org.br (D.D.S.)

[3] Shell Petróleo Brasil Ltda., Av. República do Chile, no 330, 25o andar, Torre Oeste, Centro, CEP 20031-170 Rio de Janeiro-RJ, Brazil; diego.juliano@shell.com

* Correspondence: elisakimus@metalmat.ufrj.br; Tel.: +55-21-2290-1544/233

Received: 14 September 2018; Accepted: 29 September 2018; Published: 2 October 2018

Abstract: The ultrasonic inspection of thick-walled welded joint with austenitic weld metal has proven to be a challenge due to its anisotropic microstructure that can promote ultrasonic waves attenuation. This work aimed to optimize the phased array ultrasonic inspection of the thick-walled joint of a 9% Ni steel pipe welded with Ni-based superalloy 625. The development was carried out by CIVA numeric simulation to preview the beam behavior during the inspection of GTAW (Gas Tungsten Arc Welding)/SMAW (Shielded Metal Arc Welding) joint with anisotropic weld metal. To validate the simulation results, experimental tests were performed with a phased array transducer using longitudinal waves on a calibration block withdrawn from the joint. The configuration of low frequency (2.25 MHz), 16 active elements and a scanning angle of 48° ensured the inspection of the entire joint and the computational simulation proved to be essential for the success of the inspection.

Keywords: phased array ultrasonic; dissimilar weldments; anisotropy; longitudinal wave; simulation

---

1. Introduction

Cryogenic 9% Ni steel has been used almost exclusively in the storage and transport of liquefied gases in the form of sheets of small and medium thicknesses. Nevertheless, at the beginning of the decade of 2010, 9% Ni steel found a new application as CO_2 injection equipment in Brazil pre-salt oil and gas wells. Due to the high pressure in these wells (approximately 550 bar/55 MPa), thick 9% Ni steel pipes are being used for the first time in the world [1].

The ultrasonic inspection with phased array of the thick-walled girth welded joints with an austenitic weld metal, such as a 9% Ni steel pipe welded with nickel superalloy 625 as the filler metal, has been a huge challenge for the oil and gas industry. The weld metal of this type of dissimilar welded joint has an anisotropic structure, which attenuates the ultrasonic waves and diverts the beam, making the establishment of the best inspection conditions extremely difficult and laborious. Many research works [2–6] have been carried out to describe the influence of the solidification structures and its effect on the propagation of ultrasonic wave. Since the critical defects in this type of joint are mainly found in the root of the weld, due to the corrosive fluid contact in the inner surface of the pipe, it is very important for the ultrasonic inspection to ensure the sizing of defects in this region.

Faced by this situation, our research team began an investigation using numerical simulation to preview the ultrasonic behavior during the inspection of welded joints of 9% Ni high thickness steel pipes with austenitic alloy as the filler metal. The major aim of this work is to establish the best conditions in which to optimize the phased array ultrasonic inspection of welded dissimilar steel tubular joints by drawing up an ultrasonic numeric simulation study and using ES Beam Tool software 4th version from the Eclipse Scientific Company (Canada) and the 11th version of CIVA from the Extende® Company (France). The simulation was essential, since it would be difficult, costly and time-consuming to obtain the best ultrasonic inspection technique with only experimental procedures. To validate the simulation results and ensure the effectiveness and precision of the inspection, experimental tests were performed using a phased array transducer with longitudinal linear waves on a calibration block, withdrawn from a joint fabricated by GTAW/SMAW welding processes.

The grains in austenitic weld metals grow parallel to the direction of the heat flow, thus forming a dendritic microstructure during cooling. In spite of the heat promoted by the successive welding passes deposition, it is not sufficient to ensure a modification in the structure of the elongated and crystallography-oriented grains formed in the previous welding pass deposition. These grains are approximately vertical along the center of the weld metal and are perpendicular to the fusion line, giving this region an anisotropic structure and a particular symmetry of long austenitic columnar grains [6].

During welding thermal cycles, austenitic weld metals alloys, such as nickel superalloy 625, do not experience phase transformation, which promotes the extensive growth of columnar grain. These anisotropic grains of the weld metal may promote beam deflection and consequently interfere with defect detection. There are many works [7–9] that tried to model the deflection that the austenitic weld metal can promote in the propagation of ultrasonic waves. Hirsekorn [10] compared the effect of ultrasonic longitudinal and transversal waves in austenitic weld metals and concluded that the sonic speed and the beam deflection are different according to the scanning incidence angle. For longitudinal waves, the scanning incidence angles, which represent the minimum beam deflection, are between 45° and 52°.

The attenuation coefficient α is a parameter composed by the sum of the absorption coefficient α_a and the scattering coefficient α_s. The energy absorption is affected by the interaction of the ultrasonic wave with the imperfections of the material lattice, while the scattering relies on the grain structure of the material [11].

The relation between the grain size and wavelength of the weld metal is an important feature that influences the ultrasonic attenuating effect. According to this relation, the beam scattering may be divided into three regimes: Rayleigh ($\lambda \gg D$), Stochastic ($\lambda \approx D$) and Diffusive ($\lambda \ll D$), where λ is the wavelength and D the grain size. As will be shown in the Results and Discussions, for the settings used in this work, the Rayleigh regime was the best fit. The scattering in a Rayleigh regime is described by Equation (1) [12]:

$$\alpha_s(f) = aD^3f^4 \tag{1}$$

where a = scattering coefficient; f = frequency.

As the attenuation is directly proportional to the frequency of the four powers, decreasing the frequency results in a decrease of the attenuation effect. Flotté and Bittendiebel [13] suggested the use of a frequency that would guarantee the defect sensibility without compromising the attenuation necessary for the inspection. In his work, a 2.25 MHz transducer was used.

In an anisotropic material, the average diameter of the grain will depend on the wave propagation direction and Equation (1) cannot be generalized for all directions in anisotropic media. This observation was also cited by Hirsekorn [10], who said that the scattering coefficient is also dependent on the beam to grain angle.

In addition to the scattering, the inspection might also be affected by reflection and refraction. When the weld metal fiber texture is anisotropic and the base metal has an isotropic fine grain, beam reflections, beam refractions and mode conversion may occur at the fusion line. Volkov et al. [14]

described that for scanning incidence angles larger than ~30°, the reflection coefficient of the longitudinal waves is smaller than 0.05, while for vertically polarized transverse waves, this value of reflection was only obtained for angles larger than ~60°.

Increasing the sound pressure is a way to improve the beam response. In this case, it is necessary to provide more energy to promote deeper penetration and a higher reliability of detection. Sound pressure, when the phased array technique is used, relies on the number of active elements, focal distance, propagation angle, ultrasonic speed and transducer frequency. As the number of active elements grows, the sound pressure increases, although it is important to take into consideration the signal-to-noise ratio coming from the inspection of austenitic metals [15]. When the number of active elements decreases, the beam focusing gets diffused and the sonic pressure is reduced, promoting a lower concentration at the analyzed point. On the other hand, when the number of elements is very big, the noise also increases and can hide real defects.

The directivity of the sonic beam and the angle of incidence can be adjusted to maximize the beam convergence (focusing). Lee and Choi [16] described the focalization can reduce the area of dispersion while the echo of discontinuities remains constant, thus resulting in a higher sound pressure and signal-to-noise ratio. The directivity relies on the probe design, which is composed of the number and width of active elements, pitch, frequency and bandwidth.

The results showed that with the proper combination of phased array ultrasonic parameters, it is possible to inspect the HAZ (Heat Affected Zone) and the whole weld metal of thick-walled 9% Ni welded joints, thus providing security for this new application in the Brazilian pre-salt industry.

2. Materials and Methods

2.1. Materials

To develop this work, a 9% Ni steel pipe was used with an 8″ (203.2 mm) nominal internal diameter and $1\frac{1}{4}$″ (31. 5 mm) thickness, quenched and tempered as recommended by ASTM A333 grade 8. The pipe was gas tungsten arc welded (GTAW) for root and hot passes and shielded metal arc welded (SMAW) for fill and cap passes. The adopted filler metals were Ni-based superalloy 625 (ERNiCrMo-3/AWS A5.14 and ENiCrMo-3/AWS A5.11 for GTAW and SMAW, respectively). Table 1 shows the specified and analyzed mechanical properties of the base metal and typical values of the mechanical properties of the filler metals and Table 2 shows the specified and analyzed chemical compositions (wt.%) of the base, filler and weld metals, respectively.

Table 1. Specified and analyzed mechanical properties-yield strength (YS), ultimate tensile strength (UTS), elongation (El.) and Charpy V energy at −196 °C (CVEn) and lateral expansion (CVEx)-of the base metal and typical values of filler metals mechanical properties.

Mechanical Properties	Base metal ATSM A333 Gr.8 (9% Ni Steel)		Filler Metal Ni-Based Superalloy 625	
			GTAW	SMAW
	Specified (min.) [17]	Anal.	Typical Values	
YS (MPa)	515	693	510	530
UTS (MPa)	690	740	770	770
El. (%)	22	25	42	30
CVEn (J)	N.A.	146	70	45
CVEx (mils/mm)	14.96/0.38	70/1.78	N.A.	N.A.

Where: N.A. means not applicable and Anal. means analyzed.

Table 2. Specified and analyzed chemical compositions (wt.%) of the base, filler and weld metals.

| Element | Base Metal (9% Ni) | | Filler Metal Ni-Based Superalloy 625 | | | | Weld Metal |
| | | | GTAW | | SMAW | | |
	Specified ASTM A333 (max.) [17]	Anal.	Specified AWS A5.14 [18]	Anal.	Specified AWS A5.11 [19]	Anal.	Anal.
C	0.13	0.061	<0.1	0.008	<0.1	0.034	0.0326
Mn	0.9	0.607	<0.5	0.02	<1.0	0.68	0.887
Si	0.13/0.32	0.279	<0.5	0.05	<0.75	0.35	0.494
P	0.025	<0.005	<0.02	0.006	<0.03	0.010	0.0035
S	0.025	<0.005	<0.015	0.000	<0.02	0.003	0.0088
Ni	8.4/9.6	9.70	>58	65.10	>55	62.8	63.1
Cu	N.A.	<0.005	<0.5	0.03	<0.5	0.00	0.0061
Ti	N.A.	0.0061	<0.4	0.183	N.A.	0.05	0.0694
Cr	N.A.	0.0565	20/23	21.66	20/23	21.83	20
Mo	N.A.	0.0142	8/10	8.73	8/10	8.80	8.95
Fe	Balance	89.200	<0.5	0.430	<7.0	1.84	2.49
Pb	N.A.	<0.005	<0.5	0.000	N.A.	N.A.	0.0504
Al	N.A.	0.0218	<0.4	0.140	N.A.	N.A.	<0.0005
Nb + Ta	N.A.	<0.005	3.15/4.15	3.66	3.15/4.15	3.497	3.478
Co	N.A.	N.A.	N.A.	N.A.	0.12	N.A.	N.A.

Where: N.A. means not applicable and Anal. means analyze.

2.2. Welding Procedure

The whole welding procedure was monitored by data acquisition equipment (IMC SAP 4.0 System). Figures 1 and 2 show the joint design and the illustration of butt joint weld passes distribution, respectively; Table 3 shows the adopted welding parameters.

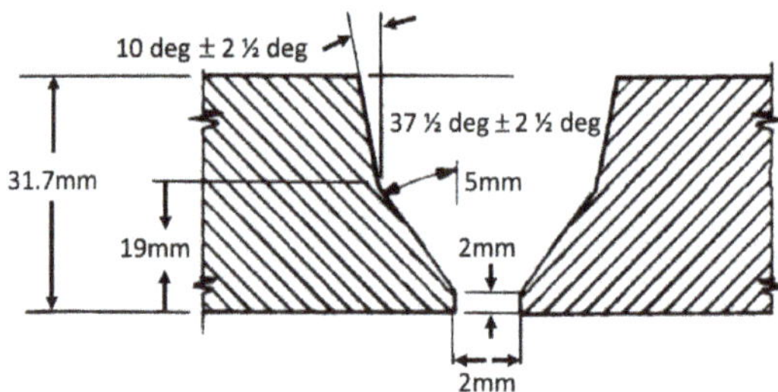

Figure 1. Design of the 9% Ni steel circumferential pipe butt joint.

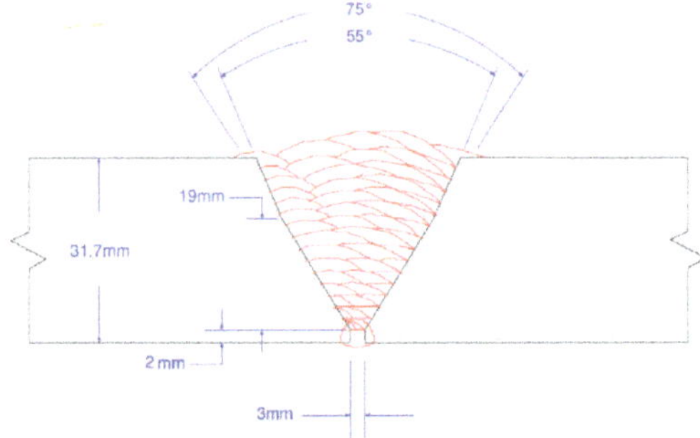

Figure 2. Illustration of the butt joint weld passes distribution.

Table 3. Processes and parameters adopted to girth weld the 9% Ni quenched and tempered steel tubular butt joint with Ni-based superalloy 625 as the filler metal.

Parameters		Welding			
		Process			
		GTAW		SMAW	
		Root	Hot	Fill	Cap
Amperage (A)		126	126	90	89
Voltage (V)		11	11	26	26
Welding speed (cm/min)		5.6	7.3	12.6	18
Heat input (kJ/mm)		1.5	1.1	1.2	0.9
Gas type* and flow rate (l/min)	Shield	Ar/12	Ar/12	N.A.	N.A.
	Purge	Ar/25	Ar/25	N.A.	N.A.

* 99.995% purity. Where: N.A. means not applicable.

2.3. Simulation Procedure

The software CIVA 11th version (Extende®, Massy, France) was used to simulate the path of the ultrasonic beam through the welded joint. To perform the scanning simulations using linear longitudinal waves, the weld reinforcement was removed to allow direct incidence promoted by the wedge passing over the weld metal, since longitudinal waves work only with a half skip because most of the energy of the incident sonic beam is lost to a converted transverse wave after reflection.

The geometry made in the software was based on a cross-sectional macrograph of the weld joint. Figure 3 shows the weld passes, grouped by blocks with the same characteristic cooling directions and highlighted by a CAD (Computer-Aided Design) drawing overlap. The arrow directions represent the average growth of the dendrites of the weld metal and the arrow senses represent the direction of the temperature gradient. This information is important to define the scanning incidence angles that allow for the least attenuation effect. The root region presented a refined microstructure and it was considered isotropic similar to the base metal. The weld metal was set with the assumption of an orthotropic symmetry and the elastic constant was extracted from the software database, which is presented in Table 4. Figure 3 also shows a SEM (Scanning Electron Microscope) micrograph of the HAZ near to the fusion line.

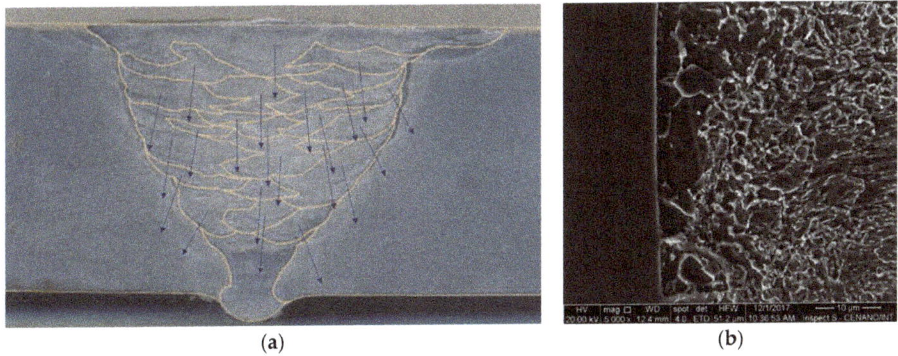

(a)　　　　(b)

Figure 3. Macrograph of the 9% Ni steel pipe welded butt joint showing the weld passes highlighted by CAD drawing overlay and the directions of dendrite growth ((**a**), arrows) and SEM micrograph of the HAZ near the fusion line (**b**).

Table 4. Elastic constants (in GPA) for austenitic stainless steel extracted from the CIVA database.

C11	C22	C33	C23	C13	C12	C44	C55	C66
250	250	250	138	180	112	117	91	70

After determining the scanning angles (45°, 48° and 52°), the software ES Beam Tool was to used preview the ultrasonic beam scanning with different indexes and number of elements, aiming to verify whether these parameters would ensure the coverage of the whole target region (weld metal). The beam opening was projected through the transversal welded joint section. The scanning previews for 45°, 16 active elements and indexes 10, 23, 35 and 48 mm are presented in Figure 4. The same analysis was made for angles of 48° and 52°.

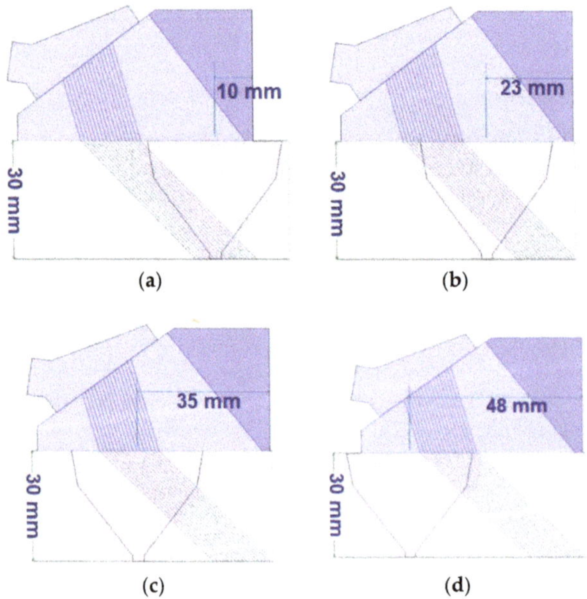

Figure 4. Simulation of inspection with scanning angle of 45° and 16 active elements. Indexes: (a) 10 mm, (b) 23 mm, (c) 35 mm and (d) 48 mm.

The final definition of the scanning incidence angles, number of active elements and indexes used in the CIVA simulation are described in the Table 5.

Table 5. Number of active elements, scanning angles and indexes of the CIVA simulation.

Scanning Angle (°)	No. of Active Elements	Index (mm)
45	16, 24, 32	23
	16	10, 35, 48
48	16, 24, 32	22
	16	8, 35, 49
52	16, 24, 32	18
	16	3, 32, 46

2.4. Inspection Procedure

For inspection procedures, a phased array device, Omniscan MX2 32-128 (Olympus, Tokyo, Japan) [20], a wedge SA32-N60L-IHC (Olympus, Tokyo, Japan) and a phased array transducer

2.25L32-A32 with frequency of 2.25 MHz and 32 active elements (maximum capacity) (Olympus, Tokyo, Japan) were used. The sonic speed and the sonic attenuation of the material were obtained through the calibration blocks. The transducer was positioned over the analyzed region and the speed was adjusted until the material's real thickness was found. For attenuation, the gain between two consecutive echoes was measured. The speed and the attenuation were evaluated in the middle of the weld metal and in the base metal. These values were obtained using a standard ultrasound transducer with no angulation (0°).

Since the CIVA simulation highlighted the best results for 16 active elements, taking into consideration the attenuation and the beam coverage, the other conditions (24 and 32 active elements) were dismissed. Table 6 presents the final configurations used in the experimental tests.

Table 6. Number of active elements, scanning angles and indexes of inspection procedures with an Olympus 2.25L-32-A32 transducer (2.25 MHz).

Scanning Angle (°)	No. of Active Elements	Index (mm)
45		10, 23, 35, 48
48	16	8, 22, 35, 49
52		3, 18, 32, 46

3. Results and Discussions

3.1. Beam Attenuation

When the grains of the weld metal are randomly oriented and small enough when compared to the wavelength, its microstructure does not interfere with its acoustic properties. However, when there are coarse crystallography-oriented grains, the sound propagation is affected by the deviation of the ultrasonic beam, resulting in a worse signal-to-noise ratio. These factors cause an increase in the difficulty of the interpretation of the ultrasonic signal, which ends up being reflected in the loss of accuracy of the location and the dimensioning of the defects, thus making it difficult to distinguish between real defects and false indications [12].

In previews tests with longitudinal and transversal waves, the sonic attenuation and sonic speed were verified in the base metal and in the weld metal. Table 7 summarizes the values obtained. Due to the less attenuation, the longitudinal waves were chosen to develop this work.

Table 7. Sonic attenuation and sonic speed in the base metal and in the weld metal.

Wave Type	Base Metal		Weld Metal	
	Sonic Speed	Attenuation	Sonic Speed	Attenuation
Longitudinal wave	5820 m/s	0.098 dB/mm	5840 m/s	0.186 dB/mm
Transversal wave	3150 m/s	0.131 dB/mm	3075 m/s	0.220 dB/mm

For longitudinal waves with a sonic speed of 5840 m/s and a frequency of 2.25 MHz, the wavelength (λ) is approximately 2595 μm, which is considerably larger than the average grain size of the weld metal (94.6 μm) and the average grain size in the fusion line (158 μm), thus agreeing with Rayleigh beam scattering ($\lambda \gg D$). Figure 5 shows the distribution, size and orientation of the grains, produced by EBSD (Electron Backscatter Diffraction) analysis in the center of the weld metal and near to the fusion boundary. These two regions were chosen to show the anisotropy along the metal and the epitaxial growth behavior of the grain next to the fusion line.

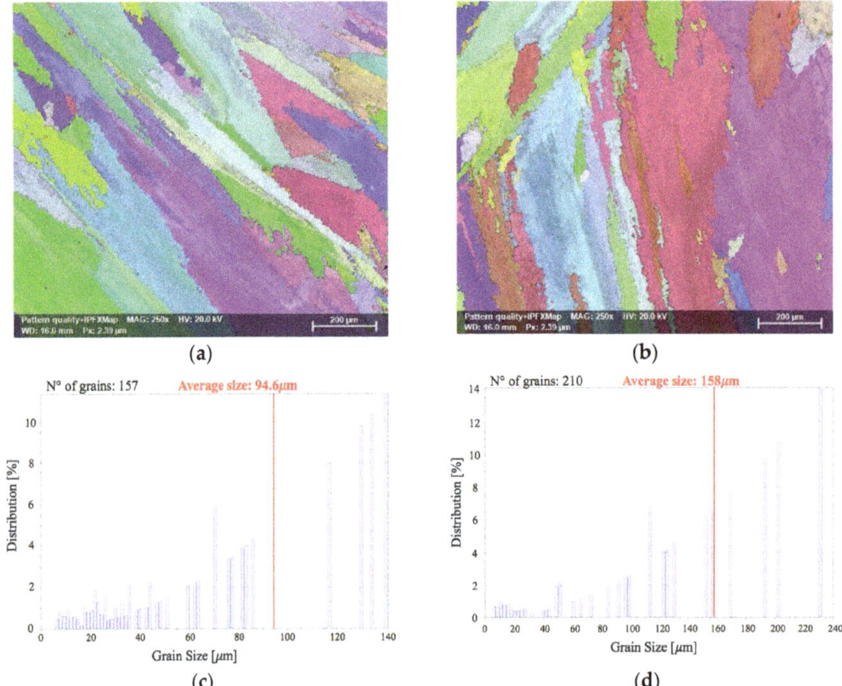

Figure 5. EBSD images of the (**a**) fusion line and the (**b**) weld metal. Below the EBSD images, are the distributions of grain size in the respective regions (**c**,**d**).

The EBSD is a microstructural and crystallographic characterization technique used to identify the grain morphology, grain orientation, phased, texture and deformation of polycrystalline materials. The analysis is conducted by an EBSD detector attached to SEM.

3.2. Simulation

The ultrasonic beam and its attenuation are represented by the pink color and the blue region depicts a collimated ultrasonic beam with high sonic pressure. The results of the linear scanning with longitudinal wave and scanning angle of 45°; the results for different active elements are presented in Figure 6.

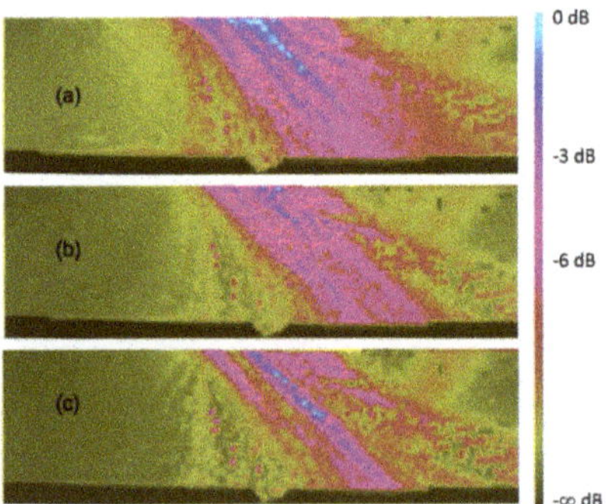

Figure 6. Longitudinal wave inspection. Scanning angle of 45°, index 23 mm. Number of elements: (**a**) 16, (**b**) 24 and (**c**) 32.

Figure 6a shows the ultrasonic beam passing through the weld metal with little attenuation and concentration near the weld cap. As the number of active elements grows, the attenuation effect intensifies, as seen in Figure 6b,c, which used 24 and 32 elements, respectively.

As expressed in theory [15], as the number of active elements grows, the sound pressure increases. However, it is important to take into consideration the signal-to-noise ratio coming from the inspection of austenitic metal. The phased array ultrasonic generates a group of delayed waves, controlled by the focal law, which propagates a maximized wave front into the austenitic microstructure. The different microstructures, grain sizes and grain orientations between the base (9% Ni steel) and weld (Ni-based superalloy 625) metals might cause a deviation of the waves generated by each activated element, which causes backscattering signals from the microstructure noise. Thus, the relation between sonic pressure and the signal-to-noise ratio must be analyzed for each focal law.

The analysis of the simulation with 16 active elements for a linear 45° angled beam demonstrates a good sound pressure and low attenuation of the beam inside the welded joint. From these results, the scanning indexes were varied, aiming to cover the whole weld metal (Figure 7).

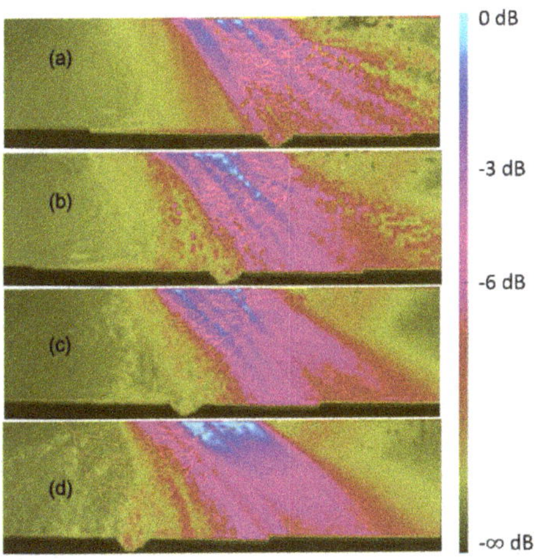

Figure 7. Longitudinal wave inspection. Scanning angle of 45° with 16 elements. Indexes: (**a**) 10 mm, (**b**) 23 mm, (**c**) 35 mm and (**d**) 48 mm.

The results of Figure 7a show high coverage on the weld bevel with low divergence and the sonic beam extends to the root region. In Figure 7b, the beam covered most of the weld metal, demonstrating a low divergence at the index of 23 mm. To cover the welded joint, 2 more indexes were made, as shown in Figure 7c,d, where the beam does not show attenuation or divergence.

The results of linear scanning with a longitudinal wave and a scanning angle of 48° are presented in Figure 8.

Figure 8a shows the 16 active elements of ultrasonic beam passing through the weld metal with little attenuation. As the number of active elements grows, the attenuation effect intensifies, as seen in Figure 8b,c, the values of which were obtained with 24 and 32 elements, respectively. As made for the scanning angle of 45°, the indexes ware varied and aimed to cover the whole weld metal. The results are shown in Figure 9. For an index of 8 mm, the sonic beam has low divergence and shows high coverage of the fusion line in the root region (Figure 9a). In Figure 9b, the beam covers most of the welded joint with low divergence for an index of 22 mm. To cover all of the welded joint, more than two indexes were made (Figure 9c,d). Using scanning indexes of 8 mm, 22 mm, 35 mm and 49 mm, it was possible to cover the whole welded joint. It may also be noted that there is excessive collimation in Figure 9b–d, which can oversize the defects.

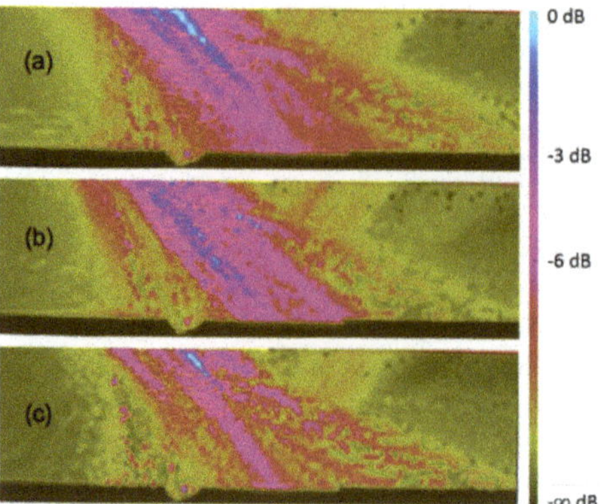

Figure 8. Longitudinal wave inspection. Scanning angle of 48°, index 22 mm. Number of elements: (**a**) 16, (**b**) 24 and (**c**) 32.

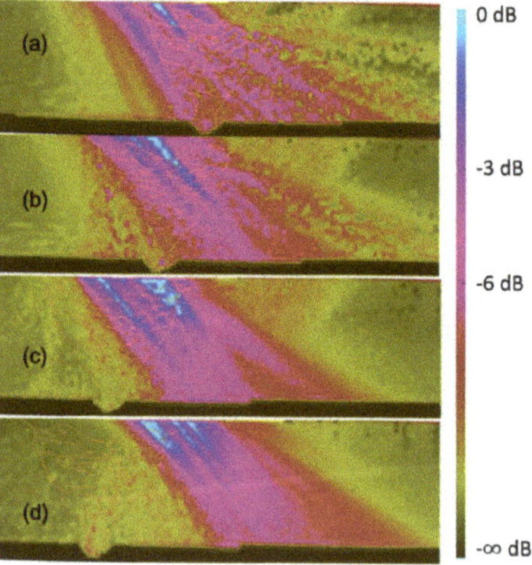

Figure 9. Longitudinal wave inspection. Scanning angle of 48° with 16 elements. Indexes: (**a**) 8 mm, (**b**) 22 mm, (**c**) 35 mm and (**d**) 49 mm.

Another linear beam angle analyzed was 52°, which was observed for the same number of active elements. Even the result of the 16 active elements showed a good coverage and the image shows a divergence of the beam angle direction and attenuation of some areas near the root. The results of the linear scanning with angle 52° are presented in Figure 10.

As the number of active elements grows from 24 to 32 (Figure 10b,c), the attenuation effect intensifies. The same methodology was made for this angle by choosing 16 active elements for the different scanning index analyses (Figure 11). More points of attenuation are visible near the centerline of the welded joint and the beam shows a divergence with indexes of 3 and 18 mm (Figure 11a,b). To cover all of the welded joint, it was necessary to insert 2 more indexes: 32 mm and 46 mm. The first one expressed a good behavior of the ultrasonic beam (Figure 11c); the second one (Figure 11d) expressed low attenuation or divergence of the ultrasonic beam.

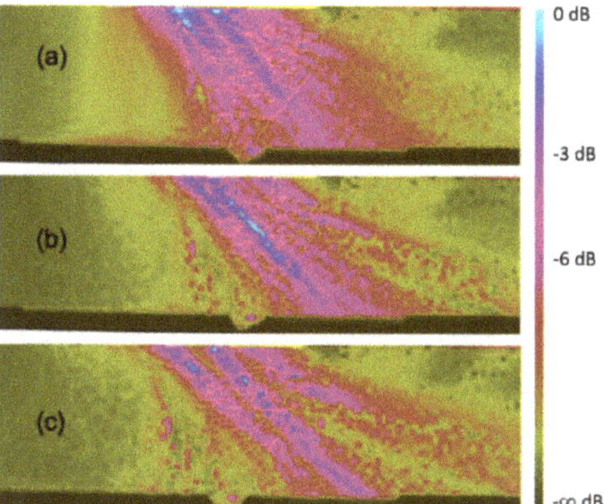

Figure 10. Longitudinal wave inspection. Scanning angle of 52°, index 18 mm. Number of elements: (**a**) 16, (**b**) 24 and (**c**) 32.

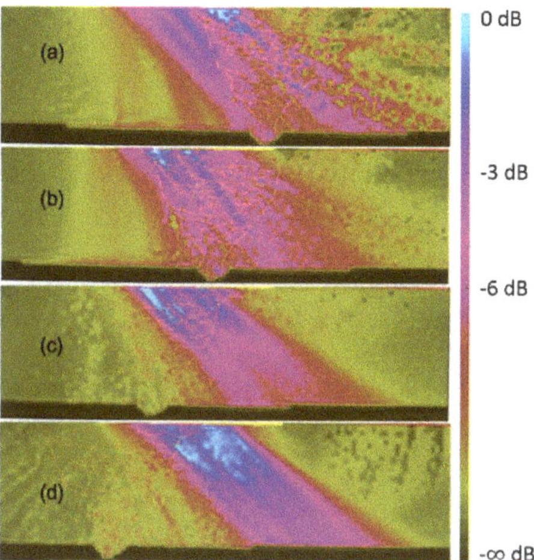

Figure 11. Longitudinal wave inspection. Scanning angle of 52°, with 16 elements. Indexes: (**a**) 3 mm, (**b**) 18 mm, (**c**) 32 mm and (**d**) 46 mm.

Angles outside the range of 40°–50° may promote significant beam attenuation, especially for austenitic metals, such as the Ni-based superalloy 625. In case of misunderstanding the metal microstructure, the scan parameters can be erroneously selected and because of that, the estimated discontinuity size detected by ultrasonic inspection may not coincide with its real size or real position in the weld metal. Table 8 shows the simulation results summarized.

Table 8. Summary of the simulation results.

Objective	Angle (°)	No. of Active Elements	Index (mm)	Simulation Results
Analyze the influence of varying the number of active elements in the ultrasonic beam propagation through the material.	45° 48° 52°	16/24/32	23 22 18	Attenuation grows as the number of active elements increases. For all the scanning angles the best results, that is, high coverage and low attenuation were the configuration of 16 elements.
Analysis of the ultrasonic beam coverage in the welded joint for different scanning indexes.	45° 48° 52°	16 16 16	10/23/35/48 8/22/35/49 3/18/32/46	All indexes showed high coverage and low attenuation on the weld bevel and on the weld metal. All indexes showed low divergence and high coverage on the fusion line and root region. Although, there was excessive beam collimation near the weld cap for the indexes of 22, 35 and 49 mm. Beam divergence was observed in the centerline of the weld for indexes of 3 and 18 mm. For 32 and 46 mm, the attenuation and divergence were low.

3.3. Experimentation

The experimental tests were performed using the results obtained in the 16 active elements of CIVA simulation, so that the results of both techniques could be compared. The calibration block was withdrawn from the GTAW/SMAW welded joint. The weld reinforcement was removed and 3 through-holes with a 2.25 mm diameter were machined, as recommended by ASME Section V-article 4. The simulation of sensitivity calibration with longitudinal waves in the calibration block is shown in Figure 12.

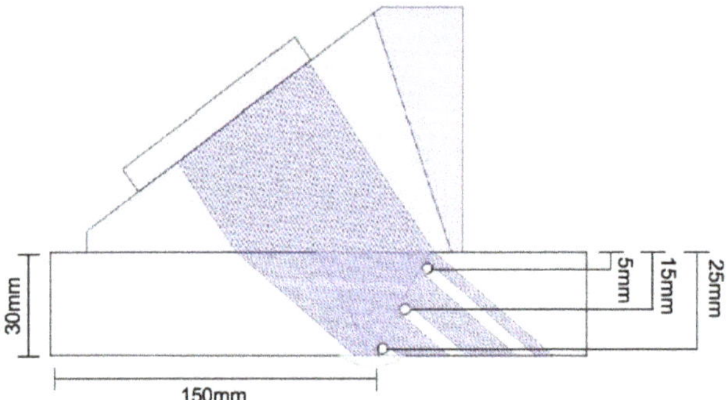

Figure 12. Simulation of sensitivity calibration with longitudinal waves in the calibration block, withdrawn from the 9% Ni quenched and tempered steel pipe joint welded with GTAW and SMAW processes using Ni-based superalloy 625 as the filler metal.

The transducer wedge was positioned on the calibration block to allow the sonic beam to strike over the third hole, near the root of the welded joint. For ultrasonic scanning with an angle of 45° and 16 active elements, the gain from the third hole reflection was 50 dB (Figure 13).

The echo from the third hole was maximized to 80% of the screen height, as indicated in Figure 14. After tracing the TCG (Time Corrected Gain) in the extension, it was necessary to inspect the weld metal with 3 holes and the primary gain was 22.8 dB.

The TCG setting was performed according to ASME V [21] using the calibration block, which was withdrawn from the 9% Ni steel joint, to equalize the sensitivity of all focal laws for defects located in different depths from the surface of the inspected piece. The TCG is corrected in time so that the reflectors (defects) have equal amplitudes for all sonic path distances, which compensates for the attenuation inside the material and inside the wedge.

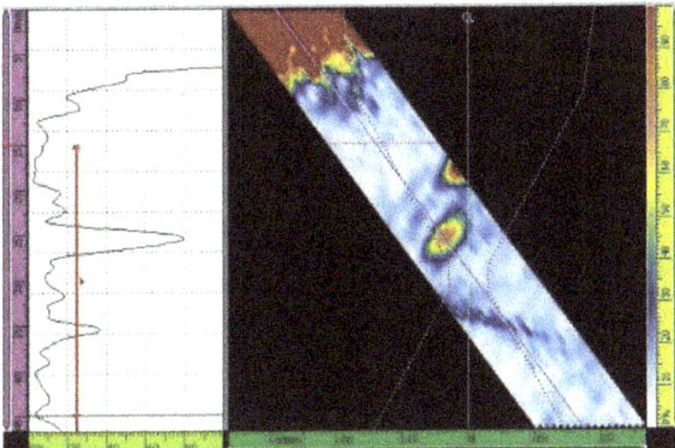

Figure 13. Echo (maximized to 80% of the screen height) from the third hole. Scanning angle of 45° and 16 active elements. Primary gain = 50 dB.

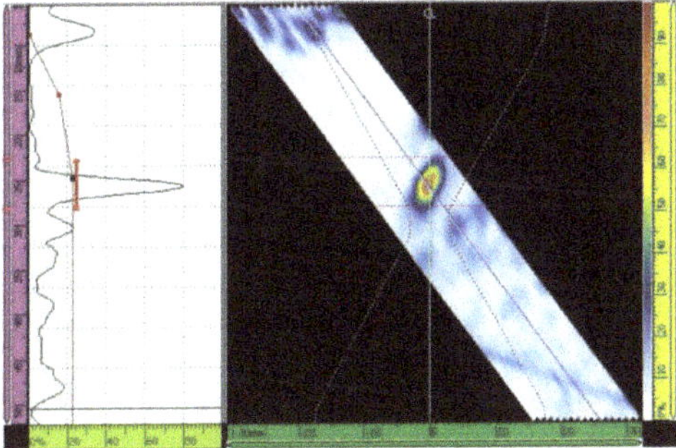

Figure 14. Time correct gain adjustment. Scanning angle of 45°, 16 active elements. Primary gain = 22.8 dB.

For the ultrasonic scanning with angle of 48° and 16 active elements, the gain from the third hole reflection was 40.1 dB and the primary gain was 22.6 dB, as shown in Figures 15 and 16.

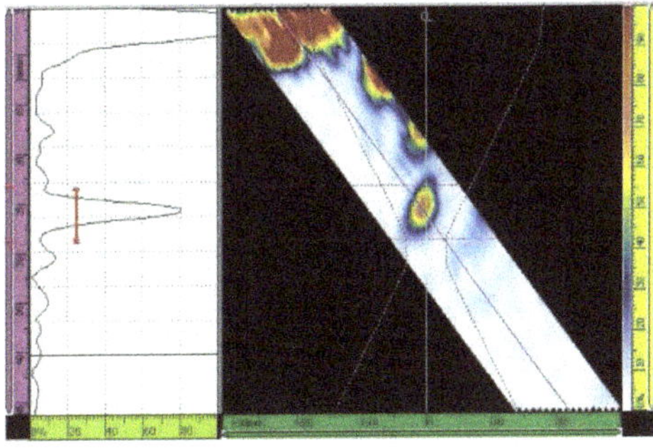

Figure 15. Echo (maximized to 80% of the screen height) from the third hole. Scanning angle of 48°, 16 active elements. Primary gain = 40.1 dB.

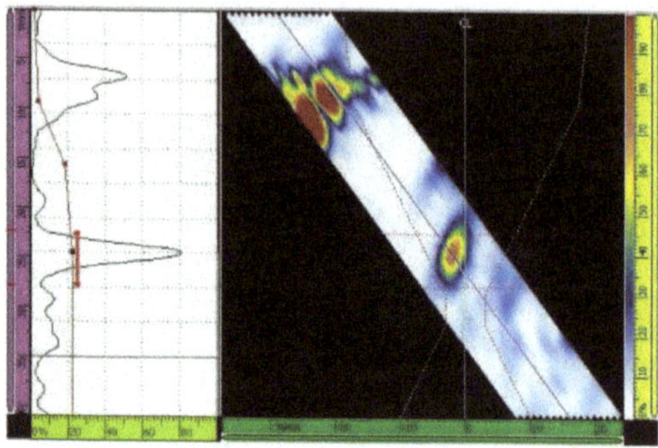

Figure 16. Time corrected gain adjustment. Scanning angle of 48°, 16 active elements. Primary gain = 22.6 dB.

Finally, for ultrasonic scanning with an angle of 52° and 16 active elements, the gain from the third hole reflection was 40.8 dB and the primary gain was 27.2 dB, as shown in Figures 17 and 18.

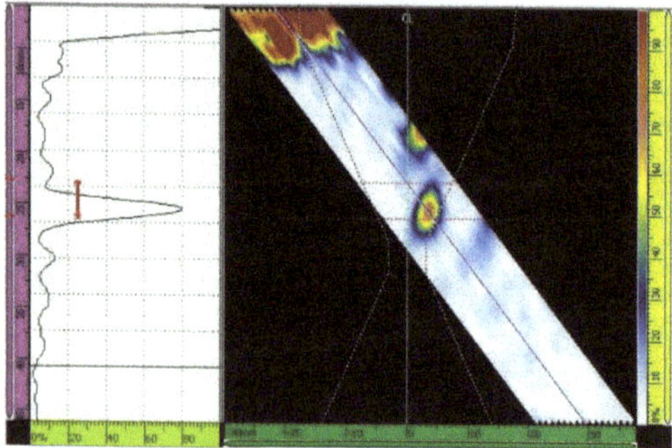

Figure 17. Echo (maximized to 80% of the screen height) from the third hole. Scanning angle of 52°, 16 active elements. Primary gain = 40.8 dB.

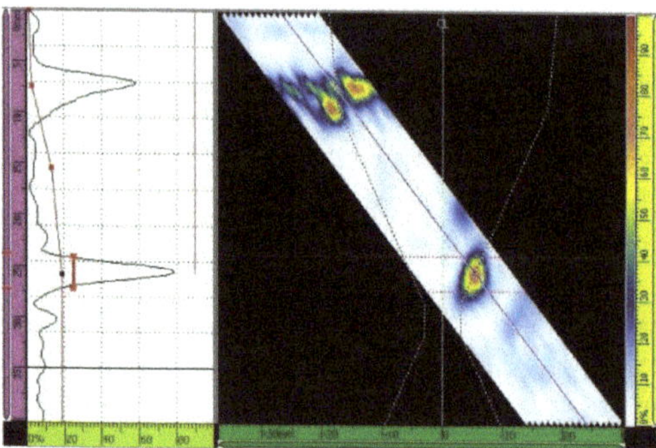

Figure 18. TCG adjustment. Scanning angle of 52°, 16 active elements. Primary gain = 27.2 dB.

During the scanning procedure, the greater the material attenuation, the greater the gain added to the signal should be to allow for a proper inspection. From that point of view, the best scanning angle for the inspection of the 9% Ni steel pipe joint SMAW welded with Ni-based superalloy 625 is 48° because the primary gains from the third hole after the TCG adjustment were lowest when compared to 45° and 52°, as seen in Table 9. Very high gain can increase the noise, thus reducing the ratio signal-to-noise and an angle of 48° presented the best signal amplitude response.

Table 9. Primary gain from the third hole after TCG adjustment for scanning angles of 45°, 48° and 52°.

Scanning Angle	Primary Gain from the Third Hole	Primary Gain after Tracing the TCG
45°	50.0 dB	22.8 dB
48°	40.1 dB	22.6 dB
52°	40.8 dB	27.2 dB

Comparing the CIVA simulations and the experimentation tests with angles of 45°, 48° and 52°, the differences among the amplitudes are clear. Greater amplitude was necessary to correct the TCG for the angle 52° because of the beam deflection and attenuation. The angles of 45° and 48° promoted minimum beam deflection during propagation, which was confirmed by experiments in the block verification for these two focal laws, as shown in Figures 13 and 15. In these figures, the maximizations of the echo to 80% show low noise levels and a clear definition of the whole amplitude.

The difference between simulations with 16, 24 and 32 active elements demonstrated that the number of active elements affects the sound pressure, the directivity, the signal-to-noise ratio and the beam attenuation, which meets the observations of Freitas [15]. The results for 16 active elements demonstrate a good sound pressure, low beam divergence and good coverage of the welded joint, which were caused by the better signal-to-noise ratio.

With respect to frequency, the experimental tests confirmed the CIVA simulations, showing that the frequency of 2.25 MHz allowed for a good penetration of the beam into the welded joint with minimal deviation, which matched the theory proposed by Neumann [3], who said that the noise generated by the grains increases with transducer frequency. The 2.25 MHz transducer used in this work has a design with a linear arrangement of 1.0 mm element pitch and 32 mm active aperture. This arrangement was reflected in a good wave front and guaranteed good sound pressure and great directivity.

As presented by Hirsekorn [10] to optimize the angle of inspection, we need to take into account the scattering of the wave in the weld metal for higher beam to grain angles. For lower angles, the effect of attenuation caused by reflection at the boundary between isotropic and anisotropic material is predominant. As the inspection was made using a longitudinal wave with the ultrasonic beam propagating directly in the weld metal, the scattering effect in attenuation has a major impact, which explains the worse results that were found when the angle of more elevated incidence was used.

As shown by Hirsekorn [10] to optimize the inspection angle, it is necessary to take into account 2 phenomena: Wave scattering in the weld metal and reflection of the wave in the fusion line. For larger angles, the scattering of the wave in the weld metal is more pronounced. For lower angles, the attenuation effect caused by the reflection at the border between isotropic and anisotropic materials is predominant. However, as the inspection was performed using a longitudinal wave with the ultrasonic beam propagating directly in the weld metal without passing through the fusion line, the scattering phenomenon has a greater effect in wave attenuation, which explains the worse results found when the greater angle of incidence was used.

In sum, the combination of the longitudinal wave, linear scanning, frequency of 2.25 MHz, transducer with 1.0 mm pitch, 16 active elements and angles of 45° or 48° made a viable austenitic metal inspection with minimum ultrasonic beam divergence and a good defect size measurement. However, the angle of 48° showed the best results. Lastly, the inspection of the welded joint between 9% Ni steel (ASTM A333 gr. 8) and Ni-based superalloy 625 is practicable but it is necessary to set up the appropriate phased array parameters.

4. Conclusions

- For linear inspection simulations with longitudinal waves, scanning angles of 45° and 48° showed the best results, with minimum beam divergence and high sonic pressure along the welded region.

- Numerical simulation presents high divergence and attenuation using an angle of 52°, for linear inspection simulations with longitudinal waves, evidencing regions where the sonic energy is absorbed by the austenitic weld metal.

- The configuration with 16 active elements proved to be the most effective scanning technique, covering a large volume of the weld, including the root region, with the same scanning index and maintaining good sonic pressure and directivity of the beam.

- According to the CIVA simulation, the optimized configuration showed that the configuration of low frequency (2.25 MHz), 16 active elements, linear scan and angles of 45° and 48° are recommended to inspect root and filler regions. In this case, at least three scanning indexes were necessary to ensure the inspection of the entire volume of the welded joint.

- Experimental tests showed that inspection with longitudinal waves, using angles of 45° and 48°, allowed the detection of the 3 holes located in the fusion line of the calibration block. However, the scanning incidence angle of 48° showed better results, since the primary gains from the third hole and after the TCG adjustment were the lowest when compared to 45° and 52°.

- The comparison between the simulation and the experimental tests demonstrated the potential of the simulation CIVA to establish the best conditions to optimize the phased array ultrasonic inspection of welded dissimilar steel tubular joints.

- The most important outcome of this work is to enable the inspection of the HAZ and the whole weld metal of thick-walled 9% Ni welded joints by choosing the proper combination of phased array ultrasonic parameters. Because of the groundbreaking application, an ultrasonic inspection procedure had not been established before the one obtained in this work.

Author Contributions: Formal analysis, E.K.D.P., R.F.F. and D.D.S.; Funding acquisition, D.R.J.; Investigation, R.F.F. and D.D.S.; Methodology, R.F.F. and D.D.S.; Project administration, J.d.C.P.F.; Supervision, J.d.C.P.F.; Validation, D.R.J. and R.S.G.; Visualization, E.K.D.P.; Writing-original draft, E.K.D.P.; Writing-review & editing, J.d.C.P.F.

Funding: This research was funded by Empresa Brasileira de Pesquisa e Inovação Industrial-Embrapii, PCCOP1512.0004" and Shell Brasil Ltda., TO821.

Acknowledgments: This work has been conducted with financial support from Shell Brasil Petróleo Ltda. (Shell Brasil), Empresa Brasileira de Pesquisa e Inovação Industrial (Embrapii) and Agência Nacional de Petróleo, Gás Natural e Biocombustíveis (ANP) and was carried out in partnership with Vallourec Soluções Tubulares do Brasil S.A. (Vallourec) and Serviço Nacional de Aprendizagem Industrial do Rio de Janeiro (Senai-RJ). Special thanks to engineer Vinícius Pereira Maia.

Conflicts of Interest: The authors declare no conflict of interest. The founding sponsors had no role in the design of the study; in the collection, analyses, or interpretation of data; in the writing of the manuscript or in the decision to publish the results.

References

1. Rodrigues, R.C. Avaliação das Transformações de Fase do Aço com 9% de Níquel e das Zonas Termicamente Afetadas Simuladas in situ com Difração de raios-X Síncrotron. Ph.D. Thesis, Universidade Federal Fluminense, Rio de Janeiro, Brasil, 2016.

2. Hudgell, R.J.; Gray, B.S. *The Ultrasonic Inspection of Austenitic Materials: State of the Art Report*; United Kingdom Atomic Energy Authority, Northern Division: Abingdon-on-Thames, UK, 1985.

3. Neumann, A.W.E. On the state of the art of the inspection of austenitic welds with ultrasound. *Int. J. Press. Vessel. Pip.* **1989**, *39*, 227–246. [CrossRef]

4. Papadakis, E.P. Influence of preferred orientation on ultrasonic grain scattering. *J. Appl. Phys.* **1965**, *36*, 1738–1740. [CrossRef]

5. Kupperman, D.S.; Reimann, K.J.; Yuhas, D. Visualization of ultrasonic beam distortion in anisotropic stainless steel. *Quant. NDE Nucl. Ind.* **1983**, *18*, 172–176.

6. Wagner, S.; Dugan, S.; Stubenrauch, S.; Jacobs, O. Modification of the grain structure of austenitic welds for improved ultrasonic inspectability. *Australas. Weld J.* **2013**, *58*, 42–48.

7. Ogilvy, J.A. Computerized ultrasonic ray tracing in austenitic steel. *NDT Int.* **1985**, *18*, 67–77. [CrossRef]

8. Apfel, A.; Moysan, J.; Corneloup, G.; Fouquet, T.; Chassignole, B. Coupling an ultrasonic propagation code with a model of the heterogeneity of multipass welds to simulate ultrasonic testing. *Ultrasonics* **2005**, *43*, 447–456. [CrossRef] [PubMed]

9. Moysan, J.; Apfel, A.; Corneloup, G.; Chassignole, B. Modelling the grain orientation of austenitic stainless steel multipass welds to improve ultrasonic assessment of structural integrity. *Int. J. Press. Vessel. Pip.* **2003**, *80*, 77–85. [CrossRef]

10. Hirsekorn, S. Directional dependence of ultrasonic propagation in textured polycrystals. *J. Acoust. Soc. Am.* **1986**, *79*, 1269–1279. [CrossRef]

11. Willems, H. A new method for the measurement of ultrasonic absorption in polycrystalline materials. In *Review of Progress in Quantitative Nondestructive Evaluation*; Springer: Boston, MA, USA, 1987; pp. 473–481.

12. Bouda, A.b.; Lebaili, S.; Benchaala, A. Grain size influence on ultrasonic velocities and attenuation. *NDT E Int.* **2003**, *36*, 1–5. [CrossRef]

13. Flotté, D.; Bittendiebel, S. A Phased Array Ultrasonic Testing of a Manual Thick Austenitic Weld-*Feeedback*. In Proceedings of the 19th World Conference on Nondestructive Testing 2016, Munich, Germany, 13–17 June 2016.

14. Volkov, A.S.; Ermolov, I.N.; Basatskaya, L.V.; Vyatskov, I.A.; Grebennik, V.S. Ultrasonic transmission by the boundary in an austenitic weld. *Sov. J. Nondestr. Test.-Ussr* **1984**, *20*, 134–137.

15. Freitas, M.I. Development of an ultrasonic Phased Array system to inspect welded joints of low thickness austenitic steel. Master's Thesis, Universidade de Lisboa, Lisbon, Portugal, May 2016.

16. Lee, J.H.; Choi, S.W. A parametric study of ultrasonic beam profiles for a linear phased array transducer. *IEEE Trans. Ultrason. Ferroelectr. Freq. Control* **2000**, *47*, 644–650. [PubMed]

17. ASTM International. *ASTM A333/A333M-16 Standard Specification for Seamless and Welded Steel Pipe for Low-Temperature Service and Other Applications with Required Notch Toughness*; ASTM International: West Conshohocken, PA, USA, 2016.

18. American Welding Society. *A5.14/A5.14M:2018: Specification for Nickel and Nickel-Alloy Bare Welding Electrodes and Rods*; American Welding Society, Inc. (AWS): Miami, FL, USA, 2018.

19. American Welding Society. *A5.11/A5.11M:2018: Specification for Nickel and Nickel-Alloy Welding Electrodes for Shielded Metal Arc Welding*; American Welding Society, Inc. (AWS): Miami, FL, USA, 2018.

20. *Ultrasonic Transducers Technical Notes*; Tech Broch Olympus NDT: Waltham, MA, USA, 2016.

21. American Society of Mechanical Engineers. *ASME Boiler and Pressure Vessel Code*; American Society of Mechanical Engineers, Boiler and Pressure Vessel Committee: New York, NY, USA, 1900.

 *metals* 

Article

Thermal Spray Coatings as an Adhesion Promoter in Metal/FRP Joints

Thomas Lindner *, Erik Saborowski, Mario Scholze, Benjamin Zillmann and Thomas Lampke

Materials and Surface Engineering Group, Institute of Materials Science and Engineering, Chemnitz University of Technology, D-09107 Chemnitz, Germany; erik.saborowski@mb.tu-chemnitz.de (E.S.); mario.scholze@mb.tu-chemnitz.de (M.S.); Benjamin.Zillmann@de.bosch.com (B.Z.); thomas.lampke@mb.tu-chemnitz.de (T.L.)
* Correspondence: th.lindner@mb.tu-chemnitz.de; Tel.: +49-371-5313-8287

Received: 17 September 2018; Accepted: 25 September 2018; Published: 27 September 2018

Abstract: In this study, various structuring methods for creating adhesion by mechanical interlocking in the interface of metal/FRP (fiber-reinforced polymer) joints are investigated. A novel processing route using thermal spray coatings as additive structure is presented. Different coating systems are first assessed by axial loading tests with spray-coated plungers for the evaluation of the additive layer adhesion on the metallic base material. Additional microstructures, produced by different abrasive processes (corundum blasting, laser structuring, and fine milling) are compared with the additive structures. All surface structures are characterized by electron microscopy for two sheet materials: DC06 and AA6016-T4. The abrasive structures show a significant material dependence, while the selected coating system offers the adjustment to different base materials by an independent surface layer. The structured metal sheets were further joined to glass-fiber-reinforced polyamide 6 (PA6) by hot pressing to evaluate the interface properties in tensile shear tests. The results confirm a suitability of thermal spray coatings for providing a high bonding strength in metal/FRP joints for both investigated metallic substrate materials.

Keywords: hybrid joining; surface structuring; thermal spraying; coating; FRP; hot pressing; bonding strength; adhesion; mechanical interlocking

---

1. Introduction

Hybrid material compounds, consisting of metal and fiber-reinforced polymer (FRP), are very common for lightweight structures in the aviation field, as well as in the automotive industry, due to their high strength-to-weight ratio. A key challenge is the joining of those dissimilar materials, especially metal and polymer. Currently, hybrid material compounds are generated by

- adhesive bonds by Huang et al. [1],
- mechanical joints with rivets shown by Di Franco et al. [2] or bolts by Matsuzaki et al. [3],
- mechanical joining by plastic deformation of the metal and polymer partner (e.g., clinching showed by Lambiase and Di Ilio [4] and Friedrich et al. [5]),
- friction spot or lap welding by Amancio-Filho et al. [6] and Liu et al. [7],
- ultrasonic spot welding by Wagner et al. [8] and Mitschang et al. [9].
- direct joining by hot pressing with adhesion promoters by Yulinova et al. [10] and without by Sickert and Haberstroh [11], and
- laser-assisted joining by Katayama and Kawahito [12].

Adhesive bonding is the state of the art in joining metal and FRP. Huang et al. [13] studied adhesively bonded joints between aluminum and FRP, which resulted in low tensile shear strength

(≈4 MPa). Huang et al. [13] also applied a plastic deformation during the joining process, which improved the strength by about 17%. Other studies by Velthuis et al. [14] and by Molitor and Young [15] showed that an adhesive bond, using advanced adhesive and an additional pretreatment of the surfaces, can reach a maximum tensile shear strength value of around 16 MPa. However, adhesively bonded joints are sensitive to several environmental conditions, like humidity (Lee et al. [16]), ultraviolet radiation (Nguyen et al. [17]) or variations in temperature (Nguyen et al. [18]). Mechanical joints with rivets or bolts increase the total weight due to the additional element. Furthermore, fibers are cut and exposed, which causes local stress concentrations that can lead to a premature failure of the component. Balle et al. [19] showed a very promising joining approach by ultrasonic spot welding. A maximum tensile shear strength of 23 MPa between an aluminum sheet and a carbon-fiber-reinforced polymer was achieved by this technology. Mitschang et al. [9] used an induction spot welding technology and gained tensile shear strength, from 15 MPa to 23 MPa, depending on the pretreatment of the metallic and polymer material. The method of friction lap welding was used for a hybrid aluminum-nylon join by Liu et al. [7], achieving a maximum tensile shear strength of 5–8 MPa. Sickert and Haberstroh [11] studied the process of direct joining of FRP to metal by hot pressing using multiple joining steps. A strength improvement of about 10% could be achieved with multiple joining steps and different pressures, compared to a constant pressure. Yulinova et al. [10] used twin polymers as an adhesion promoter on a non-treated metal surface in the hot joining process. The achieved maximum shear strength was 13 MPa, which was about 50% higher compared to the non-promoted corundum blasted surface.

Consequently, the achieved maximum strength is a function of the joining process itself. Furthermore, the surface preparation prior to the joining process has a huge influence on the final properties. Mitschang et al. [9] studied the joining strength by applying different pretreatments on the metallic and polymer partners. The pretreatment can be done by chemical, mechanical, or physical processes as listed by Velthuis et al. [14]. A major influence was found in using corundum blasting and acidic pickling, which increased the tensile shear strength by 60% compared to the other pretreatment steps. In general, the main goal of structuring the surface of the metal component are the increase in the surface area and the generation of undercuts for a mechanical clamping of the polymer. The corundum blasting can induce a certain amount of residual stresses in the material (Tosha and Iida [20]), causing deformations of thin-walled components. Another promising method is the mechanical structuring by using a laser showed by Schulze et al. [21]. The advantages are the high flexibility of the work piece geometry and process design, while area output is limited. The laser creates a microstructure with undercuts, which also allows a mechanical clamping of the dissimilar materials. Moreover, Roesner et al. [22] found an increase in tensile shear strength with a reduction of the laser line distance.

The present study introduces thermal spraying technology as a new process to create an advanced integration zone for metal/FRP joints. Thermal spray coatings can especially offer porous structures and high surface roughness. Consequently, a large specific surface area is achieved, which is beneficial for a higher joining strength. The thermal spray coatings are, furthermore, compared to three different abrasive mechanical structuring methods (corundum blasting, laser, and milling applications) in terms of the resulting strength between metal and FRP, as well as the microstructure of the integration zone after joining. The aim of this work is to produce an integration zone with high bonding strength and an easy implementation in an industrial environment.

2. Materials and Methods

2.1. Materials Characterization

The investigated hybrid metal/FRP compounds consist of a glass-fiber reinforced PA6 (2-layered with a 0°/90° fiber orientation) connected to DC06 as a low carbon deep drawing steel and an AA6016-T4 aluminum alloy, respectively. The material properties are shown in Table 1. A differential scanning calorimetry (DSC) measurement of the FRP has been carried out for determining the process window for hot pressing of the compound. Figure 1 shows two endothermic peaks for the melting and decomposition temperatures of the PA6 at 496 K and 730 K, respectively.

Table 1. Material properties.

Material	Thickness (mm)	Elongation to Failure (%)	Ultimate Strength (MPa)	Yield Strength (MPa)	Hardness (HV 10)
FRP	2	0.9	440	-	-
AA6016-T4	1	25	170	120	73
DC06	1	50	270	160	84

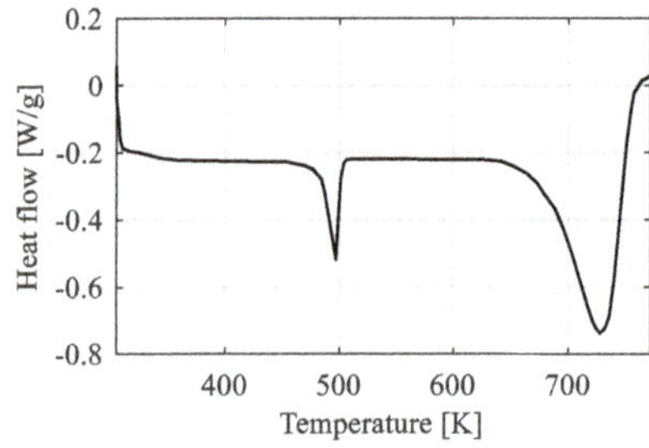

Figure 1. DSC measurement of the FRP with two endothermic peaks indicating the melting and decomposition temperatures.

2.2. Surface Pretreatment

For the specific design of an integration zone, a suitable surface treatment is necessary. During the conducted investigations, micro- and macrostructures were generated on the DC06 and AA6016-T4 surface. This leads to a higher specific surface area, and provides good conditions for an optimal mechanical interlocking of the PA6. Thermal spraying, corundum blasting, and laser structuring were chosen as microstructuring methods. The macrostructures were generated by milling. The corundum blasting was carried out with aluminum oxide EK-F-24 at a pressure of 2 bar with a blasting angle of 70° and a distance of 200 mm. Arc wire spraying has been chosen for the thermal spray coating process, due to its advantageous properties in terms of processability, productivity, and cost efficiency. The arc wire spray system VisuArc 350 (Sulzer, Winterthur, Switzerland) was used to apply the spray layers, Figure 2.

Figure 2. Sulzer VisuArc 350 arc wire spraying system.

NiAl5, NiAl20, and NiCr20 (cored wire thickness: 1.6 mm) have been chosen as spraying materials, as these alloys are typical bond coat materials in thermal spraying, referring to Davis [23] and American Welding Society [24]. The used alloys, as well as the corresponding parameters of the thermal spraying process, can be seen in Table 2.

Table 2. Coating materials and parameters used for electric arc wire spraying.

Alloy	Current (A)	Voltage (V)	Feed Velocity (m/s)	Spraying Distance (mm)	Air Pressure (bar)	Line Distance (mm)	Growth by Step (μm)	Thermal Expansion (10^{-6} K^{-1})
NiAl5	200	25	1.0	130	3.5	5	50	13.9
NiAl20	200	25	1.0	130	3.5	5	50	15.3
NiCr20	240	24	1.0	130	3.5	5	40	11.7

A fiber laser (ytterbium, wavelength: 1064 nm) was used for laser structuring the metallic joining partners. The laser parameters (Table 3) had to be adjusted to the different materials, due to the lower absorption coefficient of the aluminum, which results in a lower coupling of the laser beam and an associated restriction of the processing options.

Table 3. Parameters for laser structuring depending on the used materials.

Material	Power (W)	Velocity (mm/min)	Pulse Width (ns)	Spot (μm)	Line Distance (μm)	Penetration Depth (μm)
AA6016-T4	1300	150	200	50	250	20–50
DC06	1800	200	200	50	500	75–90

The macrostructures were produced by milling. The cutting speed was set to 150 m/min, and the feed speed to 0.3 mm/tooth. The resulting macrostructure consists of pyramid-shaped elements with a height of 0.5 mm and an angle of 45°.

Additionally, the surface roughness was measured by tactile incision technique using a T8000 system (Hommel-Etamic, Jena, Germany). All measurements were performed with a TKU300 tip using uniform parameters for all structures (measurement length = 1.5 mm, velocity = 0.5 mm/s, point distance = 0.5 μm).

2.3. Mechanical Testing of Adhesive Strength

Axial loading tests according to DIN EN 582 for the preselection of the spray system were performed utilizing adhesive pads FM1000 (Cytec Solvey Group, Woodland Park, CO, USA) in combination with the spray-coated metal plungers (Figure 3a). The specimens were joined to the adhesion pads by hot pressing. Metal plungers with a length of 50 mm and a diameter of 25 mm were used. During the joining process for the axial loading tests, the samples were treated at 473 K and fixed for a constant joining pressure in a preheated furnace for 90 min. Afterwards, cooling down was conducted directly in the furnace for approximately 300 min. After preselection of the coating system, the feedstock material was applied to the plungers of both metallic materials, DC06, and AA6016-T4. Furthermore, the bonding strength was examined in dependency of the pretreatment state of the surface using FRP as a joining partner (Figure 3b).

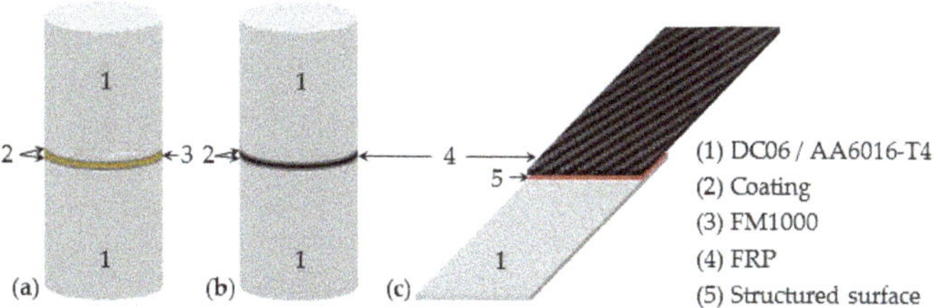

Figure 3. Setup for evaluating (**a**) the axial bonding strength of the thermal spray coating to the metal sheet; (**b**) between the coating and the FRP; and (**c**) the tensile shear strength of different surface structures to the FRP.

In order to optimize the joining process, the two plungers were inductively heated to 558 K within 20 s, followed by insertion of the FRP between the plungers and applying a joining pressure of 0.3 MPa. As a result of the optimized joining process, the samples are completely cooled down within 5 min. Axial loading tests were performed at 1 mm/min traverse speed in a tensile testing machine with a standard load cell Z020 (Zwick/Roell, Ulm, Germany). The bonding strength was determined using the calculated circular contact area of 491 mm^2.

Tensile shear tests were carried out in accordance to DIN EN 1465 to evaluate the shear strength of a single lap joint, Figure 3c. Metal and FRP stripes with a length of 100 mm and a width of 25 mm were bonded to each other with an overlapping length of 5 mm. In contrast to the standardized geometry, the overlap length was reduced from 12.5 mm to 5 mm, in order to enable a meaningful testing of high shear strength, due to a more homogeneous stress distribution among the interface (Saborowski et al. [25]). For tensile shear strength testing, the specimens were produced utilizing a die heating process (see Figure 4a for the lap shear specimen production tool). According to the production parameters determined by Haberstroh and Sickert [26], a joining pressure of 0.3 MPa and a maximum joining temperature of 558 K were chosen. During the joining process, the die is heated until the maximum joining temperature inside of the integration zone is reached, and immediately afterwards cooled down via an air-cooling system until the temperature drops below 373 K (Figure 4b). The pressure is maintained constant over the whole production time to prevent the formation of cavities in the polymer during the cooling process (Flock [27]). The temperature in the integration zone is observed by a thermocouple placed inside a drill-hole slightly below the metal surface. Finally, the specimens were tested in a tensile testing machine Z020 (Zwick/Roell, Ulm, Germany) under a constant crosshead speed of 1 mm/min. The tensile shear strength was determined using the nominal value of the rectangular contact area of 125 mm^2.

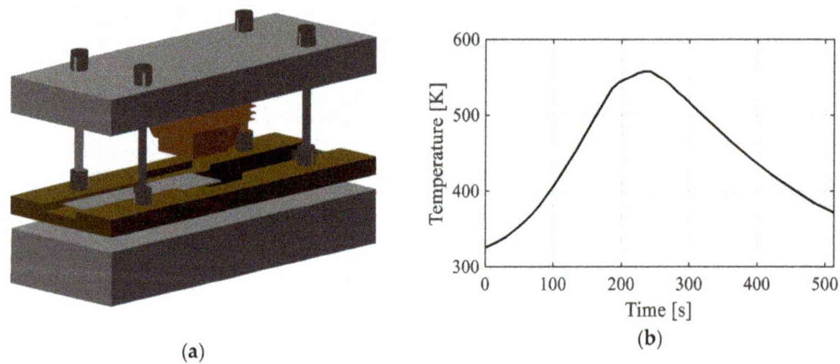

(a)   (b)

Figure 4. Joining fixture and temperature profile. (**a**) Apparatus with a heatable die for the production of tensile shear specimens; (**b**) Temperature inside the joining zone during the hot pressing process.

3. Results and Discussion

3.1. Preselection of Thermal Spray Coating Material

To choose a suitable thermal spray feedstock material, the adhesion between the coating materials and the DC06 plunger was investigated, Table 2. The coating thickness as well as the surface pretreatment were investigated, Figure 5a. The lowest coating thickness, in combination with the corundum-blasted surface, indicates the highest bonding strength (about 60 MPa), which is independent of the coating material. Moreover, the bonding strength decreases with increasing layer thickness.

NiCr20 shows a drop of only 25% when comparing the bonding strength at 40 µm and 320 µm coating thickness. The pretreatment of surface strongly affects the achievable bonding strength. A non-blasted surface state reduces the bonding strength, since the mechanical interlocking between the substrate and coating material is less pronounced. The achieved higher values for NiAl5 can be explained by micro metallurgical bonding resulting from the exothermic reaction between the alloy components [28]. The determined bonding strength of the coatings, with the exception of NiCr20 by non-blasted substrate, were far above 25 MPa. The high adhesion strength of the spray coatings is one of the key requirements for the suitability of thermal spray coatings as an adhesion promoter in metal/FRP compounds. For all further experiments, NiAl5 was chosen as suitable representative material for this feedstock class.

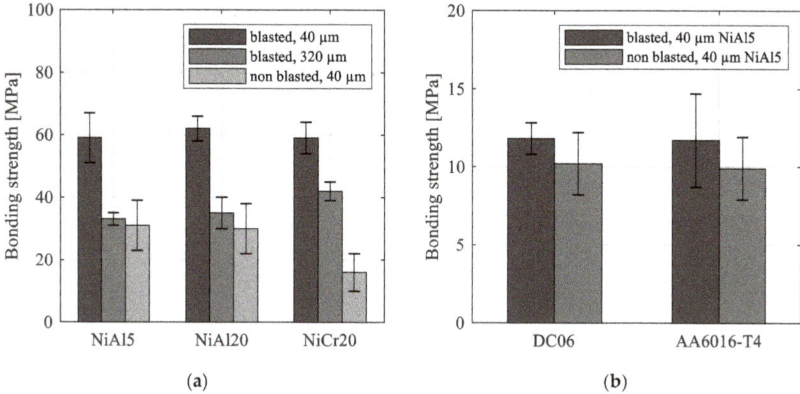

Figure 5. Bonding strength under axial tension (**a**) thermal spray layer to DC06 substrate; (**b**) FRP to thermal spray layer (NiAl5).

The results for the specimen with FRP as joining partner are presented in Figure 5b. The tensile strength is in a range of 10–12 MPa for both metallic plunger materials. A roughening of the surface by corundum blasting only affects the bonding strength insignificantly. Consequently, it is appropriate to apply the thermal spray coating to the non-blasted surface.

3.2. Characterization of Structured Metal Surfaces

Scanning electron microscope (SEM) surface images for the two sheet materials are shown in Figure 6. The corundum blasted surfaces exhibit a random distribution of the surface structure with R_z = 36 μm, whereby the aluminum material shows a higher roughness (R_z = 81 μm) because of its lower hardness. The surface structure of the NiAl5 coating offers similar surface characteristics for both substrate materials. The surface roughness is randomly distributed and has similar peak values (DC06 NiAl5: R_z = 85 μm, Al6016-T4 NiAl5: 88 μm) compared to the corundum-blasted AA6016-T4. The laser structured surfaces show a homogenous grid with height peaks of about 80 μm for DC06, and 20 μm for the aluminum material. Additionally, the edges are covered by melted substrate material, which increases the specific surface area again. The differences in the laser structure of DC06 and AA6016-T4 are presumably caused by different absorption coefficients. The milled surfaces exhibit pyramidal structures. Comparing the different milled surfaces, a homogeneous structure was achieved in case of AA6016-T4, while DC06 shows several residues, which could, for example, promote better mechanical clamping to FRP. In summary, a considerable number of undercuts, as well as a high specific surface area, ensures sufficient potential for a high joining strength between metal and FRP.

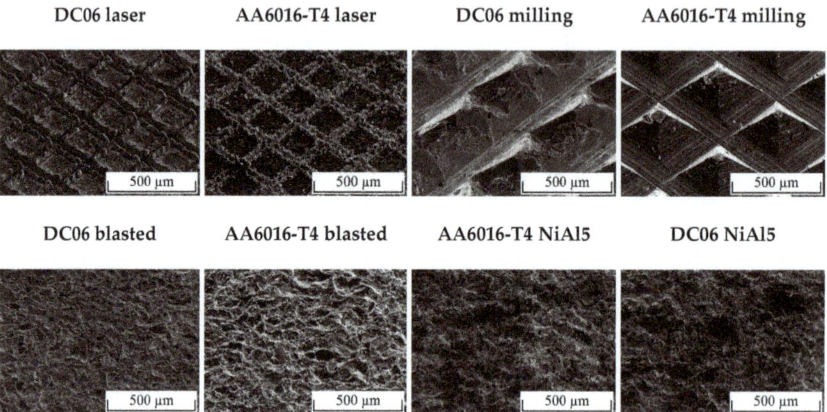

Figure 6. SEM images of the differently structured surfaces for DC06 and AA6016-T4.

3.3. Tensile Shear Properties of FRP/Metal Joints

Figure 7 shows the results of the tensile shear tests for the metal/FRP specimens with the differently structured metal surfaces. There is a significant influence of the surface pretreatment on the shear strength. The most promising methods in terms of high shear strength are laser treatment and the thermally sprayed NiAl5 coating. Thermal spraying ensures mechanical clamping by several undercuts, and reduces the influence of the substrate material through adjustment of the coating material.

Consequently, a high bonding strength between the metallic sheet and FRP is ensured. The corundum blasting and milling treatments show significant differences in surface structures as well as in the resulting shear strength, depending on the used sheet material. When comparing the surface topography (Figure 6) of the different structures with the bonding strength, assessment of the determined bonding strength becomes possible. Corundum blasting of AA6016-T4 results in a higher roughness in contrast to DC06. By using equal parameter conditions, the relatively soft aluminum substrate shows a significantly higher abrasive effect than steel. The higher roughness is presumably the main reason for the higher bonding strength caused by enhanced mechanical interlocking of the polymer melt. Laser structuring shows only a minor material influence and a good mechanical clamping is provided in tensile shear testing. The resulting structure using the milling process differs significantly between steel and aluminum. A high number of slats in the pre-milled lines of the DC06 substrate builds undercuts for the polymer melt. AA6016-T4 does not show such structures, while shear strength is the lowest of all compared structures.

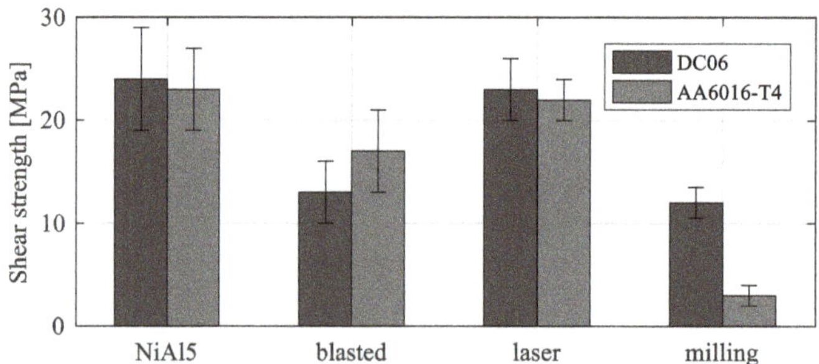

Figure 7. Tensile shear strength of metal/FRP specimen joined by die heating.

Presumably, the main influence on bonding strength of the investigated structures is mechanical clamping. A combination of undercuts and randomized surface structures with a certain roughness lead, especially in the case of spray coated and laser structured interfaces, to high values in bonding strength. When comparing the time of manufacturing for a specific surface area, the potential of thermal spray coating becomes obvious. While laser structuring offers approximately the same interface strength, the required manufacturing time for these abrasive structures is significantly higher. Considering the differences in processing parameters used in this study (which depend on the treated metal sheet) e.g., velocity, line distance and the additional secondary overrun (rotated under 90°), laser structuring takes about 6000–16,000 times longer for the same specific surface area in comparison to thermal spraying. Furthermore, processing parameters for all abrasive treatments, including laser structuring, need to be adjusted in a complex way when changing the type of base material. In contrast, thermal spraying delivers a reproducible and similar bonding strength for different substrate materials and the combination of different materials for FRP/metal compounds is facilitated.

4. Conclusions

An experimental study of different surface structuring methods as adhesion promoter for joining hybrid metal/FRP compounds has been carried out in this work. Thermal spray technology is presented as a new approach for additive structuring of the metallic surface.

NiAl5 as feedstock material exhibits a good adhesion even on smooth metallic surfaces. A material adjustment is implemented by the coating, which provides a similar bonding strength in metal/FRP compounds regardless of the metallic part. Especially, the high surface roughness with undercuts is suitable for mechanical clamping. With a shear strength of 23–24 MPa, the thermal spray coating achieved the best results of all presented structuring methods. Laser structuring showed similarly good results (−4.2% DC06/−4.3% AA6016-T4), whereas corundum blasting achieves a considerably lower adhesion (−45.8% DC06/−29.2% AA6016-T4) and the investigated milling structure is not useful at all (−51.1% DC06/−87.0% AA6016-T4). Taking into account the different surface topographies, the results can be explained in the following way:

- The adhesion is mainly achieved by mechanical clamping when just using a structured surface without chemical adhesion promoters
- A surface profile containing a considerable number of undercuts delivers the best results
- The bonding strength is not influenced by the substrate material when using thermal spraying due to the almost identical surface characteristics
- The time for manufacturing the surface structure is significantly reduced when using thermal spraying in comparison to laser structuring

Author Contributions: Thomas Lindner and Benjamin Zillmann planned and designed the experiments. Benjamin Zillmann, Thomas Lindner, Erik Saborowski and Mario Scholze conducted the experiments, analysed the data and drafted the manuscript. Thomas Lampke directed the research and contributed to the discussion and interpretation of the results.

Funding: This study was supported by the Federal Cluster of Excellence EXC 1075 'MERGE Technologies for Multifunctional Lightweight Structures'. Funding by the German Research Foundation (Deutsche Forschungsgemeinschaft, DFG), Germany is gratefully acknowledged.

Conflicts of Interest: The authors declare no conflict of interest.

References

1. Huang, Z.; Sugiyama, S.; Yanagimoto, J. Hybrid joining process for carbon fiber reinforced thermosetting plastic and metallic thin sheets by chemical bonding and plastic deformation. *J. Mater. Process. Technol.* **2013**, *213*, 1864–1874. [CrossRef]
2. Di Franco, G.; Fratini, L.; Pasta, A. Analysis of the mechanical performance of hybrid (SPR/bonded) single-lap joints between CFRP panels and aluminum blanks. *Int. J. Adhes. Adhes.* **2013**, *41*, 24–32. [CrossRef]

3. Matsuzaki, R.; Shibata, M.; Todoroki, A. Improving performance of GFRP/aluminum single lap joints using bolted/co-cured hybrid method. *Compos. Part A Appl. Sci. Manuf.* **2008**, *39*, 154–163. [CrossRef]

4. Lambiase, F.; Di Ilio, A. Mechanical clinching of metal-polymer joints. *J. Mater. Process. Technol.* **2015**, *215*, 12–19. [CrossRef]

5. Friedrich, S.; Georgi, W.; Gehde, M.; Mayer, P. Hybrid joining technology—A new method for joining thermoplastic-metal-mixed components. *AIP Conf. Proc.* **2014**, *1593*, 121–127.

6. Amancio-Filho, S.T.; Bueno, C.; dos Santos, J.F.; Huber, N.; Hage, E., Jr. On the feasibility of friction spot joining in magnesium/fiber-reinforced polymer composite hybrid structures. *Mater. Sci. Eng.* **2011**, *528*, 3841–3848. [CrossRef]

7. Liu, F.; Liao, J.; Nakata, K. Joining of metal to plastic using friction lap welding. *Mater. Des.* **2014**, *54*, 236–244. [CrossRef]

8. Mitschang, P.; Velthuis, R.; Didi, M. Induction spot welding of metal/CFRPC hybrid joints. *Adv. Eng. Mater.* **2013**, *15*, 804–813. [CrossRef]

9. Wagner, G.; Balle, F.; Eifler, D. Ultrasonic welding of aluminum alloys to fiber reinforced polymers. *Adv. Eng. Mater.* **2013**, *15*, 792–803. [CrossRef]

10. Yulinova, A.; Göring, M.; Nickel, D.; Spange, S.; Lampke, T. Novel adhesion promoter for metalplastic composites. *Adv. Eng. Mater.* **2015**, *17*, 802–809. [CrossRef]

11. Sickert, M.; Haberstroh, E. Thermal direct joining for hybrid plastic metal structures. In Proceedings of the Euro Hybrid Materials and Structures, Stade, Germany, 10–11 April 2014; pp. 42–45.

12. Katayama, S.K.Y. Laser direct joining of metal and plastic. *Scr. Mater.* **2008**, *59*, 1247–1250. [CrossRef]

13. Huang, Z.; Sugiyama, S.; Yanagimoto, J. Applicability of adhesive embossing hybrid joining process to glass-fiber-reinforced plastic and metallic thin sheets. *J. Mater. Process. Technol.* **2014**, *214*, 2018–2028. [CrossRef]

14. Velthuis, R.; Kotter, M.; Geiss, P.L.; Mitschang, P.; Schlarb, A.K. Lightweight structures made of metal and fiber-reinforced polymers. *Kunststoffe Int.* **2007**, *11*, 22–24.

15. Molitor, P.; Young, T. Adhesives bonding of a titanium alloy to a glass fibre reinforced composite material. *Int. J. Adhes. Adhes.* **2002**, *22*, 101–107. [CrossRef]

16. Lee, D.G.; Kwon, J.W.; Cho, D.H. Hygrothermal effects on the strength of adhesively bonded joints. *J. Adhes. Sci. Technol.* **1998**, *12*, 1253–1275.

17. Nguyen, T.; Bai, Y.; Zhao, X.; Al-Mahaidi, R. Effects of ultraviolet radiation and associated elevated temperature on mechanical performance of steel/CFRP double strap joints. *Compos. Struct.* **2012**, *94*, 3563–3573. [CrossRef]

18. Nguyen, T.; Bai, Y.; Al-Mahaidi, R.; Zhao, X. Time-dependent behaviour of steel/CFRP double strap joints subjected to combined thermal and mechanical loading. *Compos. Struct.* **2012**, *94*, 1826–1833. [CrossRef]

19. Balle, F.; Wagner, G.; Eifler, D. Ultrasonic spot welding of aluminum sheet/carbon fiber reinforced polymer joints. *Mater. Und Werkst.* **2007**, *38*, 934–938. [CrossRef]

20. Tosha, K.; Iida, K. Residual stress on the grit blasted surface. *Metal Behav. Surf. Eng.* **1989**, 323–328. Available online: https://www.researchgate.net/profile/Katsuji_Tosha/publication/266463084_Residual_Stress_on_the_Grit_Blasted_Surface/links/563c0c5b08ae405111a78398/Residual-Stress-on-the-Grit-Blasted-Surface.pdf (accessed on 26 September 2018).

21. Schulze, K.; Hausmann, J.; Wielage, B. The stability of different titanium-peek interfaces against water. *Procedia Mater. Sci.* **2013**, *2*, 92–202. [CrossRef]

22. Roesner, A.; Olowinsky, A.; Gillner, A. Long term stability of laser joined plastic metal parts. *Phys. Procedia* **2013**, *41*, 169–171. [CrossRef]

23. Davis, J.R. *Handbook of Thermal Spray Technology*; ASM International: Russell Township, OH, USA, 2004.

24. American Welding Society. *Thermal Spraying: Practice, Theory, and Application*; American Welding Society: Miami, FL, USA, 1985.

25. Saborowski, E.; Scholze, M.; Lindner, T.; Lampke, T. A numerical and experimental comparison of test methods for the shear strength in hybrid metal/thermoplastic-compounds. *IOP Conf. Ser. Mater. Sci. Eng.* **2017**, *181*, 012031. [CrossRef]

26. Haberstroh, E.; Sickert, M. Thermal Direct Joining of Hybrid Plastic Metal Components. *KMUTNB Int. J. Appl. Sci. Technol.* **2014**, *7*, 29–34. [CrossRef]

27. Flock, D. *Heat Conduction Bonding of Plastic-Metal Hybrid Parts*; RWTH Aachen: Aachen, Germany, 2011; pp. 81–86.
28. Sheppard, J.A. Sprayed coatings of exothermically formed nickel aluminide. *Br. Weld. J.* **1963**, *3*, 603–606.

Article

Determining Material Data for Welding Simulation of Presshardened Steel

Jonny Kaars [1,*], Peter Mayr [1] and Kurt Koppe [2]

[1] Welding Engineering, Faculty of Mechanical Engineering, Chemnitz University of Technology, 09126 Chemnitz, Germany; peter.mayr@mb.tu-chemnitz.de

[2] Production Engineering, Faculty 6, Anhalt University of Applied Sciences, 06366 Koethen, Germany; kurt.koppe@hs-anhalt.de

* Correspondence: jonny.kaars@mb.tu-chemnitz.de; Tel.: +49-371-531-33731

Received: 4 September 2018; Accepted: 18 September 2018; Published: 20 September 2018

Abstract: In automotive body-in-white production, presshardened 22MnB5 steel is the most widely used ultra-high-strength steel grade. Welding is the most important faying technique for this steel type, as other faying technologies often cannot deliver the same strength-to-cost ratio. In order to conduct precise numerical simulations of the welding process, flow stress curves and thermophysical properties from room temperature up to the melting point are required. Sheet metal parts made out of 22MnB5 are welded in a presshardened, that is, martensitic state. On the contrary, only flow stress curves for soft annealed or austenitized 22MnB5 are available in the literature. Available physical material data does not cover the required temperature range or is not available at all. This work provides experimentally determined hot-flow stress curves for rapid heating of 22MnB5 from the martensitic state. The data is complemented by a comprehensive set of thermophysical data of 22MnB5 between room temperature and melting. Materials simulation methods as well as a critical literature review were employed to obtain sound thermophysical data. A comparison of the numerically computed nugget growth curve in spot welding with experimental welding results ensures the validity of the hot-flow stress curves and thermophysical data presented.

Keywords: welding; 22MnB5; flow stress; thermophysical property; numerical simulation

---

1. Introduction

The finite-element analysis (FEA) of welding processes like spot and arc welding has a key role in the design of the faying process. The part design can be significantly influenced or even determined by requirements of the welding technology. Weldment properties eventually will determine the mechanical performance of the welded structure. Tolerances in industrial production are critical influences on the choice of welding processes and parameters, as pointed out by Podržaj et al. [1] as well as Schlosser and Jüttner [2] and Häßler and Füssel [3]. While Brauser et al. [4] emphasized the effect of gaps on spot welding quality, Moos and Vezzetti [5] for the first time employed FEA to further investigate the effect of tolerances in complex assemblies. Bi et al. [6] employed FEA to examine shunting at a challenging three-sheet joint, whereas van der Aa et al. [7] optimized the welding parameters of a new automotive steel grade using FEA. In the course of product development, FEA can therefore greatly reduce the probability of faulty designs and the necessity to carry out preliminary tests. This helps to save design costs and time, releasing resources to develop better-performing, more-lightweight welded structures.

The numerical simulation of welding processes and welding results is a challenging task. Reasons are the strong multiphysical couplings of phenomena and the steep temperature gradients at the weld site. With increasing material temperature, significant nonlinear changes in the material properties occur. As a result, a spatial field of local material properties, with steep gradients, will be present around the weld site, affecting the physics and the result of the welding process, and in turn, the numerical solution process. This necessitates a very careful selection of FEA boundary conditions, as shown by Raoelison et al. [8], and material data suitable for the specific temperature profiles of welding, as Schwenk and Rethmeier [9] emphasized. In order to carry out precise numerical simulations of the welding process, a complete dataset of physical and mechanical material properties up to the melting point is essential. According to Schwenk and Rethmeier [9], the dataset needs to include as a function of temperature:

- flow stress k_F;
- elastic modulus E and Poisson ratio ν;
- coefficient of thermal expansion α;
- specific electrical resistance ρ_{el};
- mass density ρ;
- specific heat capacity c_p/specific entropy h; and
- specific thermal conductivity λ.

As explained by Karbasian and Tekkaya [10], automotive parts made of 22MnB5 are heat-treated before the welding process. During the heat treatment, the desired, sometimes localised, martensitic microstructure with high strength is established. Merklein et al. [11] reported on various works on the production of load-adjusted parts with locally tailored properties by means of a tailored heat treatment. In the as-delivered state, 22MnB5 has a ferritic–pearlitic microstructure as Geiger et al. [12] stated. This peculiarity of 22MnB5 has important consequences. The flow stress curves available in the literature, as provided by Ngyuen et al. [13], are intended to be used for hot-stamping simulations of the material. This means that in mechanical testing, the material is fully austenitized before it is cooled down to the respective test temperature; cf. Merklein and Lechler [14] as well as Naderi [15]. In contrary to that, in the welding process the material is being heated up from the martensitic state to the respective temperature very quickly. As reported by Gumbsch et al. [16], significant softening of the material will occur although the temperature A_{c3} is not surpassed, that is, the martensite is only tempered. Consequently, the flow stress curves available are not valid for the welding simulation. Few authors have published information on the temperature-dependent physical properties of 22MnB5: Shapiro [17]; Spittel and Spittel [18]; and Wink and Kraetschmer [19]. In those datasets, various properties are missing altogether, while others are only available for a temperature range not sufficient for the welding process simulation.

It is the goal of this work to supplement the datasets available for 22MnB5 in the martensitic state by means of measurements, numerical computation and literature review. Although the scope of this work lies in the data application in spot welding, the data can be used for other welding processes as well.

2. Materials and Methods

2.1. Experimental

2.1.1. Test Procedure

Flow stress curves of 22MnB5 were measured using a DIL805 A/D Dilatometer built by BAEHR (Thermoanalyse GmbH, Huellhorst, Germany), shown in Figure 1. The specimens depicted in Figure 2 were cut out of the as-delivered sheet of 22MnB5 with 1.5 mm thickness by a water jet, and afterwards the coating was ground off. In the dilatometer, each specimen was induction-heated under vacuum to 1223 K within 180 s and soaked for 120 s. Following a Newtonian cooling law, the specimens were then quenched to 343 K with an average cooling rate of $\dot{T} = 30\ Ks^{-1}$ using argon gas. This temperature was held for another 120 s to allow some relief of residual stresses. Additional microsections and tensile tests proved that this initial heating and cooling cycle appropriately yields the desired martensitic microstructure in the specimens, and in turn resembles the presshardening process mentioned in the previous section.

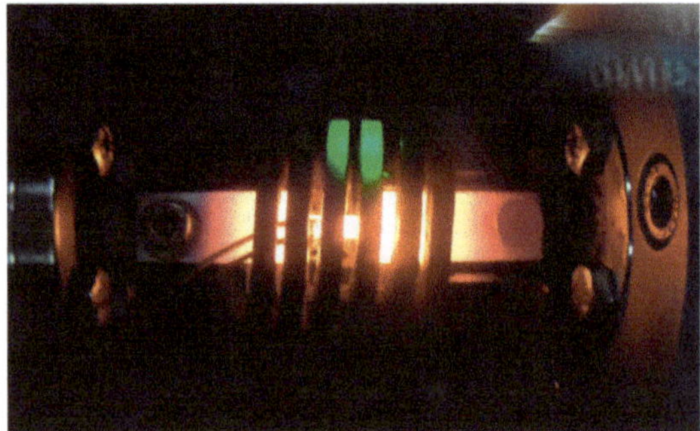

Figure 1. Austenitization of a specimen in the dilatometer.

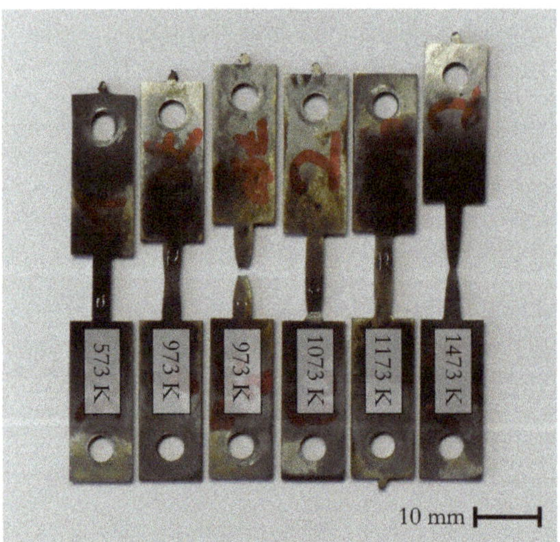

Figure 2. Some specimens after the hot tensile test in the dilatometer.

Afterwards, the specimens were heated up with $\dot{T} = 1000$ Ks^{-1} to the test temperature T_φ, followed by a soaking time of 2 s. While force and elongation were measured, the sample was then elongated with a (Hencky) strain rate of $\dot{\varphi} = 0.1$ s^{-1}. A welded-on thermocouple of the type Pt/Pt10Rh ensured correct temperature control throughout the test. The specimen was neither intentionally strained to fracture, nor was the elongation at rupture evaluated. For test temperatures up to and including $T_\varphi = 773$ K, three specimens per temperature were tested; at higher temperatures, the low scatter of the data allowed a reduction to two specimens. Due to the high strength of 22MnB5, dilatometric tests at temperatures lower than $T_\varphi = 673$ K were not possible within the machine's limitations. Additional tests on a conventional tensile testing machine at room temperature therefore supplemented the experiments. The highest temperature tested was $T_\varphi = 1473$ K.

2.1.2. Data Processing

Each measured force–elongation curve was transformed into a stress–strain curve and limited to strain values up to uniform elongation. A regression analysis then determined the parameters of the flow stress Equation (1) by Hockett and Sherby [20] to best fulfil the individual measured curve.

$$k_{f,T_\varphi}(\varphi) = k_{f,s} - \left(k_{f,s} - k_{f,0}\right) \times e^{-m \times \varphi^P} \tag{1}$$

Therein, $k_{f,0}$ is the initial flow stress and $k_{f,s}$ the saturation flow stress. m and P are dimensionless parameters. The resulting flow stress $k_{f,T_\varphi}(\varphi)$ is valid for a respective temperature T_φ. The mean of the computed Hockett–Sherby parameters within an individual test temperature then yielded the parameters of the mean flow stress curve for the respective temperature. The molten metal at the weld is described using the same structural mechanics equations as the solid material, instead of defining a multiphase approach using the Navier–Stokes equation for the molten metal. This simplification significantly eases modelling and computation of the welding process. As melt flow in most cases is little, it is allowed. Of course, molten metal will have no yield strength nonetheless. Therefore, it is formally necessary to also define flow stress curves for the melt. A virtual zero-strength, as defined below, and a finite compressibility, cf. Section 3.2.1, in this case were determined as the most reasonable material properties of the melt. In spot welding simulations by the authors, see Section 3.3 and [21], good results have been achieved by defining two additional curves fulfilling

$$k_{f,T_s=1719\ K}(\varphi) = 0{,}5 \times k_{f,T_\varphi=1473\ K}(\varphi) \tag{2}$$

$$k_{f,T_s=3273\ K}(\varphi) = 0{,}165 \times k_{f,T_\varphi=1473\ K}(\varphi). \tag{3}$$

In short, regions governed by laws of fluid flow in the FEA are represented by means of drastically reduced flow stress.

Input of the data into the FEA was required in the form of stress–Cauchy–strain curves. The strain values therein were transformed from the Hencky–strain φ applying the relation

$$\varepsilon_{pl} = e^\varphi - 1. \tag{4}$$

2.2. Numerical Material Simulation

Measurement of physical properties at elevated temperatures during rapid heating is very difficult, as shown by [22,23], which inform on the comparably large measurement errors that have to be expected. Material simulation software like JMATPRO can deliver the required data, but the reliability of the results is questionable, as the software purely relies on empirical models. Therefore, the decision was made to compute the required data using the software, and then validate the data very critically in comparison to measured data available in the literature.

Using the software, the phase composition, as well as the physical properties of the pure phases during heating of the material, were computed. Afterwards, the physical properties of the temperature-dependent phase composition were computed from the properties of the pure phases by using the lever rule. The computed austenitization temperatures for a heating rate of $\dot{T} = 1000\ \mathrm{Ks}^{-1}$ are presented in Table 1. It can be seen that the transformation delay due to the heating rate is small compared to the temperatures stated by [24].

The tempering, as well as partial or full dissolving of martensite below A_{c1}, is driven by diffusional processes. During quick heating of the material, the dwell time below A_{c1} is short. Therefore, diffusive effects are negligible. In this work, it was accordingly assumed that between A_{c1} and A_{c3}, the martensite will directly transform into austenite. Figure 3 depicts the phase composition of 22MnB5 being assumed in the thermophysical material property computation with JMATPRO.

Table 1. Computed phase-transformation temperatures of 22MnB5; quasi-static according to [24], and rapid heating.

Quantity	$\dot{T} \approx 0\ \mathrm{Ks}^{-1}$	$\dot{T} = 1\ \mathrm{k\ Ks}^{-1}$
A_{c3}/K	1153	1213
A_{c1}/K	993	1068

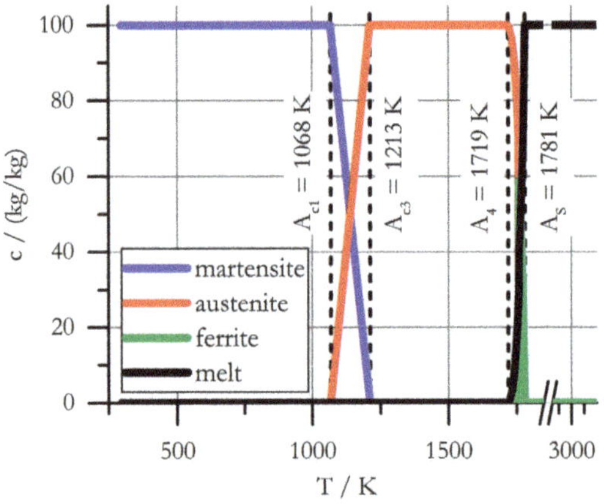

Figure 3. Phase fractions of 22MnB5 during heating with $\dot{T} = 1\ \mathrm{k\ Ks}^{-1}$.

It has to be pointed out that the assumption of neglected tempering in this work was only done for considerations regarding the thermophysical properties of 22MnB5. The flow stress curves on the contrary are derived from measurements, hence including the tempering effect of martensite as well.

The data computed with JMATPRO was compared to data published by: Bungardt and Spyra (BaS) [25], Spittel and Spittel [18] (Landolt-Boernstein), Richter [26,27], Shapiro [17], Verein Deutscher Eisenhuettenleute (VDEh) [23], Volz [28] and Wink and Kraetschmer [19]. From a chemical composition point of view, 22MnB5 is an unalloyed steel. Therefore, it is allowed to compare the order of magnitude of the computed data to data of similar unalloyed steels and pure iron, at least for the austenitic and molten phases. Accordingly, literature data on pure iron and a 23Mn6 steel were included in the comparison to ensure best verification of the data derived from JMATPRO. The data published by Volz [28] was collected from additional literature sources and is valid for structural steel. He used the data for FEAs of residual stresses resulting from arc welding and obtained very good conformity to his experimental verifications. Therefore, the data is considered to be sound.

During melting of the material, some properties considerably change their magnitude. Simulation trial runs proved that the numerical solution procedure is significantly more stable when the melting range of the material is only slightly widened. To account for this effect, the temperatures A_4 and A_s were artificially shifted about 50 K upwards and downwards, respectively; cf. the data points 'FEA' in the diagrams outlined below. An effect of this temperature shift on the computed results was not observed.

During solution of the model, the equation solver may compute very large, intermediate temperature results. In order to enhance the numerical stability, the physical data was extrapolated up to the boiling point $T = 3273$ K accordingly.

2.3. Resistance Spot Welding Simulation and Welding Experiments

The material data presented is well proven by comparing the results of a spot welding process FEA to measured results. A two-staged welding current as presented in Figure 4 along with a constant electrode force of $F_E = 6$ kN was applied in both experiment and FEA. The experiments were carried out with a pneumatic spot welding gun with high-frequency direct current (HFDC) power source. With respect to the scope of this work, details on the experimental conditions, FE model and the entirety of its boundary conditions shall be examined in [29].

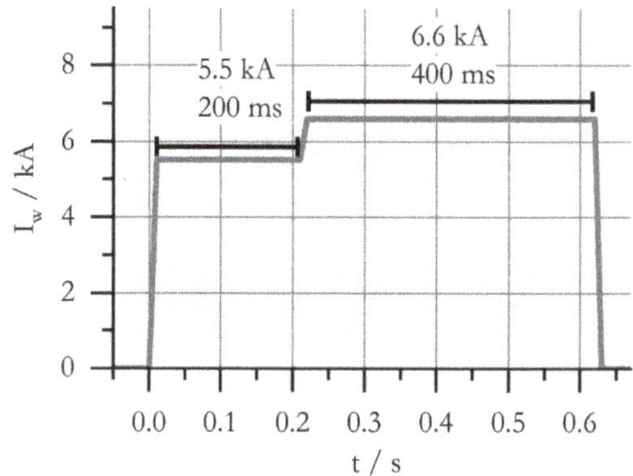

Figure 4. Welding current program used in spot welding of 22MnB5.

3. Results and Discussion

It has to be noted that the data presented is valid only for the heating phase in the welding process, as it is intended to be used in simulation-driven process development. Material property evolution in the cooling phase is not within the scope of this work. The data represented by the points denoted as 'FEA' in the diagrams is tabularly composed in Section 3.4.

3.1. Flow Stress

Figure 5 displays the measured flow stress curves, along with the two artificial curves. The curves have been extended according to (1) up to $\varphi = 1$. It can be readily seen that the work hardening effect is not pronounced in the martensitic phase. As soon as A_{c1} is surpassed, significant strain hardening becomes visible. It shall be noted that the ends of the curves do not indicate fracture of the material. At most temperatures it will fail much earlier. Due to experimental limitations outlined above, a significant gap in the data exists between $T_\varphi = 293$ K and $T_\varphi = 673$ K. As the decrease of strength in this temperature window is moderate, it is assumed that a linear interpolation between the respective curves, as FEA software modules will perform it, is reasonable.

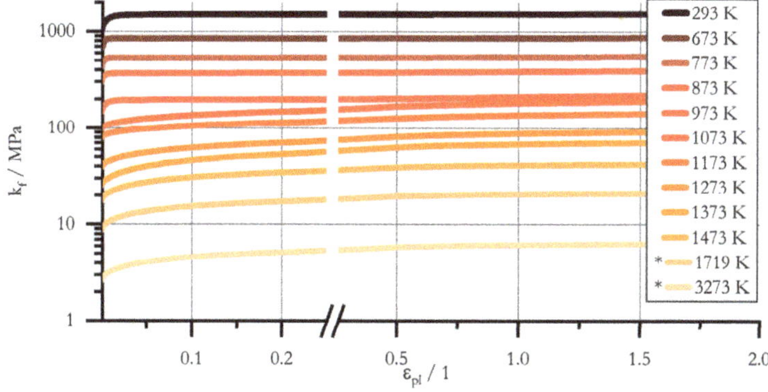

Figure 5. Flow stress curves of 22MnB5 after rapid heating from heat-treated state; * artificial curves.

For easy access, the mean of the measured flow stress parameters according to Hockett and Sherby [20] are composed in Table 2.

Table 2. Hockett–Sherby flow stress parameters of 22MnB5 after rapid heating from martensitic state.

T_φ/K	$k_{f,0}$/MPa	$k_{f,s}$/MPa	m/1	P/1
293	1066	1488	87	0.885
673	680	844	8934	1.533
773	432	529	818	1.11
873	298	368	134.5	0.82
973	147	196	90.5	0.83
1073	97	248	1	0.52
1173	79.5	180	1	0.515
1273	39	100	2	0.605
1373	25	74.3	2.6	0.663
1473	17.5	43.5	3	0.615

3.2. Physical Properties

3.2.1. Elastic Modulus and Poisson Ratio

Starting from $E = 210$ GPa, the elastic modulus decreases with temperature, while the Poisson ratio increases, as shown in Figure 6. For the melt, the software computed a very small elastic modulus along with incompressible material behaviour ($v = 0.5$), but delivered a small, finite bulk modulus K. In other words, the computed bulk modulus and Poisson ratio excluded each other. Due to lack of other data, the computed bulk modulus was then used to compute the elastic modulus of the melt using the relation:

$$E = K(3 - 6v). \tag{5}$$

As fully incompressible behaviour is unlikely for the melt, the Poisson ratio therein was set to $v = 0.45$. This prevents numerical problems resulting from incompressible material behaviour, but still reasonably represents the data computed by JMATPRO.

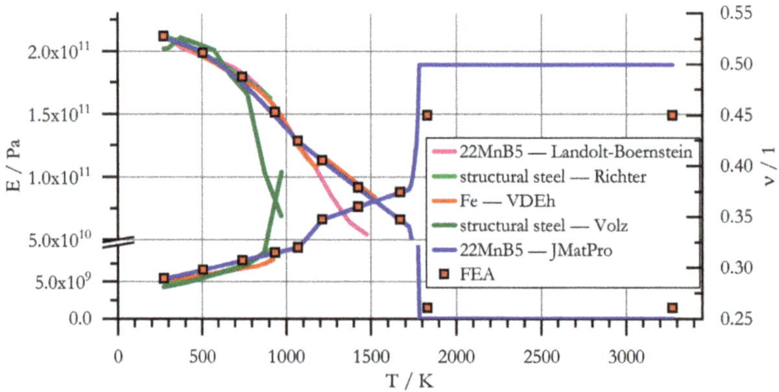

Figure 6. Elastic modulus and Poisson ratio of 22MnB5.

3.2.2. Coefficient of Thermal Expansion

The data depicted in Figure 7 represents the secant coefficient of thermal expansion with a reference temperature of $T_0 = 293$ K. Some of the literature data therein has been computed from the lattice constant taken from [23]. The coefficient of thermal expansion α strongly depends on the lattice structure, as shown in Figure 7. During austenitization, it quickly drops, followed by a steeper incline. On melting, the coefficient roughly doubles its magnitude.

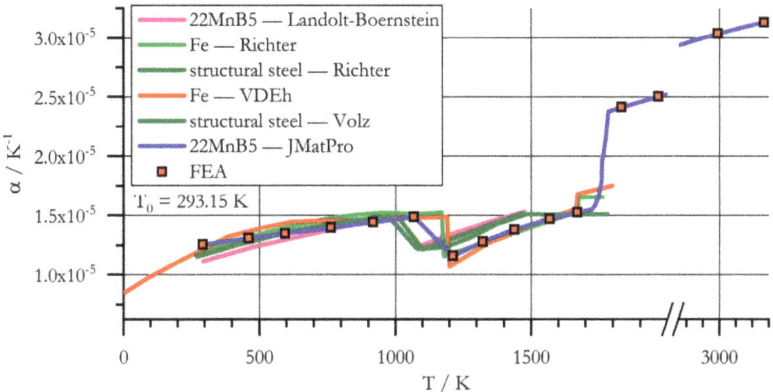

Figure 7. Coefficient of thermal expansion of 22MnB5.

3.2.3. Specific Electric Resistance

To improve readability, the specific electric resistance ρ_{el} is drawn in two separate diagrams, shown in Figures 8 and 9. Literature data on this property is particularly scarce. JMATPRO delivered data on the specific resistance only for $T > 1213$ K. To fill the gap, additional data was generated based on the thermal conductivity λ using the Wiedemann–Franz (WF) law with a Lorenz number of $L = 3.12 \times 10^{-8}$ V^{2}K^{-2} as stated by [30]:

$$\rho_{el}(T) = \frac{L \times T}{\lambda},\tag{6}$$

as well as the relation published by BaS [25],

$$\rho_{el,ferrite}(T) = \frac{2985 \times 10^{-2} \times T + 1674}{\lambda}\tag{7}$$

$$\rho_{el,austenite}(T) = \frac{2182 \times 10^{-2} \times T + 5718}{\lambda}.\tag{8}$$

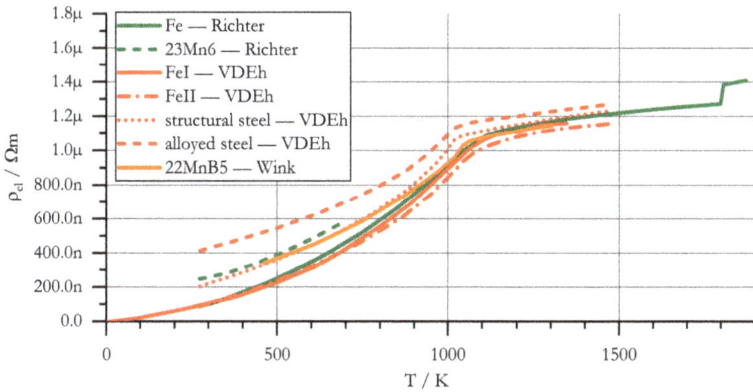

Figure 8. Specific electric resistance of 22MnB5, literature data.

Close to room temperature, the specific resistance is strongly dependent on the alloy. With increasing temperature, the curves converge towards each other. At the melting point, the resistance increases by a small percentage. For the austenite and the melt, the data delivered by JMATPRO and the WF law are in contradiction with the experimental data. The data computed with the model of BaS is in good conformity with the experimental data. Accordingly, the data computed with the BaS rule is considered to be sound. Beyond the melting point, the curve was extrapolated up to the boiling point.

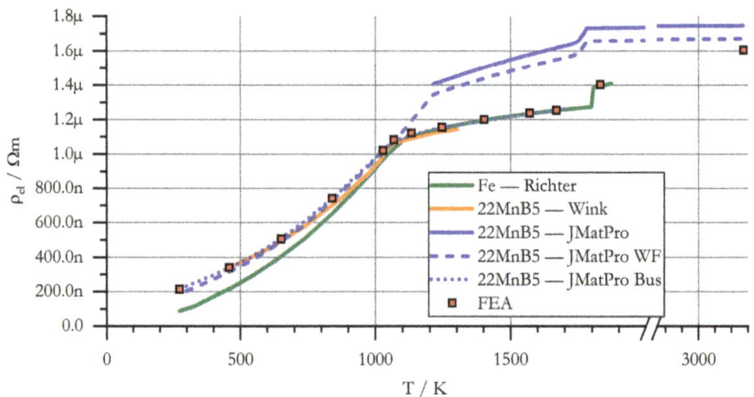

Figure 9. Specific electric resistance of 22MnB5, computed data.

3.2.4. Mass Density

In addition to the thermal expansion of the metal during heating, its mass density ρ generally decreases, as shown in Figure 10. On austenitization, the density increases by a small percentage accordingly. It is critical that the data on thermal expansion and density fit to each other. Otherwise, creation or destruction of mass in the numerical model, with adverse effects for the precision of the thermal balance, would occur. The mass density data presented here has been carefully reviewed regarding this subject, although mathematical details on the review shall not be discussed.

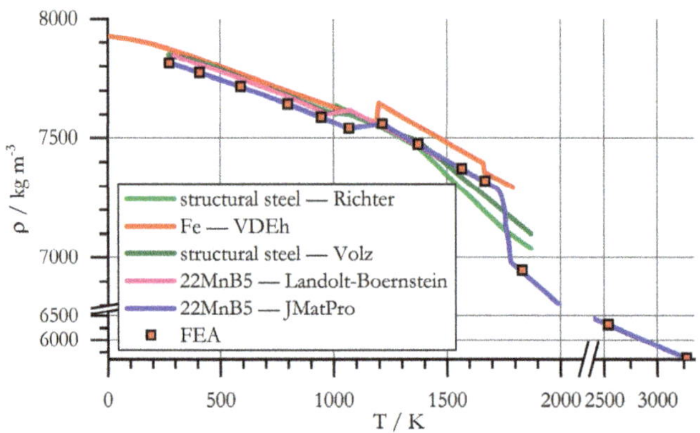

Figure 10. Mass density of 22MnB5.

3.2.5. Specific Heat Capacity

The specific heat capacity c_p is a function of temperature with discontinuities at the Curie point, during austenitization, and at the melting point, as shown in Figure 11. At the mentioned transition points, the heat capacity is theoretically infinite. To avoid the discontinuities, the specific enthalpy density h is used by the FEA following the relation

$$h(T) = \int_{T_0}^{T_1} \rho \times c_p(T)\, dT \tag{9}$$

The integrated heat capacity curves according to the relation are presented in Figure 12. Therein, an average mass density of $\rho = 7600$ kg m^{-3} was set. While the average slope of the curve is relatively constant, melting is characterised by a significant step in the curve.

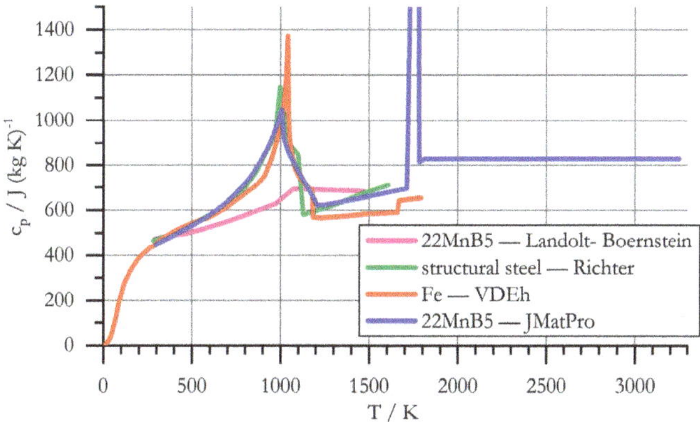

Figure 11. Specific heat capacity of 22MnB5.

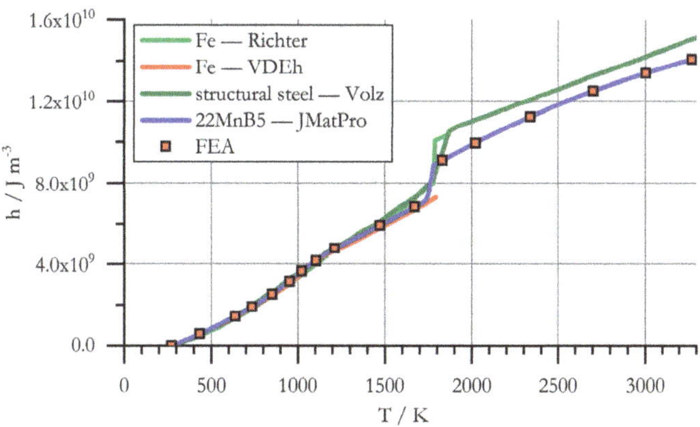

Figure 12. Specific enthalpy density of 22MnB5.

3.2.6. Specific Thermal Conductivity

The specific thermal conductivity λ is initially reduced with increasing temperature, but increases again as soon as austenitization sets in, as shown in Figure 13. The data computed by JMATPRO is in good agreement with the experimental data. The curve published by Shapiro [17] apparently refers to the cooling of the material from the austenitic state. An additional curve computed by JMATPRO for quenching of 22MnB5 being drawn in Figure 13 strongly supports this. As computed by [31,32], a hydrodynamic mixing movement will occur in the molten pool during welding. In order to account for the resulting additional convective heat transport, the specific thermal conductivity of the melt is increased by about twice the magnitude at room temperature in this analysis.

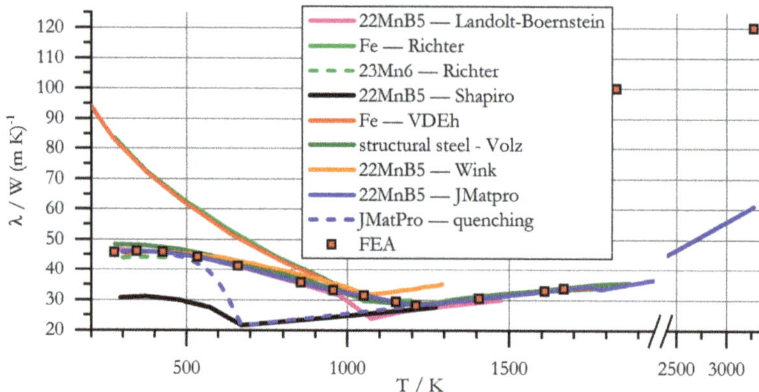

Figure 13. Specific thermal conductivity of 22MnB5.

3.3. Data Application in Resistance Spot Welding

An overview of the computed temperature distribution in the welded sheets at the end of current flow, along with a macrosection of the weld nugget, is presented in Figure 14.

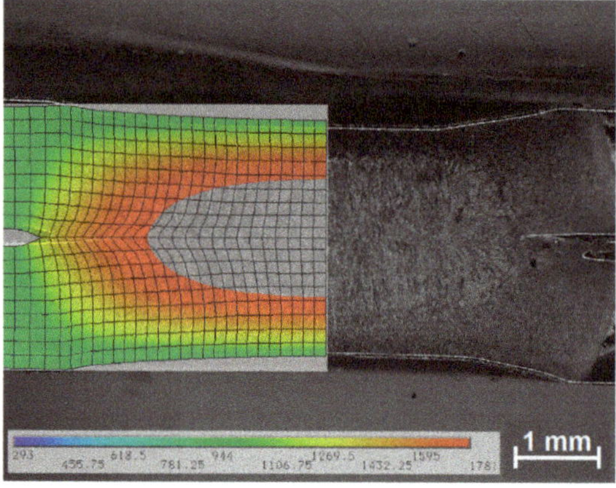

Figure 14. Macrosection of weld nugget in comparison to computed result, the latter half-translucent; temperature profile is in K, gray surface represents melt.

The computed molten zone represents the nugget shape visible in the macrosection reasonably well, although the computed nugget is slightly smaller. Quantitative examination of the nugget radius r_p in Figure 15 gives more insight: In the FEA, a decision has to be made whether the temperature A_4 or A_S is considered to be sufficient to form a joint across the sheet metals' surfaces. The experimentally observed nugget growth curve lies well in between both criteria. In Figure 14, melting is only assumed when the temperature A_S is surpassed, which explains the seemingly too-small nugget in the picture. It shall also be noted that the computed electrode indentation depth and shape conform very well with the experiment. By comparison of the sheet thicknesses on the left border of Figure 14, it is assured that the relative scaling of FEA and the macrosection is correct.

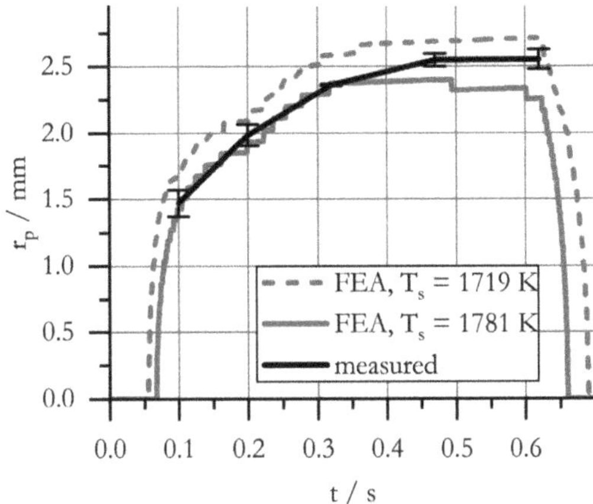

Figure 15. Nugget growth curve during the weld, measured and computed data.

3.4. Tabular Data

For convenience, the thermophysical data is composed in Tables 3 and 4.

Table 3. Physical properties of 22MnB5.

T/K	E/GPa	$v/1$	T/K	$\alpha/10^{-5}\,K^{-1}$	T/K	$\rho_{el}/\mu\Omega m$
273.15	211.804	0.290	293.15	1.256	273.15	0.214
505.05	198.693	0.298	460.98	1.307	458.96	0.340
744.51	179.279	0.307	595.73	1.348	651.12	0.506
935.80	151.383	0.315	762.56	1.398	840.70	0.742
1068.15	128.226	0.320	917.37	1.442	1027.97	1.020
1213.15	113.060	0.347	1068.15	1.485	1068.15	1.082
1426.11	91.707	0.360	1213.15	1.160	1132.22	1.122
1669.15	65.859	0.375	1322.89	1.280	1244.72	1.156
1831.15	1.470	0.450	1437.27	1.380	1402.63	1.201
3273.15	1.470	0.450	1567.51	1.469	1570.86	1.238
			1669.15	1.525	1669.15	1.254
			1831.15	2.416	1831.15	1.402
			1966.85	2.504	3273.15	1.604
			2192.90	2.646		
			2426.40	2.777		
			2708.93	2.919		
			2989.43	3.036		
			3273.15	3.133		

Table 4. Physical properties of 22MnB5.

T/K	ρ/kgm^{-3}	T/K	h/MJm^{-3}	T/K	$\lambda/Wm^{-1}\,K^{-1}$
273.15	7815	273.15	0	273.15	45.738
408.21	7775	433.92	591	343.92	46.153
589.77	7717	637.68	1441	425.87	45.807
798.48	7643	735.55	1911	534.21	44.244
946.10	7588	851.72	2529	662.40	41.172
1068.15	7541	951.01	3150	854.26	35.678

Table 4. *Cont.*

T/K	ρ/kgm^{-3}	T/K	h/MJm^{-3}	T/K	$\lambda/\text{Wm}^{-1}\text{K}^{-1}$
1213.15	7560	1019.39	3652	956.06	33.161
1372.11	7475	1103.25	4163	1051.18	31.402
1566.58	7373	1213.15	4795	1150.47	29.447
1669.15	7322	1470.95	5921	1211.79	28.152
1831.15	6947	1669.15	6838	1407.03	30.473
2092.03	6722	1831.15	9110	1608.95	32.917
2539.46	6320	2022.07	9952	1669.15	33.700
3273.72	5625	2336.66	11,239	1831.15	100.000
		2701.79	12,521	3273.15	120.000
		3001.71	13,405		
		3273.15	14,090		

4. Conclusions

Measurements of flow stress curves, as well as physical material property computations and comprehensive literature reviews, were employed to compose a set of material data necessary for welding process simulations on presshardened 22MnB5 steel. Due to the lack of literature data and/or specific requirements, a few simplifications and assumptions were made in the process of the data collection. The assumptions are described in detail in this work, so the simulation expert may carefully reconsider if the data is still valid in the specific case. The data provided is suitable to describe the materials' behaviour during the heating phase in the welding process. The materials' behaviour during cooling or quenching is not in the scope of this work.

The following results were obtained:

1. Data on the flow stress of 22MnB5 was measured and converted to stress–strain data for test temperatures ranging from $T_\varphi = 293$ K to $T_\varphi = 1473$ K.
2. Flow stress data is provided by means of flow parameters for the tested temperatures according to the Hockett–Sherby model.
3. Physical material property data of 22MnB5 as a function of temperature has been computed using material simulation software. The data was critically reviewed considering literature data.

All datasets provided have been linearly extrapolated up to a temperature of $T = 3273$ K in order to allow immediate use for the numerical process simulation procedure.

Author Contributions: Conceptualization, J.K.; Data curation, J.K., P.M. and K.K.; Formal analysis, J.K. and K.K.; Investigation, J.K.; Methodology, J.K.; Project administration, J.K.; Resources, P.M.; Software, P.M.; Supervision, P.M. and K.K.; Validation, J.K. and K.K.; Visualization, J.K.; Writing–review & editing, J.K. and P.M.

Funding: The publication costs of this article were funded by the German Research Foundation/DFG and Chemnitz University in the funding programme Open Access Publishing.

Conflicts of Interest: The authors declare no conflict of interest.

References

1. Podržaj, P.; Jerman, B.; Simončič, S. Poor fit-up condition in resistance spot welding. *J. Mater. Process. Technol.* **2016**, *230*, 21–25. [CrossRef]
2. Schlosser, B.; Jüttner, S. Indikatoren zur Beurteilung der Qualität von MSG-Schweißnähten. In *35. Assistentenseminar Füge- und Schweißtechnik*; DVS Media: Magdeburg, Germany, 2016; Volume 314.
3. Häßler, M.; Füssel, U. Fügen hochbeanspruchter Stähle durch kathodenfokussiertes WIG-Löten (CF-TIG). In *35. Assistentenseminar Füge- und Schweißtechnik*; DVS Media: Magdeburg, Germany, 2016; Volume 314.
4. Brauser, S.; Pepke, L.-A.; Weber, G.; Rethmeier, M. Influence of Production-Related Gaps on Strength Properties and Deformation Behaviour of Spot Welded Trip Steel HCT690T. *Weld. World* **2012**, *56*, 115–125. [CrossRef]

5. Moos, S.; Vezzetti, E. Compliant assembly tolerance analysis: Guidelines to formalize the resistance spot welding plasticity effects. *Int. J. Adv. Manuf. Technol.* **2012**, *61*, 503–518. [CrossRef]

6. Bi, J.; Song, J.; Wei, Q.; Zhang, Y.; Li, Y.; Luo, Z. Characteristics of shunting in resistance spot welding for dissimilar unequal-thickness aluminum alloys under large thickness ratio. *Mater. Des.* **2016**, *101*, 226–235. [CrossRef]

7. Van der Aa, E.; Amirthalingam, M.; Winter, J.; Hanlon, D.N.; Hermans, M.J.M.; Rijnders, M.; Richardson, I.M. Improved Resistance Spot Weldability of 3rd Generation AHHS for Automotive Applications. *Math. Model. Weld Phenoma* **2016**, *11*, 175–193.

8. Raoelison, R.N.; Fuentes, A.; Pouvreau, C.; Rogeon, P.; Carré, P.; Dechalotte, F. Modeling and numerical simulation of the resistance spot welding of zinc coated steel sheets using rounded tip electrode: Analysis of required conditions. *Appl. Math. Model.* **2014**, *38*, 2505–2521. [CrossRef]

9. Schwenk, C.; Rethmeier, M. Material Properties for Welding Simulation—Measurement, Analysis and Exemplary Data. *Weld. J.* **2011**, *90*, 220–227.

10. Karbasian, H.; Tekkaya, A.E. A review on hot stamping. *J. Mater. Process. Technol.* **2010**, *210*, 2103–2118. [CrossRef]

11. Merklein, M.; Johannes, M.; Lechner, M.; Kuppert, A. A review on tailored blanks—Production, applications and evaluation. *J. Mater. Process. Technol.* **2014**, *214*, 151–164. [CrossRef]

12. Geiger, M.; Merklein, M.; Hoff, C. Basic Investigations on the Hot Stamping Steel 22MnB5. *Adv. Mater. Res.* **2005**, *6–8*, 795–804. [CrossRef]

13. Ngyuen, D.-T.; Banh, T.-L.; Jung, D.-W.; Yang, S.-H.; Kim, Y.-S. A Modified Johnson-Cook Model to Predict Stress-Strain Curves of Boron Steel Sheets at Elevated and Cooling Temperatures. *High Temp. Mater. Proc.* **2012**, *31*, 37–45.

14. Merklein, M.; Lechler, J. Investigation of the thermo-mechanical properties of hot stamping steels. *J. Mater. Process. Technol.* **2006**, *177*, 452–455. [CrossRef]

15. Naderi, M. A Numerical and Experimental Investigation into Hot Stamping of Boron Alloyed Heat Treated Steels. *Steel Res. Int.* **2008**, *79*, 77–84. [CrossRef]

16. Gumbsch, P.; Roos, E.; Hahn, O. *Charakterisierung und Ersatzmodellierung des Bruchverhaltens von Punktschweißverbindungen an ultrahochfesten Stählen für die Crashsimulation unter Berücksichtigung der Auswirkung der Verbindung auf das Bauteilverhalten*; Fraunhofer Institut IWM: Freiburg, Germany; MPA Stuttgart: Stuttgart, Germany; Universität Paderborn: Paderborn, Germany, 2012.

17. Shapiro, A. Finite Element Modeling of Hot Stamping. *Steel Res. Int.* **2009**, *80*, 658–664. [CrossRef]

18. Spittel, M.; Spittel, T. Steel symbol/number: 22MnB5/1.5528. In *Metal Forming Data of Ferrous Alloys—Deformation Behaviour*; Warlimont, H., Ed.; Landolt-Börnstein—Group VIII Advanced Materials and Technologies; Springer: Berlin/Heidelberg, Germany, 2009; Volume 2C1.

19. Wink, H.-J.; Krätschmer, D. Charakterisierung und Modellierung des Bruchverhaltens von Punktschweißverbindungen in pressgehärteten Stählen—Simulation des Schweißprozesses. In Proceedings of the 11. LS-DYNA Forum, Ulm, Germany, 9–10 October 2012.

20. Hockett, J.E.; Sherby, O.D. Large strain deformation of polycrystalline metals at low homologous temperatures. *J. Mech. Phys. Solids* **1975**, *23*, 87–98. [CrossRef]

21. Kaars, J.; Mayr, P.; Koppe, K. Simple Transition Resistance Model for Spot Welding Simulation of aluminized AHSS. *Math. Model. Weld Phenoma* **2016**, *11*, 685–702.

22. Byl, C.; Bérardan, D.; Dragoe, N. Experimental setup for measurements of transport properties at high temperature and under controlled atmosphere. *Meas. Sci. Technol.* **2012**, *23*, 035603. [CrossRef]

23. *Werkstoffkunde Stahl*; Eisenhüttenleute, V.D. (Ed.) Springer: Berlin/Heidelberg, Germany, 1984.

24. Naderi, M.; Saeed-Akbari, A.; Bleck, W. The effects of non-isothermal deformation on martensitic transformation in 22MnB5 steel. *Mater. Sci. Eng. A* **2008**, *487*, 445–455. [CrossRef]

25. Bungardt, K.; Spyra, W. *Wärmeleitfähigkeit von Stählen und Legierungen bei Temperaturen Zwischen 20 °C und 700 °C*; Verlag Stahleisen: Düsseldorf, Germany, 1965.

26. Richter, F. Die wichtigsten physikalischen Eigenschaften von 52 Stahlwerkstoffen. *Stahleisen Sonderberichte* **1973**, *8*, 1–32.

27. Richter, F. Physikalische Eigenschaften von Stählen und ihre Temperaturabhängigkeit. In *Stahleisen Sonderberichte*; Verlag Stahleisen: Düsseldorf, Germany, 1983.

28. Volz, M. Die Rissentstehung in Statisch Beanspruchten Stahlkonstruktionen unter Berücksichtigung von Schweißeigenspannungen. Ph.D. Thesis, Universität Karlsruhe (TH), Karlsruhe, Germany, 2009.

29. Kaars, J.; Mayr, P.; Koppe, K. Generalized dynamic transition resistance in spot welding of aluminized 22MnB5. *Mater. Des.* **2016**, *106*, 139–145. [CrossRef]

30. *Thermal Conductivity of Pure Metals and Alloys*; Madelung, O.; White, G.K. (Eds.) Landolt-Börnstein—Group III Condensed Matter; Springer: Berlin/Heidelberg, Germany, 1991; Volume 15c.

31. Li, Y.B.; Lin, Z.Q.; Lai, X.M.; Chen, G.L. Electromagnetic Phenomena in Resistance Spot Welding and its Effects on Weld Nugget Formation. In Proceedings of the PIERS Proceedings, Moscow, Russia, 18–21 August 2009.

32. Na, S.J.; Cheon, J.H.; Kiran, D.V.; Cho, D.W. Heat and Mass Flow in Arc Welding Processes and its Application to Mechanical Analysis of Welded Structures. *Math. Model. Weld Phenoma* **2016**, *11*, 5–22.

Article

Characteristics of Welding and Arc Pressure in the Plasma–TIG Coupled Arc Welding Process

Bo Wang [1,2], Xun-Ming Zhu [3], Hong-Chang Zhang [4,*], Hong-Tao Zhang [1,2,*] and Ji-Cai Feng [1]

[1] State Key Laboratory of Advanced Welding and Joining, Harbin Institute of Technology, Harbin 150001, China; wangvbo@hit.edu.cn (B.W.); fengjc@hit.edu.cn (J.-C.F.)
[2] Institute for Shipwelding Technology, Shandong Institute of Shipbuilding Technology, Weihai 264209, China
[3] Weihai Wanfeng Magnesium Science and Technology Development Co., Ltd., Weihai 264209, China; xunming.zhu@wfjt.com
[4] Department of Electrical Engineering, Harbin University of Science and Technology, Rongcheng, 264300, China
* Correspondence: hongchangzhang123@163.com (H.-C.Z.); zhanght@hitwh.edu.cn (H.-T.Z.); Tel.: +86-178-6308-2611 (H.-C.Z.); +86-187-6910-2227 (H.-T.Z.)

Received: 25 May 2018; Accepted: 29 June 2018; Published: 3 July 2018

Abstract: In this article, a novel hybrid welding process called plasma-TIG (Tungsten Inert Gas welding) coupled arc welding was proposed to improve the efficiency and quality of welding by utilizing the full advantage of plasma and TIG welding processes. The two arcs of plasma and TIG were pulled into each other into one coupled arc under the effect of Lorentz force and plasma flow force during welding experiments. The arc behavior of coupled arc was studied by means of its arc profile, arc pressure and arc force conditions. The coupled arc pressure distribution measurements were performed. The effects of welding conditions on coupled arc pressure were evaluated and the maximum coupled arc pressure was improved compared with single-plasma arc and single-TIG arc. It was found that the maximum arc pressure was mainly determined by plasma arc current and plasma gas flow. Compared with traditional hybrid welding method, the efficiency was obviously higher and the welding heat-input was lower. The epitaxial solidification of the weld was inhibited, the tensile strength of the welded joints was higher. According to the results, the proposed coupled arc welding process has both advantages of plasma arc and TIG method, and it has a broad application prospect.

Keywords: plasma-TIG; coupled arc; arc profile; pressure distribution

1. Introduction

Given the continuous development of processing and manufacturing technologies, traditional arc welding techniques such as plasma arc welding (PAW) [1] and tungsten inert gas (TIG) [2,3] welding process, have been improved to meet the requirements of enterprises for high-quality and low-consumption welding technology [4]. Based on the situation above, the hybrid welding methods with multiple heat sources are proposed. In recent years, hybrid welding methods have also been successfully applied in shipbuilding and spacecraft. Certain hybrid welding processes, including plasma–gas metal arc welding [5–7], plasma–TIG [8,9], laser–GMAW [10], laser–TIG [11] and TIG–MIG [12], have been successfully used in the manufacturing industry. In comparison with ordinary single-arc welding, hybrid welding technology can significantly improve welding efficiency and joint quality because it is simultaneously provided with multiple heat sources that complement one another [5].

PAW produces arc plasma with high energy density and high arc pressure under the thermal pinch effect of the water-cooled copper nozzle; thus, PAW has strong fusion and penetration abilities [13,14]. However, identifying the appropriate parameters in PAW process is difficult and the process is prone to undercutting [15]. The gas tungsten arc welding with DCEN (Direct Current Electrode Negative Epolarity) is a high efficiency welding process [16,17], and has a good surface forming ability and remarkably stable welding process, but its penetration capacity is poor. Therefore, researchers also proposed several hybrid welding processes that utilize the PAW and TIG arcs, which mainly include the double-sided arc welding (DSAW) and PAW–TIG welding processes [9]. The DSAW process can increase penetration and significantly improve welding productivity [18], in this process, a TIG torch is placed on the opposite side of welds to guide the PAW arc into the keyhole, thereby permitting the current to flow from the PAW torch through the welds to the TIG torch directly instead of the conventional welding current loop [8,19]. However, the welding accessibility of DSAW is poor compared with PAW welding process. The PAW–TIG welding method is realized by the PAW and the subsequent TIG on the movable slide rail, no correlation exists between the plasma and TIG arcs. This process can also enhance the welding productivity. However, the heat-affected zone is large and the grain is coarse because the weld is reheated by the TIG arc. The plasma–TIG coupled arc welding (PTCAW) process is a novel hybrid welding method for overcoming the above deficiencies in order to achieve efficient welding process and high-quality welds.

In this study, the PTCAW process was initially constructed and the coupled arc profile was then analyzed by a charge-coupled device (CCD) camera. The effects of welding conditions on the coupled arc pressure distribution was also studied to reveal the characteristics of the coupled arc welding process. The results provided the basis to recognize the PTCAW process and expanded the application range of PAW and TIG welding processes.

2. Materials and Methods

The schematic diagram of our experimental set-up is shown in Figure 1. The PAW and TIG torches were fixed on one fixture to realize the interaction between the two processes (i.e., PTCAW process). An experimental system was established according to the above designed principle. In the present system, the TIG arc was established by using a direct current–constant current (DC–CC) power supply (REHM InverTIG.Pro digital 240 AC/DC, Blaubeuren, Germany) under the direct current electrode negative (DCEN) condition. The PAW arcs were established by using an alternating current–direct current AC/DC (LORCH V50 AC/DV) and DC–CC (LORCH Handy TIG 180 DC, Auenwald, Germany) power supplies. The Xiris XVC-1000 weld camera (55 frames/s, Burlington, ON, Canada) was applied to observe and capture the arc profile and behavior under the different welding condition in real time [20]. The anode constricting nozzle of PAW made of red cooper with excellent thermal conductivity is cooled by forced recirculation cooling water to prevent the nozzle from being burned.

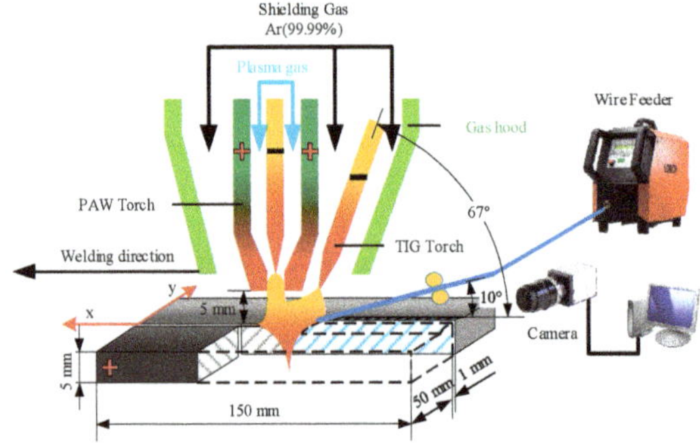

Figure 1. Proposed PTCAW processing system.

In all experiments, the PAW torch nozzle has a 3-mm orifice diameter, 3.2-mm tungsten diameter, and 3-mm tungsten setback. Pure argon (99.99%) was used as the plasma gas and shielding gas. The platform diameter of the TIG terminal electrode is 1.25 mm, and the shielding gas flow rate is 15 L/min. The electrode spacing is 7 mm, that is the distance from the plasma nozzle axis to the electrode tip of TIG. The nozzle height is the distance from the end of PAW nozzle to the workpiece surface. The detailed parameters for the designed experiments are shown in Table 1.

Table 1. Welding parameters for designed experiments.

Group	Plasma Arc Current (A)	TIG Arc Current (A)	Nozzle Height (mm)	Plasma Gas Flow (L/min)
#1	60, 80, 100, 120	100	5	4
#2	60	60, 90, 120, 150	5	4
#3	120	140, 160, 180, 200	5	4
#4	60	100	3, 4, 5, 6	4
#5	60	100	5	3, 4, 5, 6

3. Results and Discussions

3.1. TPTCAW Coupled Arc Profile

A notable characteristic of the plasma-TIG welding method is the physical coupling effect between the two arcs, which results in penetration increased and welding spatter reduced effectively compared with traditional GMAW (gas metal arc welding) hybrid welding technology. The detailed images of the plasma, TIG, and coupled arcs were acquired as shown in Figure 2. Furthermore, pseudo-color processing is implemented for converting the acquired Gray-scale images to pseudo-color images using Xiris WeldStudio$^{TM®}$ software (2.0.3, Xiris Automation, Burlington, ON, Canada), which will assists in analyzing the detailed information of arc profile. The base plate is made of red copper with forced recirculation cooling water and the welding experiment was carried out under the following conditions: a plasma welding current of 60 A while TIG current of 100 A, nozzle height of 5 mm and plasma gas flow of 15 L/min.

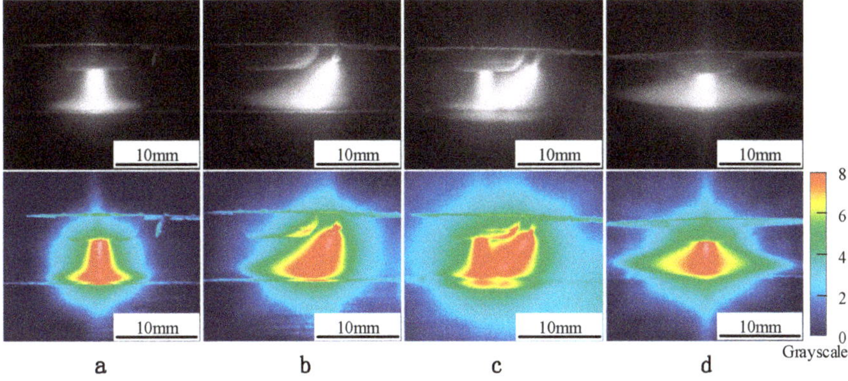

Figure 2. Arc profile under different welding methods. (**a**) Single plasma arc. (**b**) Single TIG arc. (**c**) Coupled arc perpendicular to the welding direction. (**d**) Coupled arc parallel to the welding direction.

Figure 2 illustrates typical arc profile under different welding methods. Welding arcs is steady and plasma arc seems to be trumpet-shaped while TIG arc is similar to broom-shaped during single arc welding because of the tungsten axis and the base plate have a 67 degrees angle, as shown in

Figure 2a,b. When the TIG arc is applied to plasma welding arc, it is observed that a novel coupled arc is produced between the two arcs and the arc profile has a significant change from the previous. TIG arc can be deflected to plasma arc from the view of perpendicular to welding direction, and part of TIG arc is pushed to the side of plasma arc root, as shown in Figure 2c. At the same time, the coupled arc profile seems to be bell-shaped similar to the conventional TIG arc from the view of parallel to the welding direction, as shown in Figure 2d. The reason for this behavior can be explained as follows.

Previous studies have suggested the forces play a major role in determining the arc profile [21]. Therefore, it is necessary to analyze coupled arc force condition in order to explore the interaction mechanism on coupled arc profile. It should be noted that the electromagnetic axial pressure and plasma gas flow axial pressure are recognized as the key factor of PAW arc pressure [22]. The electromagnetic axial pressure on the arc axis can be expressed as the following Equation (1):

$$p_e = \frac{\mu_0 I_P^2}{4\pi^2}\left(\frac{1}{r_2^2} - \frac{1}{r_1^2}\right) \tag{1}$$

The plasma gas flow axis pressure towards to the molten pool surface can be expressed as the following Equation (2):

$$p_g = \frac{1}{2}\rho\mu_1^2 \tag{2}$$

The plasma arc total pressure P_t on the arc axis can be expressed as the following Equation (3):

$$p_t = p_e + p_g = \frac{\mu_0 I_P^2}{4\pi^2}\left(\frac{1}{r_2^2} - \frac{1}{r_1^2}\right) + \frac{1}{2}\rho\mu_1^2 \tag{3}$$

where P_e = electromagnetic axial pressure, and it is the function of IP. P_g = plasma gas flow axial pressure, and it is the function of ρ and μ_1, simultaneously μ_1 is the function of temperature. Hence, the resultant force on the arc axis from electrode tip to the surface of base plate can be expressed as the following Equation (4):

$$F_r = p_t \cdot S = \frac{\mu_0 I_P^2}{4\pi}\left(\frac{r_1^2}{r_2^2} - 1\right) + \frac{1}{2}\pi\rho\mu_1^2 r_1^2 \tag{4}$$

Compared with the arc force condition during the single TIG welding, the Lorentz force (F_L) has a significant effect on TIG arc owing to plasma arc generates additional magnetic field acts on the welding arc, and F_L can be expressed as the following Equation (5):

$$F_L = BIL \tag{5}$$

The schematic of the arc force condition as shown in Figure 3. Considering the difference of arc force conditions on either side of TIG arc axis, it is necessary to specify four particles (location A, B, C and D) as the research objects. On the side of the TIG arc, the plasma flow force is considered as the main arc force of TIG welding arc, the plasma flow force can be expressed as the following Equation (6):

$$F_P = KI^2 \log\left(\frac{R_b}{R_a}\right) \tag{6}$$

where F_t = the plasma flow force of TIG arc at location B, C and D. and R_b are the radius of arc root surface and undersurface respectively as shown in Figure 4c.

As seen in Figures 3 and 4a, the direction of plasma flow force (F_P) on the arc axis can be deflected to plasma arc at the location B. Furthermore, the plasma flow force acts as the major driver force of the arc shift and has important influence on the weld surface forming. According to right-hand grip rule at the location A, the direction of magnetic field generated by plasma arc current is perpendicular to the paper inward on the side of TIG arc. Hence the particles on the right side of plasma arc axis move towards plasma arc along the base plate under Lorentz force (F_{L1} and F_{L2}), the motion of TIG arc can be seen in Figure 4a,c. At the same time, the plasma arc is also deflected slightly towards the location B owing to the Lorentz force (F_{L3}), as shown in Figure 4a,b. In addition, part of particles of TIG arc have a potential of coupling with that of plasma arc root, which result in transformation of TIG arc trajectory. These results indicate that the above analysis of the coupled arc force conditions is consistent with the acquired experimental results in Figure 2. It can be concluded that the TIG arc is deflected under the combination effect of plasma flow force and Lorentz force generated by the magnetic field of plasma arc column.

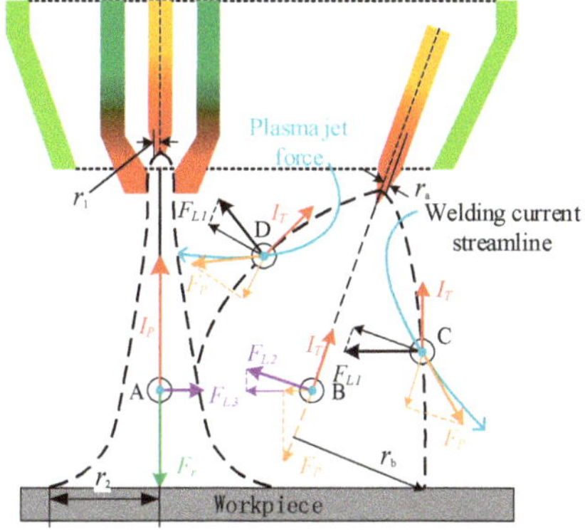

Figure 3. Schematic of the forces acting on the coupled arcs.

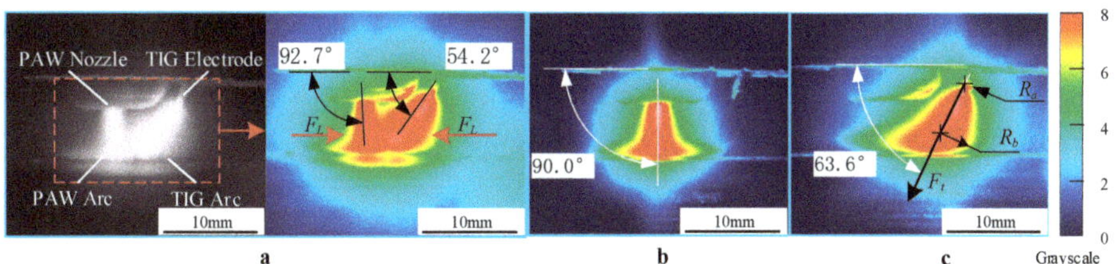

Figure 4. Schematic of the behavior on the plasma arc images of arc profile of different welding method. (**a**) Coupled arc welding. (**b**) Plasma arc welding. (**c**) TIG arc welding.

Therefore, according to the aforementioned analysis, another interesting finding has caught our attention. It is important to highlight that the coupling of two arcs and plasma flow force of arc are proportional to the welding current, this phenomenon is in good consistent with Equations (4) and (6). As can be seen from Figure 5, the coupled arc is generated only when the plasma arc current is bigger than 60 A, at the same time, the arc climbing height on plasma arc column increases with welding current (I_P) when the TIG welding current (I_T) is constant at 100 A. However, as plasma arc current is constant at 60 A, the plasma flow force (F_P) has a marked effect on welding arc profile acting on the base plate, the heat transfer area of coupled arc acting on the workpiece is apparently increased with increase of welding current (I_T), as shown in Figure 6. Therefore, it is indicated that the coupled arc profile depends on combination effect of these arcs, especially I_T and I_P, it is consistent with the above analysis of coupled arc force.

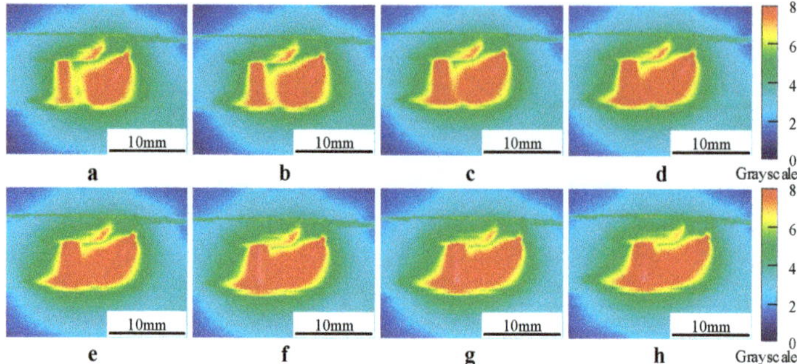

Figure 5. Acquired images of coupled arc profile with different TIG welding current: (**a**) 30 A; (**b**) 40 A; (**c**) 50 A; (**d**) 60 A; (**e**) 70 A; (**f**) 80 A; (**g**) 90 A; (**h**) 100 A.

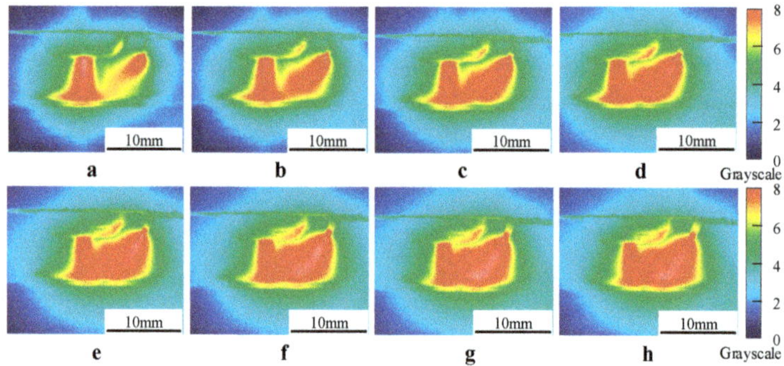

Figure 6. Acquired images of coupled arc profile with different plasma welding current. (**a**) 40 A; (**b**) 60 A; (**c**) 80 A; (**d**) 100 A; (**e**) 120 A; (**f**) 140 A; (**g**) 160 A; (**h**) 180 A.

3.2. Arc Pressure Distribution of the Coupled Arc

The arc pressure distribution on the workpiece allows the characterization of the coupled arc force on the molten pool surface and identification of arc physical properties, especially its effect on the penetration [23,24]. A differential pressure sensor (DPS: HSTL-FY01, Beijing, China) was utilized to detect the arc pressure on the surface of base metal as shown in Figure 7a. A measuring bore hole with a diameter of 0.8 mm is installed in the center of the plate, which was fixed on another water-cooled copperplate with a geometry size of $200 \times 400 \times 3$ mm^3 [25]. The real-time data acquisition of the sensor can be realized by the data acquisition card (YAV-USB2AD, Wuhan Yawei Electronic Technolgy CO., Ltd., Wuhan, China) using the technology of Labview®(National Instruments, Austin, TX, USA). The system is adjusted and calibrated through the software. To ensure accuracy of the results, the measurement point spacing is 0.1 mm and the torch moves at a constant velocity of 2 mm/s. The details of the measurement system are shown in Figure 7.

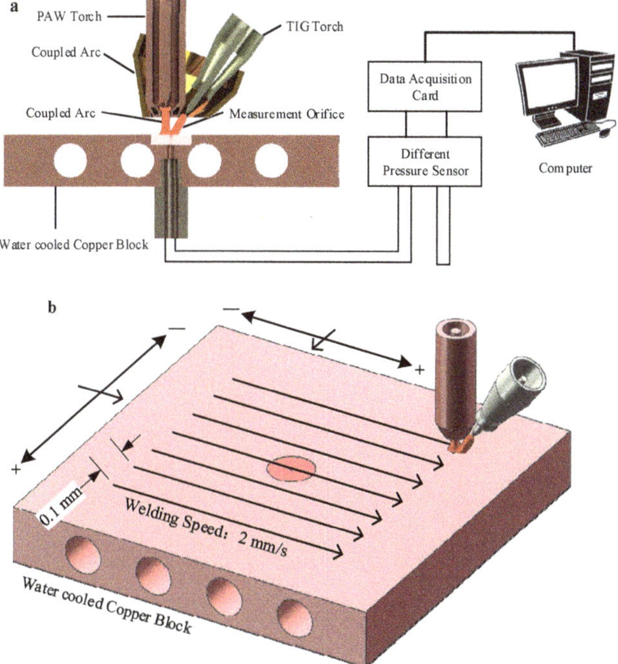

Figure 7. Measurement system of the arc pressure. (**a**) The schematic diagram of the arc pressure data acquisition system. (**b**) The schematic diagram of the acquisition methods for coupled arc pressure.

The two-dimensional pressure distribution and pressure variation along with the welding direction for each welding arc are shown in Figure 8. The surface charts for individual stagnation pressure of three welding method are drawn using Matlab® (8.3, MathWorks, Natick, MA, USA) software in order to illustrate the difference in magnitude between the values of arc pressure. According to the following experimental results, there is another interesting finding that catches our attention. It is worthwhile mentioning that the coupled arc pressure is mainly determined by plasma arc and proportion to the welding current, and the welding conditions are listed in Table 2. The contour cloud image of the coupled arc pressure distribution is shown in Figure 8e, notably, it is similar to a gourd shape and the maximum arc pressure appears between the two arcs near the side of plasma arc compared with conventional plasma arc welding method. It can be seen that the maximum value of the coupled arc is approximately 650 Pa and the maximum arc pressure decreased gradually with the increasing electrode spacing, as shown in Figure 8f. This result is significantly higher than that of plasma arc and TIG arc under the same conditions, as shown in Figure 8b,d. Therefore, it is indicated that the total arc pressure of coupled arc depends on combination effect of these arcs, and it is consistent with the above analysis of coupled arc profile.

Table 2. Experimental conditions for arc pressure measurement.

I_P (A)	I_T (A)	D_{PT} (mm)	L (mm)	D (mm)	Q_p (L/min)	Q_s (L/min)
60	100	7	5	1.25	4	15

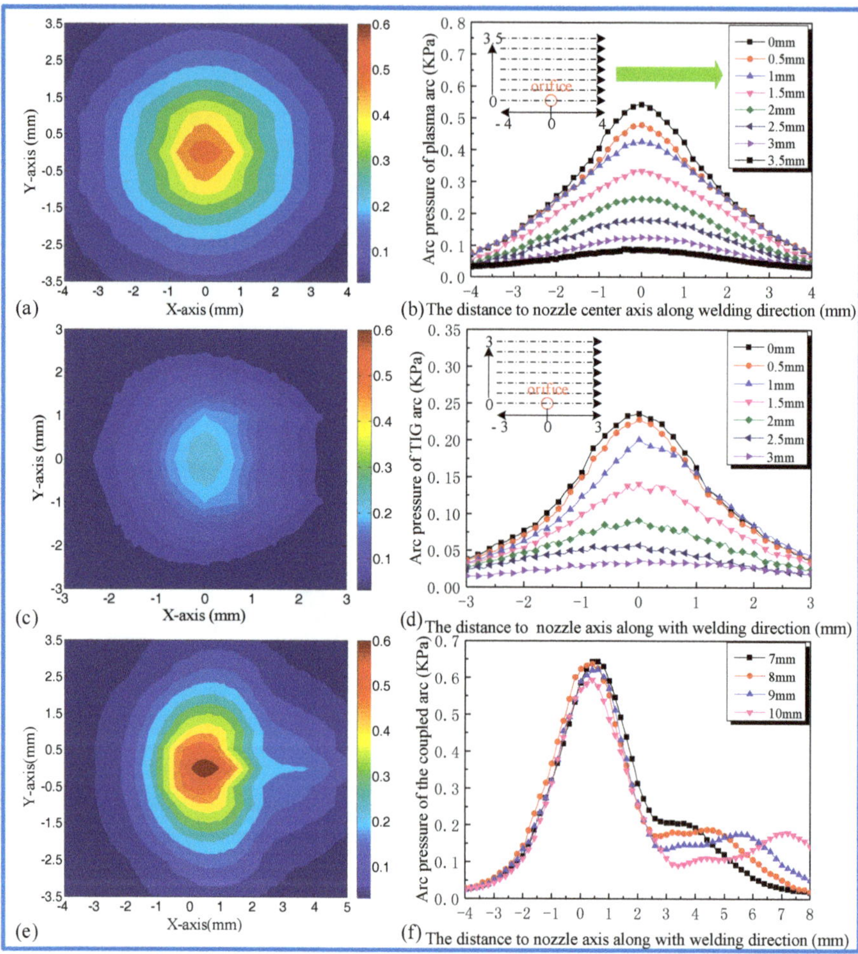

Figure 8. Stagnation pressure distribution of welding arc with the different method. (**a**) Top view of two-dimensional distribution of the plasma arc. (**b**) The plasma arc pressure along with welding direction with different distance to measurement orifice axis. (**c**) Top view of two-dimensional distribution of the TIG arc. (**d**) The TIG arc pressure along with welding direction with different distance to measurement orifice axis. (**e**) Top view of two-dimensional distribution of the coupled arc. (**f**) The coupled arc pressure along with welding direction with different electrode spacing.

3.3. The Influence Factor of Coupled Arc Pressure

Group #1 is the single-factor experiment to examine the effects of plasma current on the coupled arc pressure distribution. The plasma arc as an important part of the coupled arc [26], the variation in the current value has a significant influence on plasma flow force and mainly determines the maximum arc pressure of the coupled arc. The histogram of the coupled arc pressure is acquired by means of extracting the maximum value of distributed line type, as shown in Figure 9b. As can be seen from part a and b of Figure 9, the maximum arc pressure increases gradually with the rise in plasma current and reaches 1600 Pa as plasma current is 120 A. In addition, according to the changes of arc pressure curves at point a, it can be illustrated that the TIG arc pressure peak is overlapped with that of plasma gradually with increase of plasma current, that is due to increase in Lorentz force during the process. This result suggest that the coupling effect is strengthened between the two arcs with the increase of plasma current.

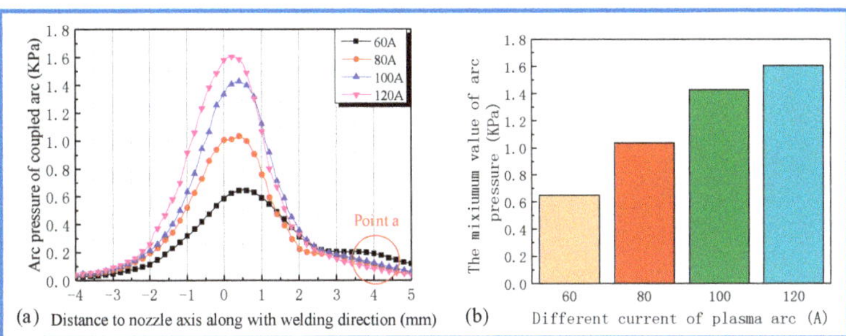

Figure 9. Coupled arc pressure with different plasma current. (**a**) Coupled arc pressure along with welding direction. (**b**) Max value of coupled arc pressure.

Groups #2 and #3 are the single-factor experiments to examine the effect of TIG current on the coupled arc pressure distribution. As previously mentioned, the change in TIG arc would has different effects on the coupled arc pressure. It is necessary to analysis the arc pressure in two situation, when the plasma arc current is 60 A, as shown in Figure 10a, as the TIG current increases, the maximum arc pressure of the coupled arc gets deviated to the side of TIG arc within a small scale. This condition was mainly attributed to plasma flow force increases with welding current, it results in a good TIG arc stiffness. On the contrary, when the plasma current increases to 120 A, as shown in Figure 10b, the maximum arc pressure of the coupled arc gets deviated to TIG arc by a large margin, the results indicate that the Lorentz force acting on plasma arc increases enough to deflect the plasma as the TIG current increases. This is consistent with the observation from the coupled arc profile.

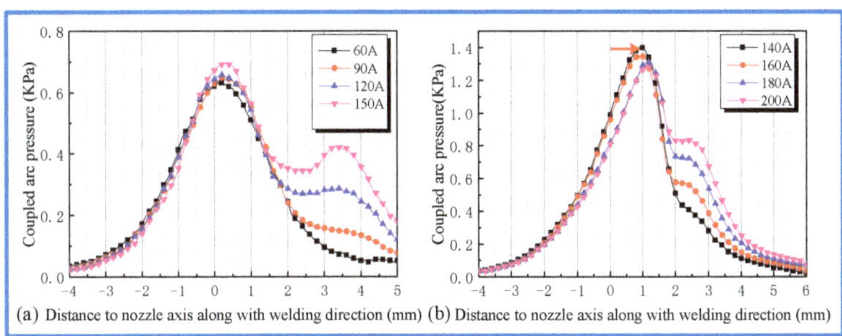

Figure 10. Coupled arc pressure along with welding direction with different current. (**a**) The plasma current is 60 A. (**b**) The plasma current is 120 A.

Group #4 is the single-factor experiment to examine the effect of the coupled arc length on the coupled arc pressure distribution. The histogram of coupled arc along with the welding direction is acquired by means of extracting the maximum value of distributed line type, as shown in Figure 11b. The maximum arc pressure initially decreased rapidly with the increase of nozzle height and then plateaued; meanwhile, another interesting finding is that the coupled arc no longer decreases as the nozzle height exceeds 5 mm. The results suggest that the horizontal component of the Lorentz force and plasma flow force cancels each other and reach equilibrium state along with welding direction, as shown in Equation (5).

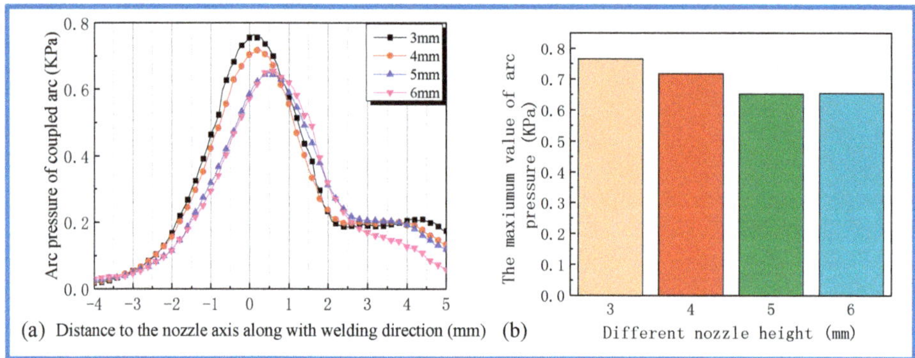

Figure 11. Coupled arc pressure with different nozzle height. (**a**) Coupled arc pressure along with welding direction. (**b**) Max value of coupled arc pressure.

Group #5 is the single-factor experiment to examine the effect of plasma gas flow rates on the coupled arc pressure distribution. Plasma gas flow rates (QP) has a significant effect on plasma movement during welding process [27]. As shown in Figure 12, the maximum arc pressure of the coupled arc evidently increased with the increase of the plasma gas flow, it can be inferred the increased plasma gas flow has led to the increase in axial component of plasma velocity and plasma gas density, as shown in Equation (2).

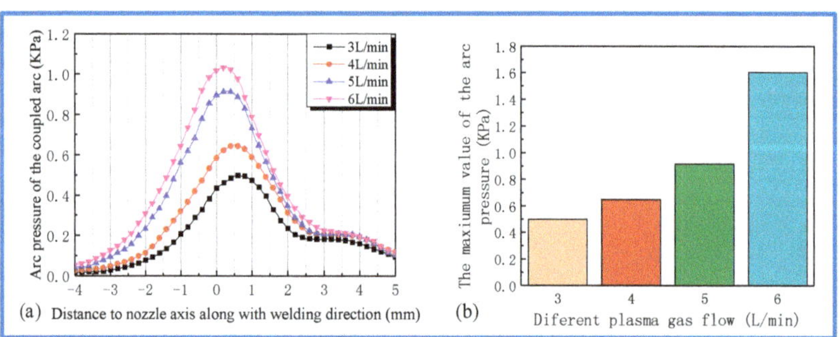

Figure 12. Coupled arc pressure with different plasma gas flow. (**a**) coupled arc pressure along with welding direction. (**b**) Max value of coupled arc pressure.

3.4. Formation and Microstructure of the Weld

The base metal is Q235B low carbon steel (with nominal chemical compositions of C-0.15, Mn-0.4, Si-0.2, S-0.03, P-0.035, balance Fe in wt.%) with dimensions of $200 \times 80 \times 5$ mm^3. MG70S-6 (with nominal chemical compositions of Cu-0.5, Mn-1.85, Si-1.15, C-0.15, Ni-0.15, Cr-0.15, Mo-0.15, S-0.025, P-0.025, balance Fe in wt %) with 1.2 mm diameter is used as the filler metal. The square groove is 1.0 mm ($\pm$ 0.1 mm). The welding positions of base metal are prepared by the conventional grinding method and then cleaned with acetone. The appearance of the surface and back of the weld using a plasma current of 150 A and a TIG current of 70 A, as shown in Figure 13. As can be seen from Figure 13a,b, a smooth and uniform weld seam with one side welding and both sides formation was successfully produced by P/TCAW. As illustrated in Figure 13c,d, a Charge Coupled Device (CCD) camera is used to acquire a clear image of the coupled arc during the welding process, the coupled arc burns stably and no obvious welding defects can be found, the weld section looks like the goblet profile and has a smaller the section dimensions. It can be suggested that the welding heat input on the base metals is limited. Therefore, P/TCAW is suitable for completing high speed and efficiency welding of mild steel plate.

The joining strength of the weld is determined by the microstructure and the typical metallurgical structure as shown in Figure 13e. The microstructure represented in the following sub-figure is consistent with the position presented in Figure 13d. The microstructure of fusion zone is finer and the upper part mainly consists of pro-eutectoid ferrite and acicular ferrite in the columnar grains, however, the lower part mainly consists of acicular ferrite. This is due to the upper part is welding under the welding heat input of the coupled arc and the lower part mainly determined by the plasma arc. Moreover, a narrow heat affected zone can be achieved using the coupled arc and the microstructure consists of fine pearlite and a small amount of acicular ferrite.

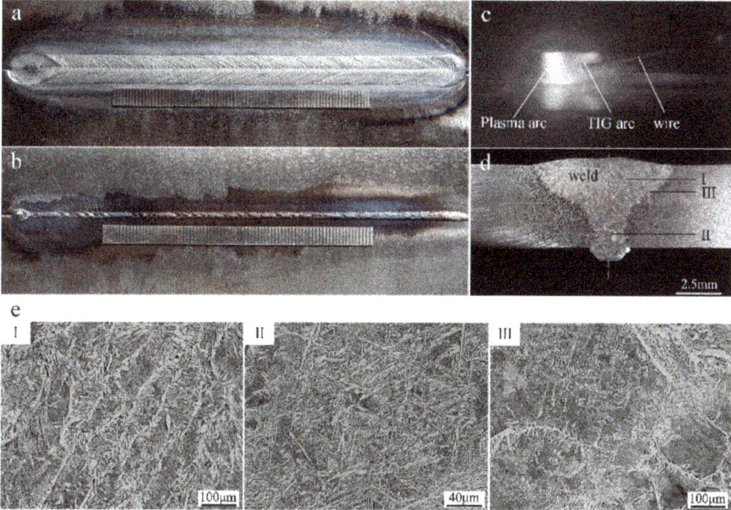

Figure 13. Appearance and microstructure of the weld. (**a**) The surface of the weld. (**b**) The back of the weld. (**c**) The coupled arc welding process with filler wire. (**d**) The cross-section of the weld. (**e**) The microstructure of the cross section.

4. Conclusions

From the current study, a novel hybrid welding method has been proposed, and the coupled arc behavior has been explored. The coupled arc profile and coupled arc pressure have been considered as ascertaining the mechanics circumstance of coupled arc behavior. The following conclusions could be obtained:

1. Under the function of Lorentz force and plasma flow force, TIG arc is deflected and couples with plasma arc. A coupled arc is generated and simultaneously involves the arc characteristics of plasma and TIG.

2. The coupled arc profile is mainly determined by the combination effect of plasma flow force and Lorentz force generated by the magnetic field of plasma arc column, which is affected by the welding current from plasma and TIG.

3. The coupled arc pressure distribution in two-dimensional surface is similar to gourd shape. The maximum arc pressure appears between plasma arc and TIG arc and it is mainly determined by the plasma current. The welding conditions that affect the coupled arc distribution and the maximum pressure could be ranked according to their effects on the experimental results, namely, plasma current, plasma gas flow, TIG current and electrode spacing.

4. The high-efficiency welding process of mild steel can be achieved without obvious defects. A smooth and uniform weld seam with on side welding both side formation was acquired at a plasma current of 150 A, a TIG current of 70 A and a welding speed of 240 mm/min.

Author Contributions: H.-T.Z. and B.W. conceived and designed the experiment; H.-C.Z. performed the experiments and contributed to the data analysis; B.W. analysed the data and wrote the article; X.-M.Z. and J.-C.F. contributed to the data analysis and discussion.

Funding: This research received no external funding.

Acknowledgments: The research was sponsored by Shandong Provincial Natural Science Foundation (ZR2018MEE027), Shandong Key research and development plan (2017GGX30132), Shandong Key research and development plan (GG201709250277), Natural Scientific Research Foundation in Harbin Institute of Technology (HIT.NSRIF.201707) and Young Taishan Scholars Program of Shandong Province (tsqn20161062).

Conflicts of Interest: The authors declare no conflict of interest.

Nomenclatures

r_1	The plasma arc radius at the place of the tungsten cathode top (mm)
r_2	The plasma arc radius of arc column on the base metal surface (mm)
μ_0	Permeability of free space (T*m/A)
I_P	Plasma arc welding current (A)
p	Plasma gas density (kg/m^3)
μ_1	Axial component of the plasma velocity (m/s)
P_e	Electromagnetic axial pressure (Pa)
P_g	Plasma gas flow axial pressure (Pa)
F_r	The plasma arc axis resultant force
S	The arc sectional area at the place of base plate on the arc radial and $S = \pi r_1^2$
B	Magnetic induction of magnetic field generated by plasma arc (T)
L	Displacement of particle per unit
F_t	The plasma flow force of TIG arc
K	A constant that equals $\mu/4\pi$
μ	Medium magnetic conductivity (T*m/A)
R_a	The radius of TIG arc root surface (mm)
R_b	The radius of TIG arc root undersurface (mm)

References

1. Liu, Z.M.; Cui, S.; Luo, Z.; Zhang, C.; Wang, Z.; Zhang, Y. Plasma arc welding: Process variants and its recent developments of sensing, controlling and modeling. *J. Manuf. Process.* **2016**, *23*, 315–327. [CrossRef]

2. Tanaka, M.; Terasaki, H.; Ushio, M.; Lowke, J.J. A unified numerical modeling of stationary tungsten-inert-gas welding process. *Metall. Mater. Trans. A* **2002**, *33*, 2043–2052. [CrossRef]

3. Wu, C.S.; Ushio, M.; Tanaka, M. Analysis of the tig welding arc behavior. *Comput. Mater. Sci.* **1997**, *7*, 308–314. [CrossRef]

4. Huang, J.; He, X.; Guo, Y.; Zhang, Z.; Shi, Y.; Fan, D. Joining of aluminum alloys to galvanized mild steel by the pulsed de-gmaw with the alternation of droplet transfer. *J. Manuf. Process.* **2017**, *25*, 16–25. [CrossRef]

5. Asai, M.S.; Ogawa, M.T.; Ishizaki, M.Y.; Minemura, M.T.; Minami, M.H.; Iyazaki, M.S.M. Application of plasma mig hybrid welding to dissimilar joints between copper and steel. *Weld. World* **2013**, *56*, 37–42. [CrossRef]

6. Yurtisik, K.; Tirkes, S.; Dykhno, I.; Gur, C.H.; Gurbuz, R. Characterization of duplex stainless steel weld metals obtained by hybrid plasma-gas metal arc welding. *Soldag. Insp.* **2013**, *18*, 207–216. [CrossRef]

7. Ton, H. Physical properties of the plasma-mig welding arc. *J. Phys. D Appl. Phys.* **1975**, *8*, 922. [CrossRef]
8. Zhang, Y.M.; Zhang, S.B. Double-sided arc welding increases weld joint penetration. *Weld. J.* **2000**, *77*, 57–61.
9. Taban, E. Joining of duplex stainless steel by plasma arc, tig, and plasma arc+tig welding processes. *Mater. Manuf. Process.* **2008**, *23*, 871–878. [CrossRef]
10. Acherjee, B. Hybrid laser arc welding: State-of-art review. *Opt. Laser Technol.* **2018**, *99*, 60–71. [CrossRef]
11. Liming, L.; Jifeng, W.; Gang, S. Hybrid laser–tig welding, laser beam welding and gas tungsten arc welding of az31b magnesium alloy. *Mater. Sci. Eng. A* **2004**, *381*, 129–133. [CrossRef]
12. Chen, J.; Wu, C.S.; Chen, M.A. Improvement of welding heat source models for tig-mig hybrid welding process. *J. Manuf. Process.* **2014**, *16*, 485–493. [CrossRef]
13. Hsu, Y.F.; Rubinsky, B. Two-dimensional heat transfer study on the keyhole plasma arc welding process. *Int. J. Heat Mass Transf.* **1988**, *31*, 1409–1421. [CrossRef]
14. Wu, C.S.; Wang, L.; Ren, W.J.; Zhang, X.Y. Plasma arc welding: Process, sensing, control and modeling. *J. Manuf. Process.* **2014**, *16*, 74–85. [CrossRef]
15. Tashiro, S.; Miyata, M.; Tanaka, M.; Shin, K.; Takahashi, K. Numerical analysis of basic property of keyhole welding with plasma arc. *Trans. Mater. Res. Soc. Jpn.* **2010**, *35*, 589–592. [CrossRef]
16. Stenbacka, N. On arc efficiency in gas tungsten arc welding. *Soldag. Insp.* **2013**, *18*, 380–390. [CrossRef]
17. Fuerschbach, P.W.; Knorovsky, G.A. A study of melting efficiency in plasma arc and gas tungsten arc welding. *Weld. J.* **1991**, *70*, 287.
18. Sun, J.S.; Wu, C.S.; Zhang, Y.M. Heat transfer modeling of double-side arc welding. *Acta Phys. Sin.* **2002**, *51*, 286–290.
19. Zhang, Y. Keyhole double-sided arc welding process. *J. Mater. Sci. Technol.* **2001**, *17*, 159–160. [CrossRef]
20. Wang, J.; Sun, Q.; Feng, J.; Wang, S.; Zhao, H. Characteristics of welding and arc pressure in tig narrow gap welding using novel magnetic arc oscillation. *Int. J. Adv. Manuf. Technol.* **2017**, *90*, 413–420. [CrossRef]
21. Qi, B.J.; Yang, M.X.; Cong, B.Q.; Liu, F.J. The effect of arc behavior on weld geometry by high-frequency pulse gtaw process with 0cr18ni9ti stainless steel. *Int. J. Adv. Manuf. Technol.* **2013**, *66*, 1545–1553. [CrossRef]
22. Dai, D.S.; Song, Y.L.; Zhang, L.P. Study on the pressure in plasma arc. *Chin. J. Mech. Eng.* **2003**, 34–36. [CrossRef]
23. Cheng, L.; Hu, S.; Wang, Z. Arc pressure analysis in variable polarity tig welding. *Trans. China Weld. Inst.* **2014**, *35*, 101–104.
24. Huang, Y.; Huaiyu, Q.U.; Fan, D.; Liu, R.; Kang, Z.; Wang, X. Arc pressure measurement and analysis of coupling arc aa-tig. *Trans. China Weld. Inst.* **2013**, *34*, 33–36.
25. Ham, H.S.; Oh, D.S.; Cho, S.M. Measurement of arc pressure and shield gas pressure effect on surface of molten pool in tig welding. *Sci. Technol. Weld. Join.* **2013**, *17*, 594–600. [CrossRef]
26. Murphy, A.B.; Tanaka, M.; Yamamoto, K.; Tashiro, S.; Sato, T.; Lowke, J.J. Modelling of thermal plasmas for arc welding: The role of the shielding gas properties and of metal vapour. *J. Phys. D Appl. Phys.* **2009**, *42*, 194006. [CrossRef]
27. Murphy, A.B. Thermal plasmas in gas mixtures. *J. Phys. D Appl. Phys.* **2001**, *34*, R151–R173. [CrossRef]

Article

Liquation Cracking in the Heat-Affected Zone of IN738 Superalloy Weld

Kai-Cheng Chen [1], Tai-Cheng Chen [2,3], Ren-Kae Shiue [3] and Leu-Wen Tsay [1,*

[1] Institute of Materials Engineering, National Taiwan Ocean University, Keelung 20224, Taiwan; k51275127@gmail.com

[2] Nuclear Fuels and Materials Division, Institute of Nuclear Energy Research, Taoyuan 32546, Taiwan; tcchen@iner.gov.tw

[3] Department of Materials Science and Engineering, National Taiwan University, Taipei 10617, Taiwan; rkshiue@ntu.edu.tw

* Correspondence: b0186@mail.ntou.edu.tw; Tel.: +886-2-24622192 (ext. 6405)

Received: 7 May 2018; Accepted: 25 May 2018; Published: 27 May 2018

Abstract: The main scope of this study investigated the occurrence of liquation cracking in the heat-affected zone (HAZ) of IN738 superalloy weld, IN738 is widely used in gas turbine blades in land-based power plants. Microstructural examinations showed considerable amounts of γ' uniformly precipitated in the γ matrix. Electron probe microanalysis (EPMA) maps showed the γ-γ' colonies were rich in Al and Ti, but lean in other alloy elements. Moreover, the metal carbides (MC), fine borides (M_3B_2 and M_5B_3), η-Ni_3Ti, σ (Cr-Co) and lamellar Ni_7Zr_2 intermetallic compounds could be found at the interdendritic boundaries. The fracture morphologies and the corresponding EPMA maps confirmed that the liquation cracking in the HAZ of the IN738 superalloy weld resulted from the presence of complex microconstituents at the interdendritic boundaries.

Keywords: IN738 superalloy; welding; HAZ cracking; microconstituent

---

1. Introduction

The superior tensile strength, creep and oxidation resistance at elevated temperature make Ni-base superalloys used extensively in industrial gas turbines [1]. A wide range of Ni-base superalloys, from solid solution-strengthened to highly-alloyed precipitation-hardened materials, have been achieved to satisfy the requirements of high-temperature performance and environmental corrosion resistance. IN738, a cast Ni-base superalloy, is strengthened by extensive precipitation of γ' ($Ni_3(Al,Ti)$) and metal carbides (MC) to achieve excellent mechanical properties at elevated temperature [2,3]. After a long term of service at elevated temperature and high stress, the turbine blades eventually suffer from fatigue and creep damages, which degrade the microstructures and/or introduce defects into the components. To lower the life cost during service, less expensive means are developed as an alternative to replace new components. Furthermore, rejuvenation treatments are reported to be able to restore the microstructures and mechanical properties of degraded superalloys [4–6]. More often, repair-welding is chosen as the major way to refurbish cracked or damaged blades in gas turbines [1].

Traditionally, repair-welding of Ni-base superalloys has been conducted by gas tungsten arc welding (GTAW) processes [7–11]. However, the precipitation-hardened Ni-base superalloys are very sensitive to hot cracking in the heat-affected zone (HAZ) and fusion zone (FZ) during welding [7,8,12–14]. IN738 contains considerable amounts of Al and Ti, which is known for its poor weldability. Liquation cracking of precipitation-hardened Ni-base superalloys is strongly dependent on forming low-melting liquid film at the grain boundaries of the HAZ. The final solidified products with low melting point at the grain boundaries of cast Ni-base superalloys, such as M_5B_3 and M_2B borides, γ-γ' ($Ni_3(Al,Ti)$) eutectics, MC carbides, and some other intermetallic compounds, are responsible for their high cracking susceptibility [15–19]. Cracking caused by the HAZ liquation in the as-welded condition is exacerbated by subsequent postweld heat treatment [20]. Preheating at sufficiently high temperatures before welding reduces thermally induced stress. Such preheating more effectively prevents cracking of IN738LC superalloy than does controlling the material ductility through preweld heat treatment [21].

To date, many efforts have been made to eliminate or reduce the HAZ cracking of precipitation-hardened Ni-base superalloys through the control of primary microstructures by preweld heat treatment [18,19,22–24], by laser [12,18,20,23,25–27] or electron beam [28,29] welding processes, or by using filler with lower Al + Ti content [30–32]. In practical applications, the use of under matched filler or low heat-input welding processes to reduce weld-shrinkage stress are able to lower cracking in those Ni-base superalloys with poor weldability. Hot isostatic pressing is effective to heal all the cracks present in FZ and HAZ of laser-repaired CM247LC superalloy weld [33]. The aim of this work was to investigate the influence of grain boundary constitutents on the hot cracking sensitivity of IN738 alloy. The microstructures and chemical compositions of the microconstituents in cast IN738 were investigated. The coverings on liquated crack surfaces were examined and their compositions were determined. Furthermore, the relationship between microstructural features and the hot cracking mechanism of the repaired specimens was correlated with the inherent solidification products at the interdendritic boundaries of the cast IN738 superalloy.

2. Materials and Experimental Procedures

The IN738 alloy ingot used in this work, which was melt and cast in vacuum, had a diameter of 75 mm. Disc samples with a thickness of 5 mm were wire-cut from the vacuum-melted ingot. The chemical compositions of the experimental material in wt. % were: 0.17 C, 3.5 Al, 3.5 Ti, 16 Cr, 8.5 Co, 1.7 Mo, 2.5 W, 0.8 Nb, 0.1 Zr, 0.01 B, and the balance Ni. To relieve the residual stress in the as-cast sample, the samples were heated at 1050 °C for 5 h in high vacuum. GTAW was used to perform bead-on-plate welding in this study. The welding parameters included a welding current of 60 A, welding voltage of 10 V, and travel speed of 100 mm/min. After welding, the samples were cut either parallel or normal to the FZ. The welds experienced a standard metallographic preparation, and great attentions were paid to the HAZ microcracks.

Microstructures of the substrate and weldment were examined using a JSM-7100F field emission scanning electron microscope (FESEM, JEOL Ltd., Tokyo, Japan). Chemical compositions of various phases at the solidified boundaries were measured using either an X-Max^N energy-dispersive spectrometer (EDS, Oxford Instruments, Abingdon, UK) or a JXA-8200 electron probe micro-analyzer (EPMA, JEOL Ltd., Tokyo, Japan) equipped in the SEM. D2 Phaser X-ray diffraction (XRD, Bruker, Karlsruhe, Germany) using Cu-Kα radiation was employed to reveal various phases in the sample. In the XRD analysis, the tested samples wire-cut from the IN-738 ingot was approximately 10 mm in width and 20 mm in length. Although the analyzed region varies with the 2θ angle, it changed not much when 2θ varied between 30° and 80° [34]. The fracture features of the cracked specimens were also inspected by a FESEM. The crystallographic analyses of various phases at solidification boundaries were investigated by using an FESEM equipped with a detector of NordlysMax^2 electron backscatter diffraction (EBSD, Oxford Instruments, Abingdon, UK).

3. Results

3.1. Microstructural Examinations

The back-scatter electron (BSE) images can be used to distinguish solidification products in a cast superalloy according to their image brightness or phase contrast. Figure 1 presents the SEM micrographs in BSE images, showing the microstructures of the cast IN738 alloy. Regarding the image brightness and phase contrast of the interdendritic microconstituents, the γ-γ′ colonies and γ matrix are revealed by high darkness in the figure. The segregation of carbide-forming (Nb, Ta, Ti) elements resulted in the formation of MC carbides, which are shown by high brightness. As shown in Figure 1a,b, MC carbides in different sizes and morphologies were the primary second phase in the cast alloy. The MC carbides were in different shapes: irregular, island-like, blocky, or Chinese-script. Moreover, lamellar eutectics were more likely to be found in the central part of the ingot. At higher magnification (Figure 1c,d), lamellar eutectics were observed nearby the γ-γ′ eutectic colonies, indicating the formation of low-melting microconstituents along solidified boundaries. A few fine white phases were found in the narrow boundaries between γ-γ′ colonies (Figure 1c,d). The lamellar eutectics were not as bright as the MC carbides, possibly due to lower concentrations of heavy elements in the former than those in the latter. In addition, the grain boundary product of gray color in Figure 1c was deduced to be boride. The fine particles at the boundaries between γ-γ′ colonies were likely fine MC carbides. The compositions and structures of those microconstituents would be confirmed below.

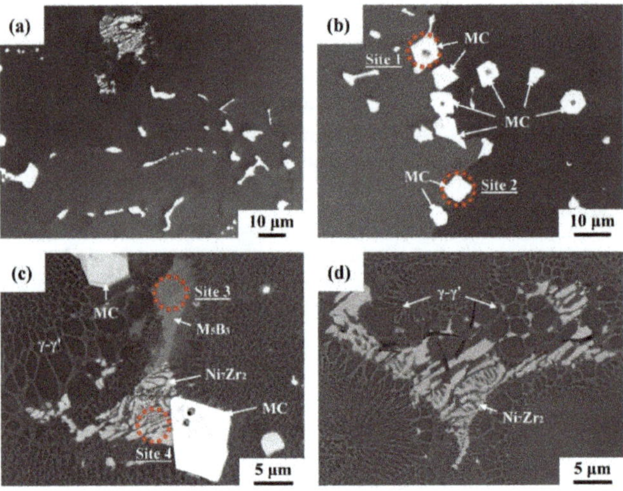

Figure 1. SEM micrographs showing (**a**) the microstructures of the IN738 superalloy; (**b**) the metal carbides (MC) carbides; (**c**,**d**) γ-γ′ eutectic colonies, lamellar intermetallics and borides around the boundaries. The electron probe microanalysis (EPMA) analysis of carbides at sites 1 and 2 in (**b**), boride at site 3 and lamellar structure at site 4 in (**c**) were performed.

EPMA was used to determine the chemical compositions of the microconstituents at the interdendritic boundaries, as shown in Figure 1. Table 1 lists chemical compositions (in at. %) of the EPMA analysis, which were obtained at various locations displayed in Figure 1b,c. As illustrated in Table 1, MC carbides (Figure 1b, sites 1 and 2) were rich in Ti and alloyed with additional Ta and Nb. Because high Nb and Ta concentrations were observed in the MC carbides, the carbides naturally exhibited higher brightness in the BSE image. EDS analysis was also used to screen the chemical compositions of the carbides of different morphologies, which could be located inter- and intragranularly. All carbides were composed of the main alloy elements of Ti, Ta, and Nb. Figure 1c illustrates complex microconstituents ahead of the γ-γ′ eutectic colony, including a gray blocky phase (site 3) and white lamellar eutectic (site 4). The EPMA analysis results showed that the gray blocky phase comprised of high boron concentration, and enriched in Cr and Mo. It was reasonable to deduce that the gray blocky phase was Cr-Mo borides. Furthermore, the bright lamellar eutectic showed the high segregation of Zr along with high Ni content. The low alloy contents of the measurement at site 4 were due to the high yield volume of the detected site relative to the size of the lamellar structure. It was deduced that the bright lamellar eutectic could be intermetallic compound.

Table 1. The chemical compositions in at. % of the microconstituents indicated in Figure 1b,c.

Element	Al	B	C	Cr	Mo	Nb	Ti	Zr	Co	Ta	Ni
Site 1	0.02	-	45.72	0.89	1.69	6.21	30.07	0.05	0.27	11.25	Bal.
Site 2	0.01	-	44.03	0.98	1.66	6.52	30.34	0.04	0.26	11.98	Bal.
Site 3	-	27.14	1.87	54.3	9.59	0.40	0.98	-	1.31	0.08	Bal.
Site 4	1.06	-	2.28	3.49	0.23	0.94	2.34	16.67	7.01	0.31	65.58

3.2. Phase Identification by XRD

Figure 2 presents the XRD patterns of various phases in the cast IN738 substrate or base metal. The results indicated that MC carbides, M_3B_2 borides, η-Ni_3Ti and Ni_7Zr_2 intermetallics associated with γ' precipitates in the γ matrix, suggesting that the segregation of alloy elements during solidification caused the formation of many solidification products in the cast superalloy. According to previous research [7,8,12,14], different grain boundary constituents, such as MC carbide, M_2SC sulphocarbide, Cr-Mo boride, Ni-Zr intermetallics, and γ-γ' ($Ni_3(Al,Ti)$) eutectic have been observed in IN738/IN738LC superalloys. In this study, η-Ni_3Ti was also detected in the alloy. The presence of these final solidification products along interdendritic boundaries significantly deteriorated the weldability of cast IN738 superalloy.

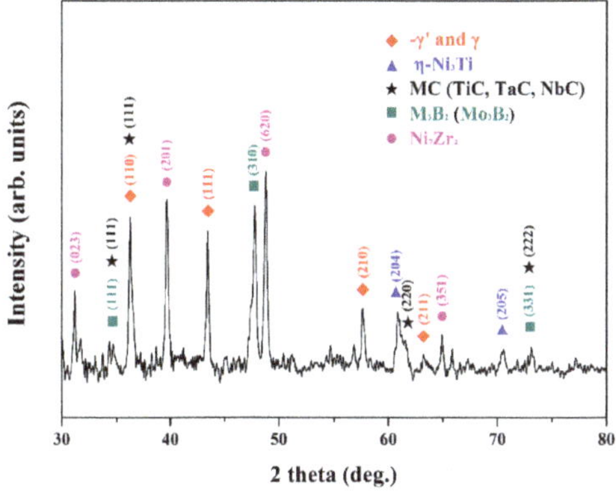

Figure 2. XRD patterns of the cast IN738 superalloy.

3.3. Microconstituents Identified by EBSD

Figure 3 illustrates BSE image, the elemental mappings of EPMA and EBSD crystallographic map in order to identify various phases at the interdendritic boundary. In the BSE image (Figure 3a), the solidification products can be distinguished by image contrast. Additionally, the elemental maps (Figure 3b) illustrate the distributions of various elements in the microconstituents determined by EPMA. A bright color indicates higher count numbers of an element than does a dark one. EBSD map (Figure 3c) provides crystallographic data of different phases in the microconstituents. As shown in Figure 3a, the BSE image showed complex products formed nearby the dendrite boundary (Figure 3a). Both γ-γ' colonies and γ matrix were in relatively high darkness in the figure. Isolated slender white phase, lamellar eutectic gray phase, and blocky gray product were also observed in Figure 3a. EPMA maps revealed that the γ-γ' eutectic colonies were significantly enriched in Al and Ti as shown in high-intensity red (Figure 3b). The strong segregation of Al and Ti into the γ-γ' eutectic colonies ahead of the final solidification products was also reported. Although the intensity of the Ta map was weak, it still showed the tendency of segregation to the γ-γ' colonies. The C map was related to the sites of MC carbides, which showed the formation of fine, isolated carbides inter-dispersed in the grain boundary products. EPMA mappings displayed the co-segregation of B, Mo, Cr and W, which was expected to promote the formation of island-like Cr-Mo-W borides as shown in Figure 3a. It was also noted that a severe segregation of Co with Cr, Mo and W into the solidification products, which occupied a great fraction in this analysis. Moreover, according to the Ti and Nb maps, the distribution of Ti in the microconstituents was coincident with that of Nb therein. In addition, high Zr intensity was only observed in certain sites in the microconstituents of a slender form, as shown in Figure 3b.

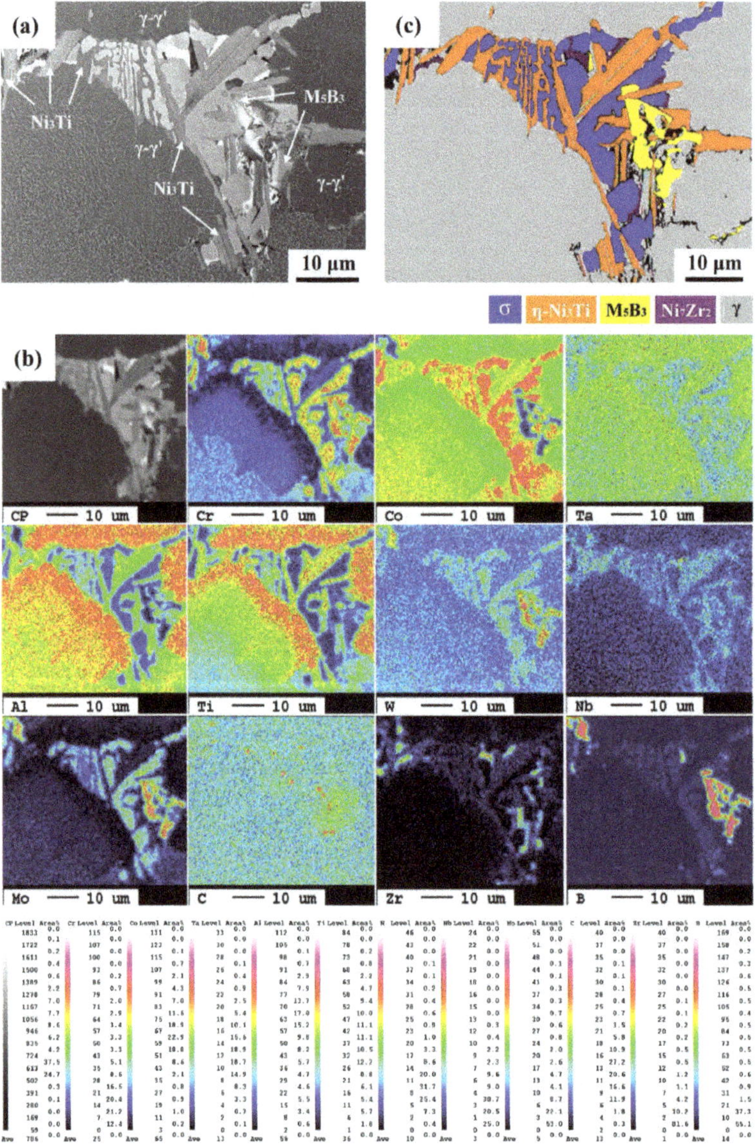

Figure 3. (**a**) Back-scatter electron (BSE) image of the microconstituents; (**b**) the elemental maps determined by EPMA; (**c**) Electron backscatter diffraction (EBSD) crystallographic map to identify various phases in the microconstituents.

The EBSD map (Figure 3c) showed the distributions of distinct phases in the solidification product. The co-segregation of B, Cr, Mo and W assisted the formation of M_5B_3 $(CrMoW)_5B_3$ borides, shown in yellow in Figure 3c. Additionally, the lamellar η-Ni_3Ti, indicated by orange, has previously been pointed out in studies of Ni-Co superalloy and IN738 [35–37]. Although the Nb and Al intensity in the microconstituents was not so intense (Figure 3b), the distribution of Nb and Al in the solidification product was consistent with that of Ti. It was deduced that Nb and Al were partitioned into η-Ni_3Ti (η-$Ni_3(TiAlNb)$) intermetallic compound. Besides, locations with high Zr intensity were associated with the formation of Ni_7Zr_2 intermetallics, indicated by purple in the EBSD crystallographic map. EPMA mappings in this work showed a very high Co intensity in the solidification products. Basically, co-segregation of Co, Cr, Mo and W, which was confirmed by EPMA mappings, enhanced the formation of σ (CrCo) phase, as shown by dark blue in the EBSD map. Nodular σ phase has also been reported to nucleate and grow from the interdendritic liquid [37]. The result indicated that the chemical compositions of every phase in the terminal solidification products were quite complex. Compounds or precipitates predicted by the binary phase diagram were frequently alloyed with the third or even fourth elements.

The solidification products with blocky and network morphologies ahead of the γ-γ' eutectic colonies were examined and subjected to further investigation (Figure 4). As listed in Table 2, the chemical compositions of the blocky gray and white phases, indicated by arrows in Figure 4a, were determined by EPMA. At sites 1 and 2, the compositions of the blocky and footprint-like gray phases were high in B and Cr mixed with extra Mo. It was deduced that such solidification products would be Cr-Mo borides. The EPMA maps of various elements (Figure 4b) also confirmed the partitions of B, Cr, Mo and W into grain boundary borides. At site 3, the white phase possessed high Ti, Nb, Ta and C concentrations; they were expected to be MC carbides. The distribution of C in the EPMA map was correlated with carbide sites. The EPMA mappings revealed that all fine white phases in Figure 4a had high C, Ti, Ta and Nb intensity. Thus, those fine white phases having different morphologies were deduced to be MC carbides. Although the Ti and Al intensities were not so strong (Figure 4b), it still showed the fact of co-segregation of both elements into the γ-γ' colonies. In addition, the lamellar structures shown in Figure 4a were more likely to coexist with borides, which were rich in Zr in the EPMA map (Figure 4b). The EBSD map showed that the blocky and footprint-like gray phases in Figure 4a were M_3B_2 borides, as shown by red in Figure 4c. A few slender M_5B_3 borides, indicated by yellow in Figure 4c, also could be seen in the solidification products. Moreover, the lamellar structures shown in Figure 4a were identified as Ni_7Zr_2 intermetallics, as indicated by purple in Figure 4c. It was obvious that the presence of borides and intermetallics at solidification boundaries would greatly depress the final solidification temperature, thus, leading to a high tendency of liquation cracking in the IN738 superalloy weld.

Table 2. The chemical compositions in at. % of the microconstituents indicated in Figure 4a.

Element	Al	B	C	Cr	Mo	Nb	Ti	Ta	Ni
Site 1	0.04	29.06	1.39	37.43	17.32	2.37	3.2	0.48	Bal.
Site 2	0.01	27.82	1.38	38.35	18.13	2.82	3.01	0.43	Bal.
Site 3	-	-	24.14	1.22	0.46	16.89	39.74	12.66	Bal.

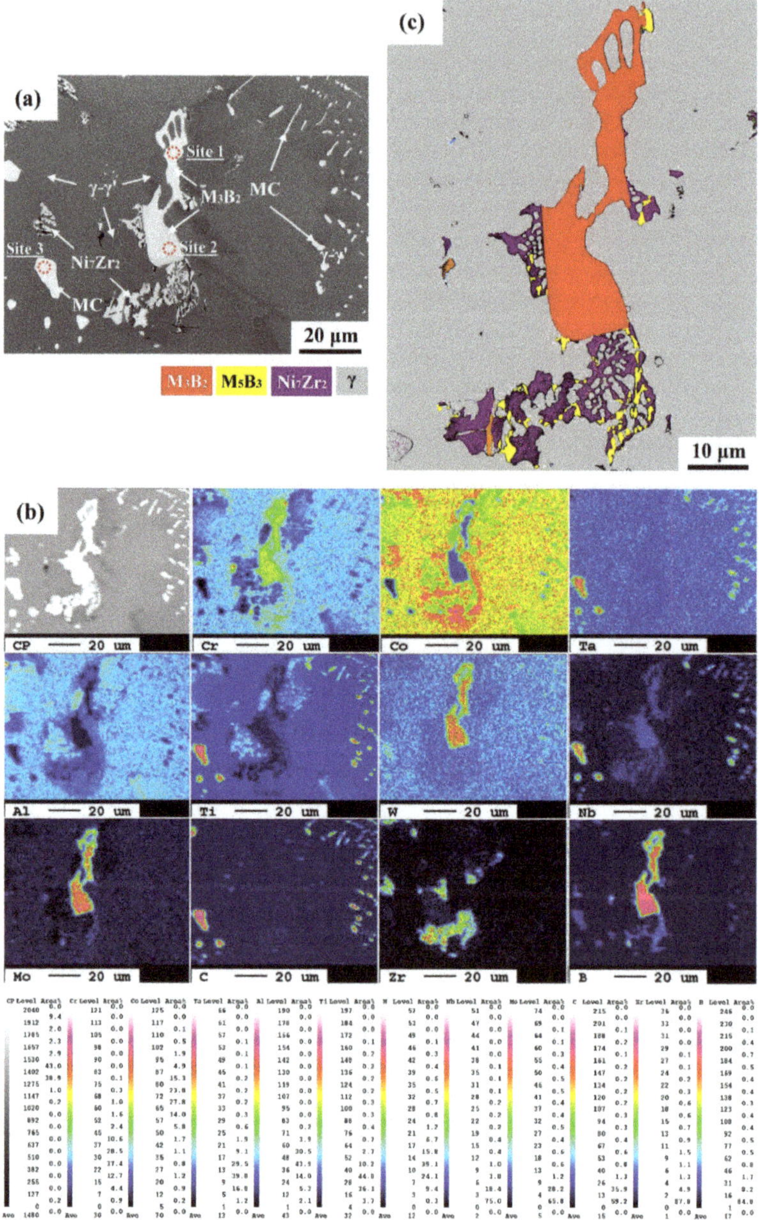

Figure 4. (**a**) BSE image of the IN738 superalloy; (**b**) the elemental maps of the solidification products determined by EPMA; (**c**) the EBSD map to identify the terminal solidification products.

3.4. HAZ Microcrack Inspections

The welds were subjected to metallurgical preparations to reveal the liquation cracks. Figure 5a displays the crack path of the HAZ microcracks adjacent to the FZ. The results indicated that the serrated crack path could be linked to the melting of low-temperature microconstituents at the boundaries. The EBSD map showed that the HAZ microcrack had a tendency to propagate interdendritically between grains of different orientations (Figure 5b). It was deduced that the imposed shrinkage stress on the liquated phases caused the occurrence of liquation cracking in the HAZ of the IN738 weld. Higher magnifications revealed that the crack was more likely to propagate along the interfaces between the MC carbides and the matrix (Figure 5c,d), due to the presence of substantial amounts of MC carbides in the IN738 superalloy. Moreover, liquation cracking also occurred along the boundaries of γ-γ' colonies, indicating the low solidus temperature of the colony boundaries or the presence of low-melting point microconstituents therein. Therefore, it was obviously that the induced HAZ microcracks of the IN738 weld clearly resulted from the inherently metallurgical issue of the IN738 superalloy itself. In material concerns, the weldability of the IN738 superalloy could be linked with the populations and distributions of those harmful microconstituents in the alloy. Advanced welding or repair-welding processes can lower the HAZ microcracks by reducing the shrinkage stress imposed on the weld, and by controlling the heat input to minimize the melting of those microconstituents. Besides, the usage of filler metals with low (Al + Ti) contents for repair-welding of superalloys is able to reduce the HAZ microcracks.

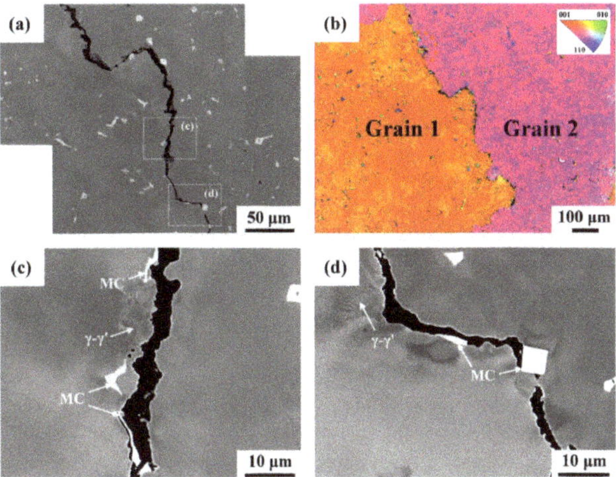

Figure 5. (**a**) The liquation crack in the HAZ of the IN738 weld; (**b**) EBSD map showing interdendritic crack between adjacent grains; (**c,d**) grain boundary crack along γ-MC and γ-γ' interfaces.

3.5. SEM Fractography

The microcracks in the HAZ of the weld were opened by using a bending fixture after immersion in liquid nitrogen for a few seconds. Figure 6 displays the fracture features of HAZ microcracks in an IN738 superalloy weld, which were inspected by SEM in BSE images. In addition, the EPMA maps were used to determine the distributions of alloy elements of the coverings on the fracture surface of the liquated crack. As shown in Figure 6a, the cobble-like fracture surface, deposited by fine solidified droplets, was rich in Al (Figure 6b). These regions with the cobble-like features, which displayed the same morphologies as the γ-γ' colonies, were associated with the liquation cracking of the γ-γ' colonies therein. Irregular bright phases of different sizes were present on the cracked surface. The EPMA maps (Figure 6b) revealed that the co-segregation of C, Ta, Ti, and Nb to those bright phases was expected to form MC carbides (Figure 6a). Both the constitutional liquation of fine carbides and the γ/MC eutectic reaction deteriorated the weldability and accounted for the HAZ microcracking of the IN738 weld. Also noted was strong segregation of Co to those zones with low concentrations of Al and carbide-forming elements (Figure 6b). As mentioned previously, σ phase (Co-Cr intermetallic compound) was found in the final solidification products. Therefore, it was deduced that the smooth fracture surface with high Co content was related to the liquation cracking of σ phase. Although the Zr intensity was weak in the EPMA map (Figure 6b), Zr tended to segregate to the lamellar structures, as shown in Figure 6a. It was deduced that the lamellar features on the fracture surface were associated with the formation of Ni_7Zr_2. Overall, the complex grain boundary microconstituents found in the cast superalloy were also observed on the fracture surface of the opened liquation cracks.

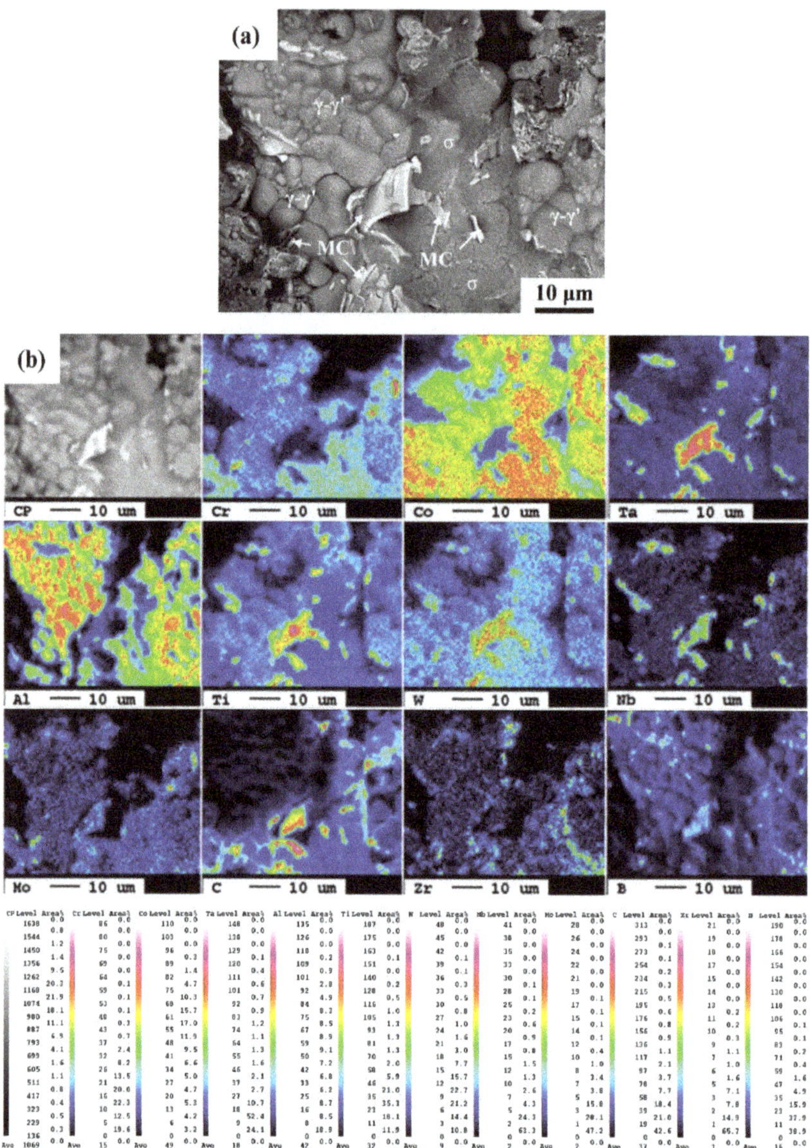

Figure 6. (a) SEM fractograph of liquation crack; (b) the associated elemental maps determined by EPMA.

4. Discussion

It is reported that the original microstructures have great influence on the HAZ microfissuring of Ni-based superalloy welds [9]. The reduction in grain boundary area per unit volume accounts for the obvious decrease in microcracks in directionally solidified Rene 80 relative to the polycrystalline IN738LC [9]. Moreover, a proprietary single crystal alloy can be avoided from cracking when build-up welds with different fillers are performed [9]. Several published literatures have elucidated the mechanism of liquation cracking in the HAZ and FZ of IN738/IN738LC welds [8,14,16,19,32]. The grain boundary constituents [8,14,16,19,32], such as MC carbide, M_2SC sulphocarbide, γ-γ' eutectic, lamellar Ni_7Zr_2 intermetallics and Cr-Mo (M_3B_2) boride, are main causes of their poor weldability. Moreover, the major second phases are MC carbides and γ-γ' eutectics formed along interdendritic boundaries [14]. Substantial microsegregation of strong carbide forming elements (Ti, Nb, Ta, Mo) along with C results in forming grain boundary carbides [8]. In addition, lamellar and dendritically shaped carbides support the $L \rightarrow \gamma + MC$ eutectic type transformation [8]. The supersaturation of Ti and Al in interdendritic liquid will assists the formation of γ-γ' eutectic [8]. The terminal solidification products also consist of Ni_7Zr_2 and M_3B_2 formed in front of γ-γ' eutectic in IN738 superalloy [14]. Furthermore, these boride particles decrease the solidus temperature at which non-equilibrium intergranular liquation is induced in the HAZ during the heating cycle of welding, thereby enhancing the cracking susceptibility of the weld [38]. In prior studies, the borides and intermetallics at the interdendritic boundaries are more likely to melt during repair-welding of Mar-M004 superalloy [39,40].

In addition to the reported Ni-Zr intermetallic, Cr-Mo boride and γ-γ' eutectic, platelet η-Ni_3Ti and nodular σ (Co-Cr) phase were also found in the terminal solidification products in this study. In fact, platelet η and nodular σ phases have been observed in cast Ni-Cr-Co superalloys like IN738 [35–37]. The σ phase consists of high Co and Cr concentrations [41], besides, the plate-like σ enriched in Co and Cr nucleates around the grain boundaries of GTD-111 superalloy [42]. Platelet and blocky η phases are also formed in IN738 [36] and in the interdendritic region of IN939 (Ni-Cr-Co) superalloy [43]. Local melting around the η phase occurs in the boundary region of IN939 at 1150 °C [43]. It seemed that local melting of those intermetallics (η, σ and Ni_7Zr_2) mixed borides at the dendrite boundaries would make the repair-welding of IN738 alloy become much difficult. Therefore, the HAZ liquation cracking of the investigated superalloy can be attributed to the presence of low-melting-temperature microconstituents at the grain boundaries, as confirmed by the coverings on the open surface of the hot cracks.

As mentioned in this work, the presence of complex solidification products in the IN738 superalloy made the welding or repair-welding very difficult. The occurrence of liquation cracking could not be avoided if the low-melting constituents were present. Hot isostatic pressing is reported to be able to heal those microcracks in the welds. To improve the weldability of this alloy, proper pre-weld heat treatments to reduce borides and intermetallic contents should be put in the first priority. Regarding the complete refurbishment of turbine blades, the machining of weld-repaired components is also a very important issue. Super Abrasive Machining is reported to be a good solution to increase machining efficiency during the production of blades and turbine disks [44]. Moreover, in case of machining IN718 superalloy, high performance cutting can be obtained in those tools with proper chemical vapor deposition (CVD) coating [45]. All those achievement in machining technologies will make the refurbishment of aeronautic components more reliable and successful.

5. Conclusions

GTAW was applied to investigate liquation cracking in the HAZ of the IN738 superalloy weld.

(1)   The microstructures of IN738 superalloy showed extensive precipitation of cuboidal γ' and coarse MC carbides in the γ matrix. Several different microconstituents were found in the interdendritic boundaries of the cast alloy. The terminal microconstituents in distinct morphologies present at

the grain boundaries included MC carbides, Cr-Mo borides, Ni-Zr intermetallics, σ (Co-Cr) and η-Ni$_3$Ti phases.

(2)   The occurrence of liquation cracking in the HAZ of IN738 weldment was interrelated with the eutectic melting of terminal solidification products along interdendritic boundaries. Liquation cracks were prone to initiate and propagate at interfaces between MC carbides and the γ matrix, due to their higher contents among the solidification products. In addition, the lamellar eutectics formed ahead of the γ-γ' colonies. They were the mixture of Cr-Mo borides, Ni-Zr intermetallics, σ and η-Ni$_3$Ti phases, which were expected to melt at much lower temperatures than the matrix.

(3)   The fracture appearance of the liquation cracks showed that the causes of liquation cracking of the IN738 superalloy weld were strongly associated with the interdendritic microconstituents in the cast structures.

Author Contributions: L.-W.T. and R.-K.S. designed and planned the experiment. K.-C.C. and T.-C.C. carried out the experiments and inspections. All authors contributed to submission and manuscript proof.

Acknowledgments: Authors acknowledge the funding support of this research by the Ministry of Science and Technology, R.O.C. (Contract No. MOST 105-2221-E-019-003). Authors also appreciate Chung-Yuan Kao at National Taiwan University for his great help in EPMA analyses.

Conflicts of Interest: The authors declare no conflict of interest.

References

1.   Henderson, M.B.; Arrell, D.; Larsson, R.; Heobel, M.; Marchant, G. Nickel Based Superalloy Welding Practices for Industrial Gas Turbine Applications. *Sci. Technol. Weld. Join.* **2004**, *9*, 13–21. [CrossRef]

2.   Balikci, E.; Raman, A.; Mirshams, R.A. Influence of Various Heat Treatments on the Microstructure of Polycrystalline IN738LC. *Metall. Mater. Trans. A* **1997**, *28*, 1993–2003. [CrossRef]

3.   Balikci, E.; Raman, A. Characteristics of the γ' Precipitates at High Temperature in Ni-Base Polycrystalline Superalloy IN738LC. *J. Mater. Sci.* **2000**, *35*, 3593–3597. [CrossRef]

4.   Qian, M.; Lippold, J.C. The Effect of Rejuvenation Heat Treatments on the Repair Weldability of Wrought Alloy 718. *Mater. Sci. Eng. A* **2003**, *340*, 225–231. [CrossRef]

5.   Yao, Z.; Degnan, C.C.; Jepson, M.A.E.; Thomson, R.C. Effect of Rejuvenation Heat Treatments on Gamma Prime Distributions in a Ni based Superalloy for Power Plant Applications. *Mater. Sci. Technol.* **2013**, *29*, 755–780. [CrossRef]

6.   Monti, C.; Giorgetti, A.; Tognarelli, L.; Mastromatteo, F. On the Effects of the Rejuvenation Treatment on Mechanical and Microstructural Properties of IN-738 Superalloy. *J. Mater. Eng. Perform.* **2017**, *26*, 2244–2256. [CrossRef]

7.   Ojo, O.A.; Richards, N.L.; Chaturvedi, M.C. Contribution of Constitutional Liquation of Gamma Prime Precipitate to Weld HAZ Cracking of Cast Inconel 738 Superalloy. *Scr. Mater.* **2004**, *50*, 641–646. [CrossRef]

8.   Ojo, O.A.; Richards, N.L.; Chaturvedi, M.C. Microstructural Study of Weld Fusion Zone of TIG Welded IN 738LC Nickel-based Superalloy. *Scr. Mater.* **2004**, *51*, 683–688. [CrossRef]

9.   Sidhu, R.K.; Ojo, O.A.; Richards, N.L.; Chaturvedi, M.C. Metallographic and OIM Study of Weld Cracking in GTA Weld Build-up of Polycrystalline, Directionally Solidified and Single Crystal Ni-based Superalloy. *Sci. Technol. Weld. Join.* **2009**, *14*, 125–131. [CrossRef]

10.   Sidhu, R.K.; Ojo, O.A.; Chaturvedi, M.C. Microstructural Response of Directionally Solidified Rene 80 Superalloy to Gas-Tungsten Arc Welding. *Metall. Mater. Trans. A* **2009**, *40*, 150–162. [CrossRef]

11.   González, M.A.; Martínez, D.I.; Pérez, A.; Guajardo, H.; Garza, A. Microstructural Response to HAZ Cracking of Prewelding Heat-Treated Inconel 939 Superalloy. *Mater. Charact.* **2011**, *62*, 1116–1123. [CrossRef]

12.   Montazeri, M.; Malek Ghaini, F.; Ojo, O.A. Heat Input and the Liquation Cracking of Laser Welded IN738LC Superalloy. *Weld. J.* **2013**, *92*, 258–264.

13.   Lachowicz, M.; Dudziński, W.; Hainmann, K.; Podrez-Radziszewska, M. Microstructure Transformations and Cracking in the Matrix of γ–γ' Superalloy Inconel 713C Melted with Electron Beam. *Mater. Sci. Eng. A* **2008**, *479*, 269–276. [CrossRef]

14. Ojo, O.A.; Richards, N.L.; Chaturvedi, M.C. Study of the Fusion Zone and Heat-Affected Zone Microstructures in Tungsten Inert Gas-Welded Inconel 738LC Superalloy. *Metall. Mater. Trans. A* **2006**, *37*, 421–433. [CrossRef]
15. Li, Q.; Lin, X.; Wang, X.H.; Yang, H.; Song, M.; Huang, W.D. Research on the Grain Boundary Liquation Mechanism in Heat-Affected Zone of Laser Forming Repaired K465 Nickel-Based Superalloy. *Metals* **2016**, *6*, 64. [CrossRef]
16. Ojo, O.A.; Richards, N.L.; Chaturvedi, M.C. On Incipient Melting During High Temperature Heat Treatment of Cast Inconel 738 Superalloy. *J. Mater. Sci.* **2004**, *39*, 7401–7404. [CrossRef]
17. Zhang, H.R.; Ojo, O.A. Cr-rich Nanosize Precipitates in a Standard Heat-Treated Inconel 738 Superalloy. *Philos. Mag.* **2010**, *90*, 765–782. [CrossRef]
18. Egbewande, A.T.; Zhang, H.R.; Sidhu, R.K.; Ojo, O.A. Improvement in Laser Weldability of INCONEL 738 Superalloy through Microstructural Modification. *Metall. Mater. Trans. A* **2009**, *40*, 2694–2704. [CrossRef]
19. Ola, O.T.; Ojo, O.A.; Chaturvedi, M.C. On the Development of a New Pre-weld Thermal Treatment Procedure for Preventing Heat-Affected Zone (HAZ) Liquation Cracking in Nickel-Base IN 738 Superalloy. *Philos. Mag.* **2014**, *94*, 3295–3316. [CrossRef]
20. Rush, M.T.; Colegrove, P.A.; Zhang, Z.; Broad, D. Liquation and Post-Weld Heat Treatment Cracking in Rene 80 Laser Repair Welds. *J. Mater. Process. Technol.* **2012**, *212*, 188–197. [CrossRef]
21. Danis, Y.; Arvieu, C.; Lacoste, E.; Larrouy, T.; Quenisset, J.M. An Investigation on Thermal, Metallurgical and Mechanical States in Weld Cracking of Inconel 738LC Superalloy. *Mate Des.* **2010**, *31*, 402–416. [CrossRef]
22. Shahsavari, H.A.; Kokabi, A.H.; Nategh, S. Effect of Preweld Microstructure on HAZ Liquation Cracking of Rene 80 Superalloy. *Mater. Sci. Technol.* **2007**, *23*, 547–555. [CrossRef]
23. Pakniat, M.; Malek Ghaini, F.; Torkamany, M.J. Effect of Heat Treatment on Liquation Cracking in Continuous Fiber and Pulsed Nd:YAG Laser Welding of Hastelloy X Alloy. *Metall. Mater. Trans. A* **2017**, *48*, 5387–5395. [CrossRef]
24. González Albarrán, M.A.; Martínez, D.I.; Díaz, E.; Guzman, I.; Saucedo, E.; Guzman, A.M. Effect of Preweld Heat Treatment on the Microstructure of Heat-Affected Zone (HAZ) and Weldability of Inconel 939 Superalloy. *J. Mater. Eng. Perform.* **2014**, *23*, 1125–1130. [CrossRef]
25. Zhong, M.L.; Sun, H.Q.; Lin, W.J.; Zhu, X.F.; He, J.J. Boundary Liquation and Interface Cracking Characterization in Laser Deposition of Inconel 738 on Directionally Solidified Ni-Based Superalloy. *Scr. Mater.* **2005**, *53*, 159–164. [CrossRef]
26. Sidhu, R.K.; Ojo, O.A.; Chaturvedi, M.C. Microstructural Analysis of Laser-Beam-Welded Directionally Solidified INCONEL 738. *Metall. Mater. Trans. A* **2007**, *38*, 858–870. [CrossRef]
27. Osoba, L.O.; Ding, R.G.; Ojo, O.A. Microstructural analysis of laser weld fusion zone in Haynes 282 superalloy. *Mater. Charact.* **2012**, *65*, 93–99. [CrossRef]
28. Lachowicz, M.; Dudzinski, W.; Radziszewska, M.P. TEM Observation of the Heat-Affected Zone in Electron Beam Welded Superalloy Inconel 713C. *Mater. Charact.* **2008**, *59*, 560–566. [CrossRef]
29. Angella, G.; Barbieri, G.; Donnini, R.; Montanari, R.; Richetta, M.; Varone, A. Electron Beam Welding of IN792 DS: Effects of Pass Speed and PWHT on Microstructure and Hardness. *Materials* **2017**, *10*, 1033. [CrossRef] [PubMed]
30. Liu, D.; Lippold, J.C.; Li, J.; Rohklin, S.R.; Vollbrecht, J.; Grylls, R. Laser Engineered Net Shape (LENS) Technology for the Repair of Ni-Base Superalloy Turbine Components. *Metall. Mater. Trans. A* **2014**, *45*, 4454–4469. [CrossRef]
31. Banerjee, K.; Richards, N.L.; Chaturvedi, M.C. Effect of Filler Alloys on Heat-Affected Zone Cracking in Preweld Heat-Treated IN-738 LC Gas-Tungsten-Arc Welds. *Metall. Mater. Trans. A* **2005**, *36*, 1881–1890. [CrossRef]
32. Sidhu, R.K.; Richards, N.L.; Chaturvedi, M.C. Effect of Filler Alloy Composition on Post-weld Heat Treatment Cracking in GTA Welded Cast Inconel 738LC Superalloy. *Mater. Sci. Technol.* **2008**, *24*, 529–539. [CrossRef]
33. Hsu, K.T.; Wang, H.S.; Chen, H.G.; Chen, P.C. Effects of the Hot Isostatic Pressing Process on Crack Healing of the Laser Repair-Welded CM247LC Superalloy. *Metals* **2016**, *6*, 238. [CrossRef]
34. Oliveira, J.P.; Miranda, R.M.; Braz Fernandes, F.M. Welding and Joining of NiTi Shape Memory Alloys: A Review. *Prog. Mater. Sci.* **2017**, *88*, 412–466. [CrossRef]
35. Cui, C.Y.; Gu, Y.F.; Ping, D.H.; Harada, H.; Fukuda, T. The Evolution of η-phase in Ni–Co Base Superalloys. *Mater. Sci. Eng. A* **2008**, *485*, 651–656. [CrossRef]

36. EI-Bagoury, N.; Nofal, A. Microstructure of an Experimental Ni Base Superalloy under Various Casting Conditions. *Mater. Sci. Eng. A* **2010**, *527*, 7793–7800. [CrossRef]

37. Long, F.; Yoo, Y.S.; Jo, C.Y.; Seo, S.M.; Jeong, H.W.; Song, Y.S.; Jin, T.; Hu, Z.Q. Phase Transformation of η and σ Phases in an Experimental Ni-based Superalloy. *J. Alloys Compd.* **2009**, *478*, 181–187. [CrossRef]

38. Montazeri, M.; Ghaini, F.M. The Liquation Cracking Behavior of IN 738LC Superalloy during Low Power Nd:YAG Pulsed Laser Welding. *Mater. Charact.* **2012**, *67*, 65–73. [CrossRef]

39. Chen, T.C.; Cheng, Y.H.; Tsay, L.W.; Shiue, R.K. Effects of Grain Boundary Microconstituents on Heat Affect Zone Cracks in a Mar-M004 Weldment. *Metals* **2018**, *8*, 201. [CrossRef]

40. Cheng, Y.H.; Chen, J.T.; Shiue, R.K.; Tsay, L.W. The Evolution of Cast Microstructures on the HAZ Liquation Cracking of Mar-M004 Weld. *Metals* **2018**, *8*, 35. [CrossRef]

41. Hou, J.S.; Guo, J.T.; Yang, G.X.; Zhou, L.Z.; Qin, X.Z. The Microstructure Instability of a Hot Corrosion Resistant Superalloy during Long-Term Exposure. *Mater. Sci. Eng. A* **2008**, *498*, 349–353.

42. Lee, H.-S.; Kim, D.-S.; Yoo, K.-B.; Song, K.-S. Quantitative Analysis of Carbides and the Sigma Phase in Thermally Exposed GTD-111. *Met. Mater. Int.* **2012**, *18*, 287–293. [CrossRef]

43. Jahangiri, M.R.; Arabi, H.; Boutorabi, S.M.A. Investigation on the Dissolution of η Phase in a Cast Ni-based Superalloy. *Int. J. Min. Metall. Mater.* **2013**, *20*, 42–48. [CrossRef]

44. Gonzalez, H.; Calleja, A.; Pereira, O.; Ortega, N.; de Lacalle, L.N.L.; Barton, M. Super Abrasive Machining of Integral Rotary Components Using Grinding Flank Tools. *Metals* **2018**, *8*, 24. [CrossRef]

45. Thakur, D.G.; Ramamoorthy, B.; Vijayaraghavan, L. Some Investigations on High Speed Dry Machining of Aerospace Material Inconel 718 Using Multicoated Carbide Inserts. *Mater. Manuf. Process.* **2012**, *27*, 1066–1072. [CrossRef]

Article

Characterizing the Soldering Alloy Type In–Ag–Ti and the Study of Direct Soldering of SiC Ceramics and Copper

Roman Koleňák [1], Igor Kostolný [1,*], Jaromír Drápala [2], Martin Sahul [1] and Ján Urminský [1]

[1] Faculty of Materials Science and Technology in Trnava, Slovak University of Technology in Bratislava, Jána Bottu č. 2781/25, 917 24 Trnava, Slovakia; roman.kolenak@stuba.sk (R.K.); martin.sahul@stuba.sk (M.S.); jan.urminsky@stuba.sk (J.U.)

[2] FMMI—Faculty of Metallurgy and Material Engineering, 17. listopadu 15, Poruba, 708 33 Ostrava, Czech Republic; jaromir.drapala@vsb.cz

* Correspondence: igor.kostolny@stuba.sk; Tel.: +421-906-068-304

Received: 13 March 2018; Accepted: 11 April 2018; Published: 16 April 2018

Abstract: The aim of the research was to characterize the soldering alloy In–Ag–Ti type, and to study the direct soldering of SiC ceramics and copper. The In10Ag4Ti solder has a broad melting interval, which mainly depends on its silver content. The liquid point of the solder is 256.5 °C. The solder microstructure is composed of a matrix with solid solution (In), in which the phases of titanium (Ti_3In_4) and silver ($AgIn_2$) are mainly segregated. The tensile strength of the solder is approximately 13 MPa. The strength of the solder increased with the addition of Ag and Ti. The solder bonds with SiC ceramics, owing to the interaction between active In metal and silicon infiltrated in the ceramics. XRD analysis has proven the interaction of titanium with ceramic material during the formation of the new minority phases of titanium silicide—SiTi and titanium carbide—C_5Ti_8. In and Ag also affect bond formation with the copper substrate. Two new phases were also observed in the bond interphase—$(CuAg)_6In_5$ and $(AgCu)In_2$. The average shear strength of a combined joint of SiC–Cu, fabricated with In10Ag4Ti solder, was 14.5 MPa. The In–Ag–Ti solder type studied possesses excellent solderability with several metallic and ceramic materials.

Keywords: solder; ceramics; copper; flux-less soldering

---

1. Introduction

The application of ultrasonic power to form joints between different materials is a frequently used method [1–3]. Ultrasonic power has numerous advantages regarding the formation of joints between different materials: The absence of flux, bond soundness, high speed joint formation, and the possibility to join metals with non-metals. These traits make the use of ultrasonic power greatly desired in the field of soldering as well as in the electronics industry. The suitability of this technology has been successfully documented by many scientific studies in the fields of brazing, soldering [1–4] and transient liquid phase (TLP) bonding [5–9].

Applications of In-based solders are mainly used to solder dissimilar materials in the electronics industry. One of the common applications of these alloys occurs when the service temperature is well below the freezing point. This mainly concerns in the fields of space and inter-planetary research. Electronic equipment suffers at very low temperatures. Advantages of using In and its alloys in cryogenic temperatures include; excellent wettability, higher toughness, and excellent conductivity when compared to the standard Sn–Pb solders [10–12].

Many researchers and research workplaces in the world are devoted to the study of solders with high indium content. A study released in 1991 [13] investigated the deformation properties of

In-based solders at both room temperature and $-196\,^\circ$C. InBiSn and InBi type solders were studied. The InBi solder showed considerably higher toughness at $-196\,^\circ$C than the InBiSn solder. Therefore, the authors demonstrated that the InBi solder is suitable for application in the electronics industry at low service temperatures. Research of the properties of In and In-based solders at low temperatures was mentioned in several studies [14–17].

All authors agreed on the excellent properties of In and In-based solders in cryogenic temperatures. However, residual stresses were observed in the soldered joints. The joint fabricated from In solder was subjected to tensile loading after cooling down and exhibited the highest residual stresses during thermal changes.

In-based solders are used in the electronics industry as a substitute for banned Pb-based solders. The authors of this work [18] examined the issues of soldering Al–Si alloys using a solder of high indium content with ultrasound assistance. They studied the formation of multi-phase reinforced bonds when soldering with a Sn51In solder. The bonds were formed from Si particles, creating a solid solution of Al–In and intermetallic phases. They found that joint strength may increase with longer periods of ultrasound assistance. The period of ultrasound activity was 0.2, 1, 15, and 25 s. At 0.2 s, the measured joint strength was 0.28 MPa, at 15 s it was 4.89 MPa, and at 25 s, the strength rose to 6.81 MPa.

The authors of Reference [19] used an In-based solder for the study of intermetallic phases formed in the interface between the Ni substrate and In49Sn solder. In that case, the soldering was performed in a vacuum furnace, with infrared heating, and the application of flux. The intermetallic phases of NiInSn were formed in the interface. The soldering time varied from 15 to 240 min. However, this process was rather time demanding, therefore, the authors preferred to use technologies that applied ultrasound, as the soldering time takes just a few seconds. This prevents the dissolution of the substrate in the liquid solder and excessive formation of intermetallic phases, both of which affect the strength of the joints.

The aim of this research is to characterize the soldering alloy type In10Ag4Ti. This solder is intended for lower temperature soldering, which Sn active solders (e.g., Sn3.5Ag2Ti solder) do not cover. Indium and titanium were selected because both are active metals. Indium has excellent wettability on many metallic and non-metallic materials. Its disadvantage as a base solder is its low tensile strength of 2–4 MPa. Therefore, the solder was alloyed with 10 wt % of Ag, in order to increase the strength of the In matrix of the solder and improve electrical conductivity. The amount of active metal Ti used varied between 2 to 6 wt %. Therefore, 4 wt % of Ti was proposed. Titanium is an active metal with a high affinity for many elements. It was examined whether the designated composition of the soldering alloy was suitable for soldering SiC ceramics and copper substrates under defined conditions. Thus, this research consisted of the study of solder proper, and its interactions with solder-substrate interfaces.

2. Experimental Section

After determining the weight proportions of the prepared alloy, weighing of individual components was performed. Materials with a purity of 4N or higher were used for solder manufacturing.

The chemical composition of the prepared alloy is shown in Table 1.

Table 1. The chemical composition of soldering alloy in wt %.

Sample	In [%]	Ag [%]	Ti [%]
In–Ag–Ti	86.0	10.0	4.0

Substrates of the following materials were used in experiments;

- ceramic SiC substrate in the form of disks Ø 15 × 3 mm;

- metallic Cu substrate with 4N purity of dimensions Ø 15 × 2 mm and 10 × 10 × 2.5 mm.

A hot-plate with thermostatic regulation was used for the fabrication of soldered joints. The SiC substrate was placed on the hot-plate, the solder was then added, and heated to soldering temperature. Soldering was performed using Hanuz UT2ultrasonic equipment with the parameters given in Table 2.

Solder activation was accomplished via an encapsulated ultrasonic transducer consisting of a piezo-electric oscillating system and a titanium sonotrode with a tip diameter of Ø 3 mm. The soldering temperature was 230 °C. Soldering temperature was checked by a continuous temperature measurement of the hot-plate by a NiCr–NiSi thermocouple. The time of ultrasonic power use was 5 s.

Soldering was performed without flux. The redundant layer of oxides on the surface of molten solder was removed. An identical procedure was repeated with the other substrate. Substrates with molten solder were then attached to each other, thus, forming a joint. A schematic representation of this procedure is shown in Figure 1.

Table 2. Soldering parameters.

Ultrasound Power	400	[W]
Working Frequency	40	[kHz]
Amplitude	2	[μm]
Soldering Temperature	230	[°C]
Time of Ultrasound Activation	5	[s]

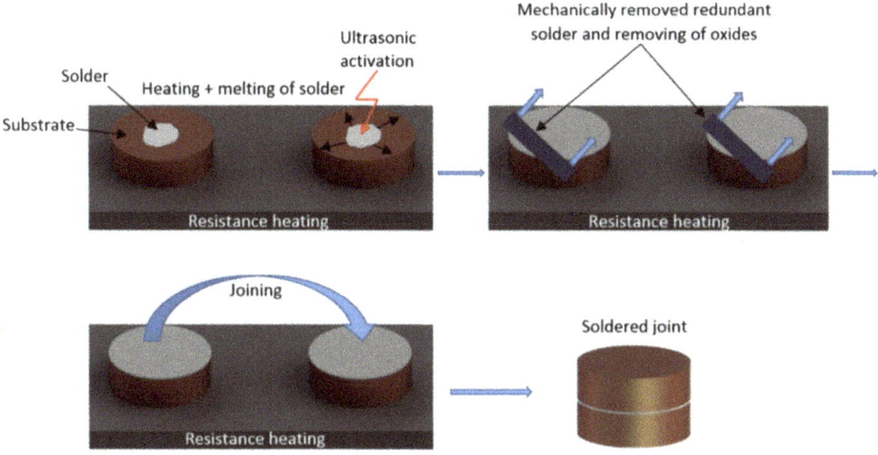

Figure 1. Schematic representation of soldering process at the presence of ultrasonic power.

Metallographic preparation of specimens from soldered joints was done using the standard metallographic procedures for specimen preparation. Grinding was performed using SiC emery papers with granularities of 240, 320, and 1200 g/cm^2. Polishing was performed with diamond suspensions of grain size: 9, 6, and 3 μm. Final polishing was performed by a OP-S (Struers, Detroit, MI, USA) polishing emulsion with 0.2 μm granularity.

The solder microstructure was studied with scanning electron microscopy (SEM) using microscope types TESCAN VEGA 3 (Brno, Czech Republic) and JEOL 7600 F (Belfast, Northern Ireland) with a X-ray micro-analyzer type Microspec WDX-3PC, used to perform both qualitative and semi-quantitative chemical analysis. The X-ray diffraction analysis was used for the identification of phase composition of the solder. X-ray diffraction measurements were carried out using a PANalytical Empyrean diffractometer in Bragg–Brentano geometry (EA Almelo, The Netherlands). Characteristic $CuK_{\alpha1,2}$ ($CuK_{\alpha1}$ = 1.540598 × 10^{-10} m, $CuK_{\alpha2}$ = 1.544426 × 10^{-10} m) was emitted at an accelerating voltage of 40 kV and a beam current of 40 mA, and was collimated using fixed slits. Diffracted radiation was collected using area-sensitive detectors operating in 1D scanning mode. XRD data were analyzed using the ICSD Inorganic Crystal Structure Database and ICDD PDF2 powder diffraction and crystal structure database. The differential scanning calorimetry (DSC) analysis of the In–Ag–Ti solder was done using Netzsch STA 409 C/CD equipment that was shielded with Ar gas of 6N purity.

A shear test was performed to determine the mechanical properties of the soldered joints. A schematic representation of the sample and a measurement scheme of shear stress are shown in Figure 2. Shear strength was measured using a versatile LabTest 5.250SP1-VM tearing machine. To alter the direction of tensile force acting upon the test piece, a special jig with the defined shape of the test piece was applied. This shearing jig ensured uniform shear loading of the specimen in the plane of the interface between the solder and substrate. The loading velocity of the sample was 1 mm·min^{-1}.

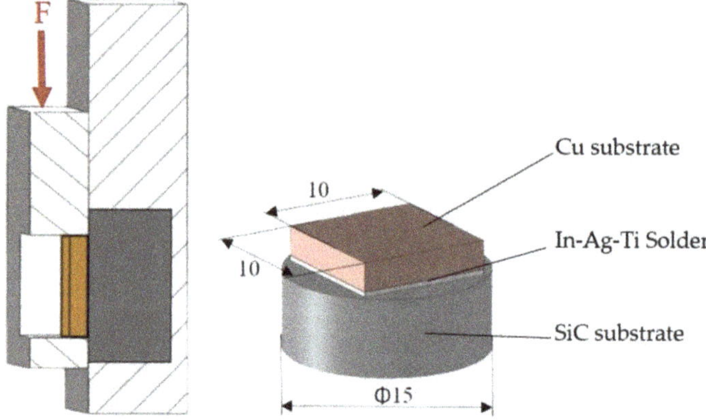

Figure 2. The scheme of shear stress measurement (unit: mm).

3. Experimental Results

3.1. DSC Analysis

DSC analysis was performed to determine the melting point of the solder. The curve of the In10Ag4Ti solder at a heating rate of 10 K·min^{-1} is documented in Figure 3. Two pronounced peaks were obtained. The first peak, of maximum temperature146.9 °C, corresponds to the temperature of eutectic transformation, after which the mechanical mixture of the solid solution (In) and intermetallic phase of $AgIn_2$—φ was formed. In accordance with the binary diagram of the authors [20], the temperature of eutectic transformation should be 144 °C, but the slight amount of Ti contained in the solder increased the melting point of eutectics to 146 °C. Approximately 65% of the solder volume was molten at the first peak.

The second peak, of maximum temperature 178 °C, represents the peritectic reaction in the Ag–In system. After this peritectic reaction, an intermetallic phase of Ag_2In was formed in the melt. At the second peak, approximately 19% of the volume of solder was molten.

The third peak, of temperature approximately 215 °C, represented the second peritectic reaction in the Ag–In system. After this peritectic reaction, the Ag_3In phase was formed in the melt.

The fourth peak, at 245.6 °C, represented the termination of melting of the components in the Ag–In system—Figure 4. The titanium phase, Ti_3In_4, which occurs in the matrix of the indium solder, was not yet fully molten at this temperature. In accordance with the binary diagram of In–Ti [21], this phase will be fully molten at 796 °C—Figure 4.

From the results of the DSC analysis it is obvious that the Ag addition decreased the melting point of the In10Ag4Ti solder, while the Ti addition slightly increased the melting point.

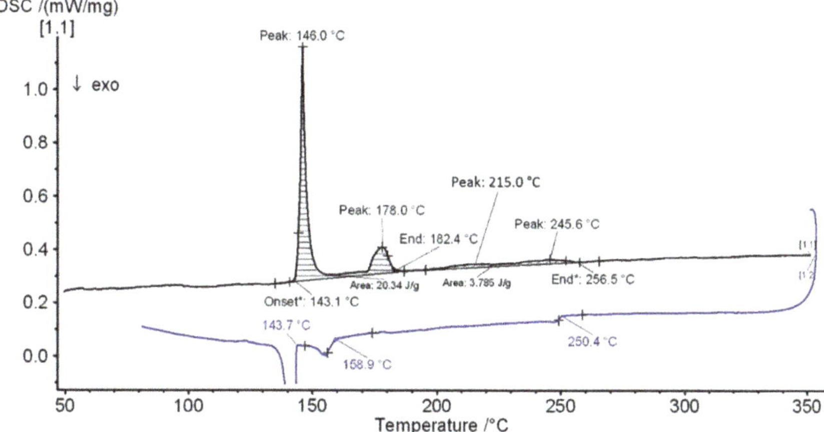

Figure 3. DSC analysis of In10Ag4Ti solder at the heating rate of 10 K·min^{-1}.

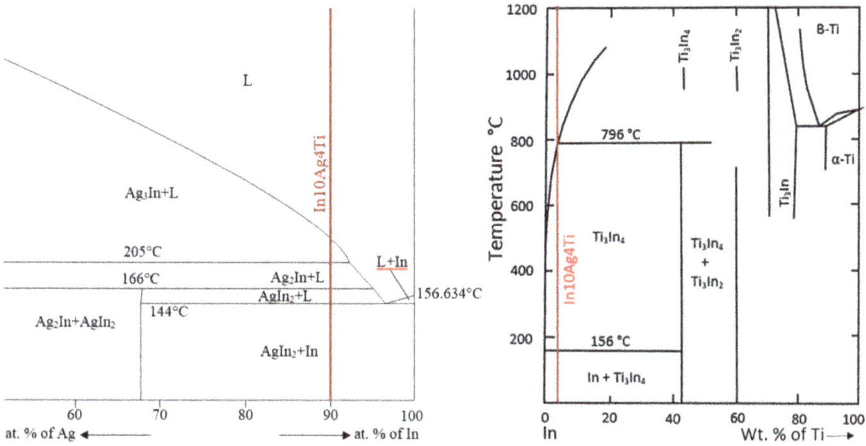

Figure 4. Incomplete binary diagrams Ag–In and In–Ti [20,21].

3.2. Microstructure of In10Ag4Ti Solder

The microstructure of the In10Ag4Ti soldering alloy, shown in Figure 5, consists of a indium solder matrix, where the intermetallic silver phases, mainly AgIn$_2$, are uniformly distributed, it also contains the non-uniformly distributed phases of the titanium solid solution, α-Ti, see Figure 5b.

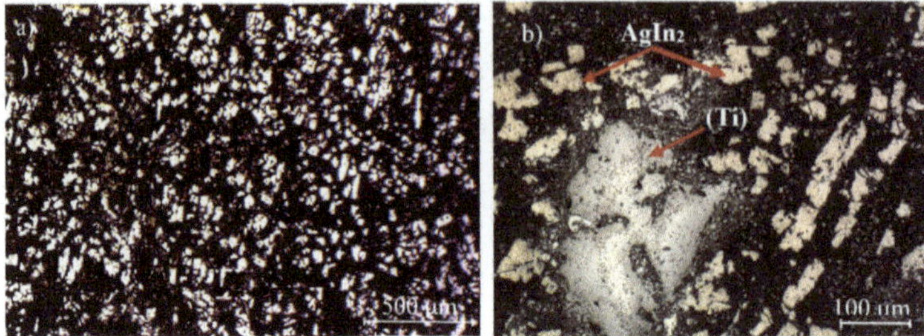

Figure 5. Microstructure of In10Ag4Ti solder from the optical microscope (**a**) in polished condition and (**b**) in etched condition.

EDX analysis was performed to determine the chemical composition of the individual components in the soldering alloy. The points of measurements are shown in Figure 6 and Table 3. These are marked from 1 to 4.

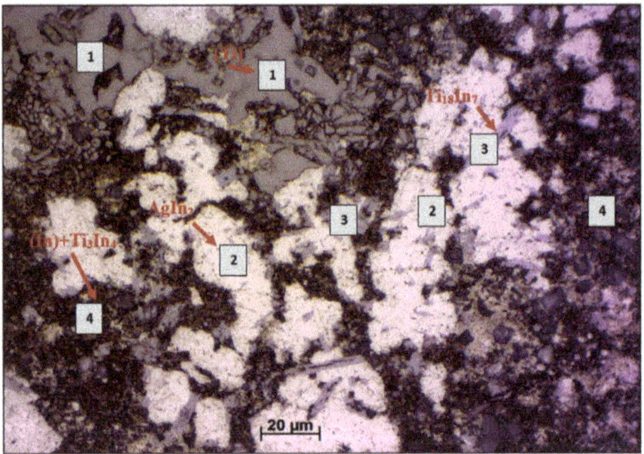

Figure 6. Point EDX analysis of In10Ag4Ti solder.

Table 3. Point EDX analysis of In10Ag4Ti solder.

Spectrum	In [wt %]	Ag [wt %]	Ti [wt %]	Solder Component
Spectrum 1	13.8	0	86.2	solid solution (Ti)
Spectrum 2	68.4	31.6	0	phase AgIn$_2$
Spectrum 2	67.7	32.3	0	phase AgIn$_2$
Spectrum 3	85.9	0	14.1	phase Ti$_{18}$In$_7$
Spectrum 4	99.2	0	0.8	matrix (In) + phase Ti$_3$In$_4$

The dark-grey phase in Figure 6, designated as Spectrum 1, is composed of a titanium (Ti) solid solution. The bright zones are due to the AgIn$_2$ phase. The solder matrix consists of a fine mechanical mixture of (In) + Ti$_3$In$_4$ phase. The dark particles in the microstructure are comprised of SiC and abrasive. The solder is very soft and the grains of abrasive are stuck to it.

Diffraction XRD analysis of the In10Ag4Ti solder has proven the presence of solid solution (In), intermetallic phase AgIn$_2$ and the presence of intermetallic phase titanium Ti$_3$In$_4$. The record of the diffraction analysis is documented in Figure 7. The solid solution (Ti) and intermetallic phase (Ti$_{18}$In$_7$) occur in the matrix only scarcely and were undetected by XRD analysis.

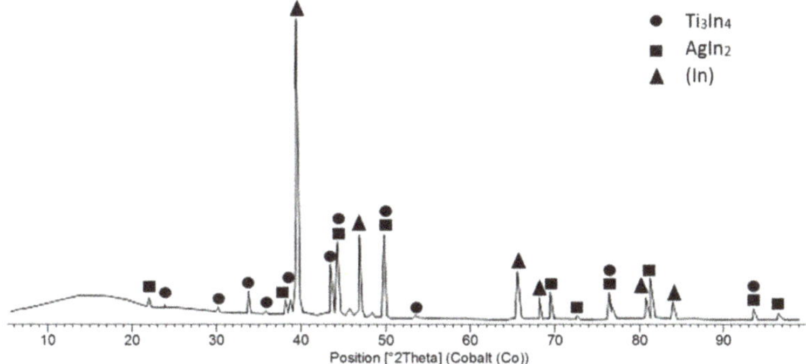

Figure 7. XRD analysis of In10Ag4Ti solder.

The planar distribution of silver phase AgIn$_2$ and titanium phases in the matrix of indium solder is documented in Figure 8. The origin of the dark particles in the solder matrix is from abrasive.

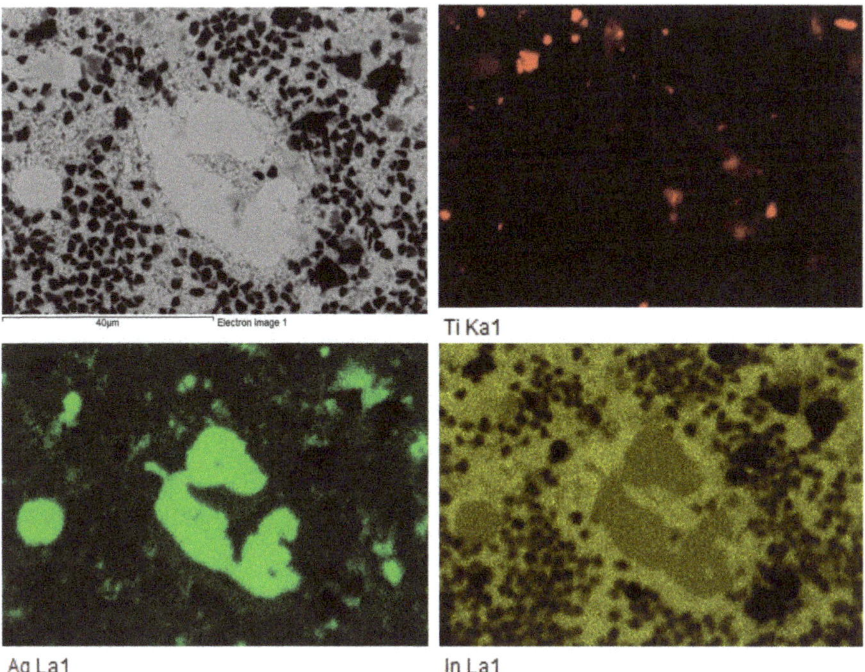

Figure 8. Map of In, Ag, and Ti elements.

3.3. Microstructure of SiC–In10Ag4Ti–Cu Joint

The SiC–In10Ag4Ti–Cu soldered joint was fabricated at 230 °C. Owing to ultrasound activation, an acceptable bond was achieved using a soldering process that did not contain cracks or other irregularities. The microstructure of the soldered joint is shown in Figure 9.

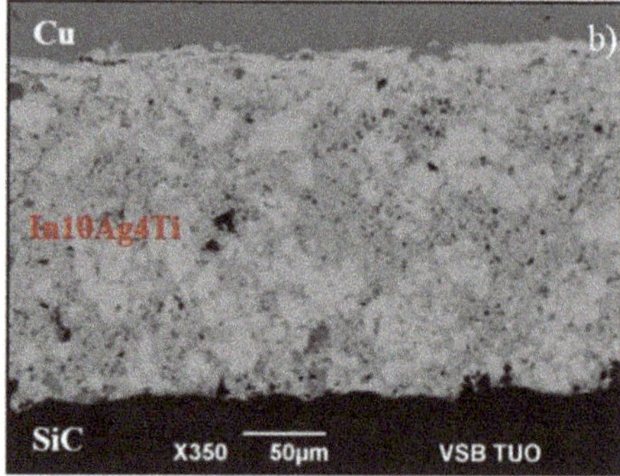

Figure 9. Microstructure of the SiC–In10Ag4Ti–Cu joint (**a**) from optical microscope; (**b**) from SEM.

Figure 9a shows that the larger particles of solid solution (Ti) remained preserved in the solder matrix after ultrasonic soldering. EDX analysis of the soldered joint was done to determine the chemical composition and identification of individual phases (Figure 10 and Table 4).

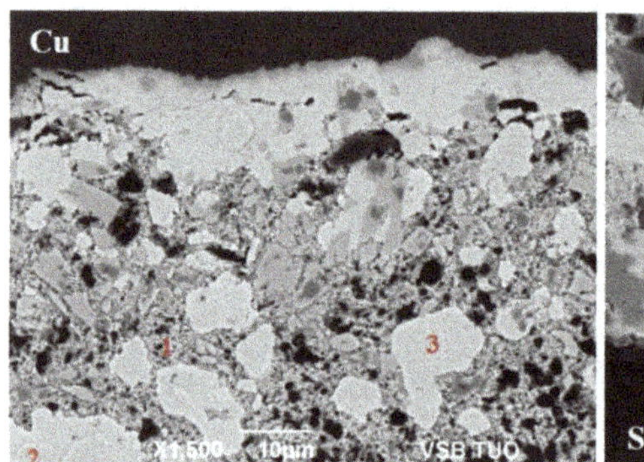

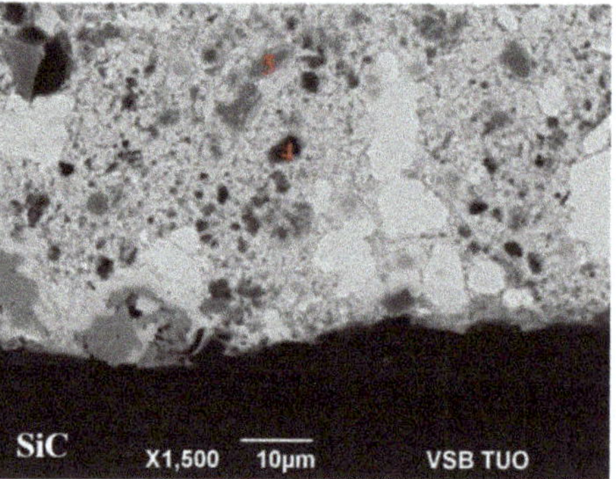

Figure 10. EDX point analysis of the SiC–In10Ag4Ti–Cu joint.

Table 4. EDX point analysis of the SiC–In10Ag4Ti–Cu joint.

Spectrum	In [wt %]	Ag [wt %]	Ti [wt %]	Solder Component
Spectrum 1	99.2	0.1	0.7	eutectic In-Ti, In-Ag
Spectrum 2	35.6	64.4	0	phase Ag_2In
Spectrum 3	70.1	29.9	0	phase $AgIn_2$
Spectrum 4	0	0	0	abrasive particle Al_2O_3
Spectrum 5	29.8	0	70.2	solid solution (Ti)—α-Ti

Silver phases of Ag_2In and $AgIn_2$ as well as titanium phases occurred in the solder matrix after soldering. The matrix was composed of In eutectics; Ti and Ag elements were present, but only in low concentrations. Zones with undiluted copper have also occurred in the solder matrix.

Based on previous studies [22,23], it was supposed that the active Ti element would concentrate in the interface with the ceramic SiC material, where it would then form new phases as a result of the interaction between the solder and the substrate, for example, the formation of Ti–C and Ti–Si. However, no interaction of titanium was accompanied with the formation of new phases in the interface of the solder-SiC ceramics, as observed by EDX analysis, in spite of Ti mapping along the entire interface of the solder-ceramics.

However, the connection of both materials was satisfactory and lacked formation pores or cracks at the joint interface. The EDX point analysis of the SiC-solder interface is shown in Figure 11 and Table 5. The planar distribution of elements is documented in Figure 12. From the planar distribution it is obvious that Ti is non-uniformly distributed across the entire solder volume, however, its local interaction cannot be excluded.

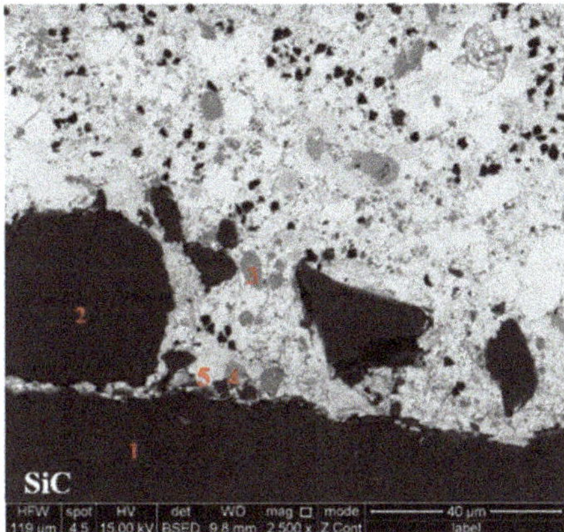
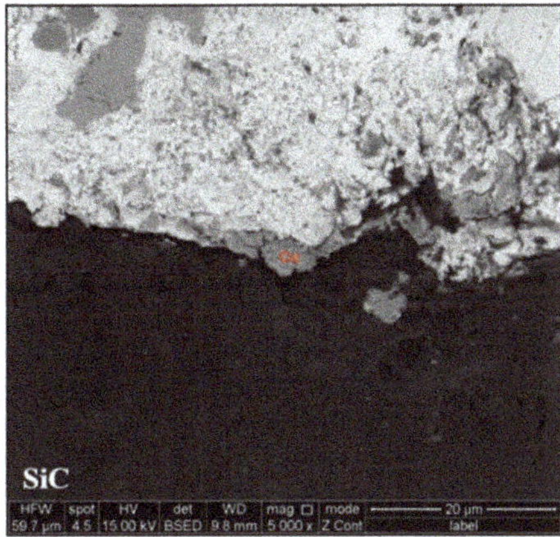

Figure 11. EDX point analysis of the SiC–In10Ag4Ti joint.

Table 5. EDX point analysis of the SiC–In10Ag4Ti joint.

Spectrum	In [wt %]	Ag [wt %]	Ti [wt %]	Cu [wt %]	Si [wt.%]	C [wt %]	Component
Spectrum 1	0	0	0	0	60.1	39.9	ceramics SiC
Spectrum 2	0	0	0	0	59.3	40.7	particle SiC
Spectrum 3	15.3	0.3	0	84.4	0	0	solid solution (Cu)
Spectrum 4	0	0	0	100	0	0	particle Cu
Spectrum 5	54.7	19.4	19.5	6.4	0	0	-

Bond formation is due to the interaction of indium with the surface of the SiC material: During the soldering process the indium particles are distributed to the interface of the SiC ceramics, due to the effect of ultrasonic activation, where they are combined with the silicon infiltrated in the SiC ceramics. The bond of indium with ceramics has an adhesive character, and does not form the new type Ti–C or Ti–Si contact phases.

Line analysis and concentration profiles of elements Ti and In (Figure 13) prove that Ti is not segregated in the interface of the ceramic SiC material, but that a significant effect is exerted one bond formation by indium.

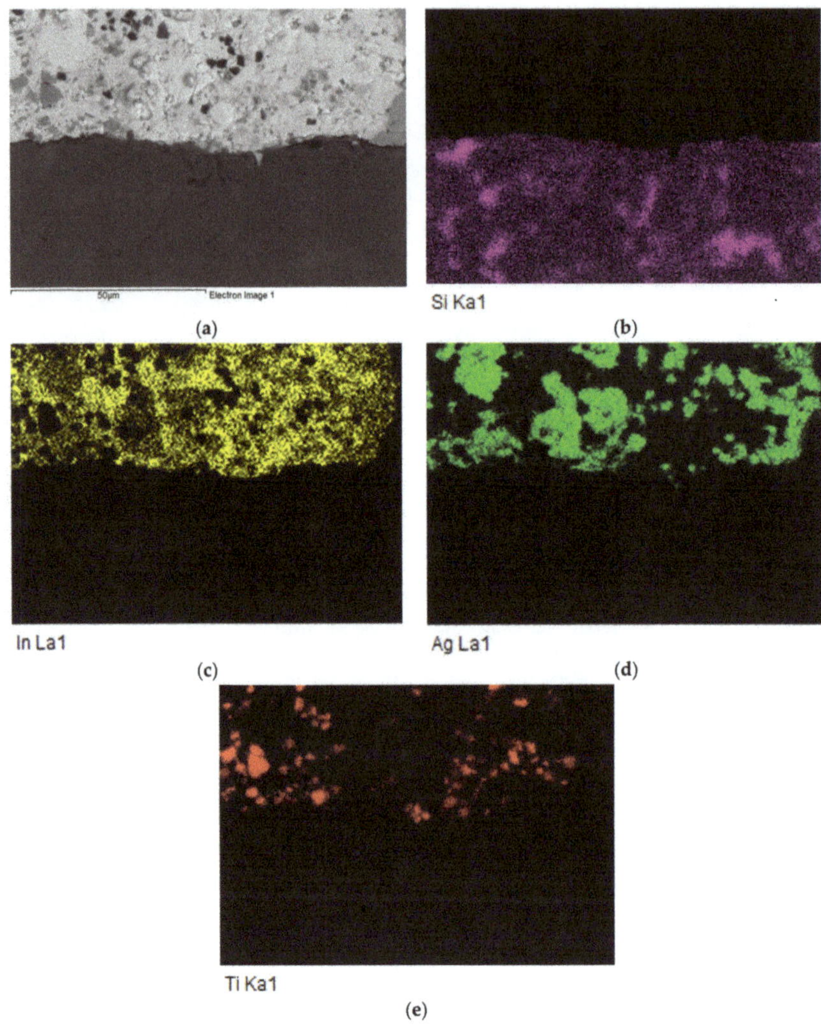

Figure 12. Planar distribution of Si, In, Ag, and Ti elements in the interface of the SiC–In10Ag4Ti joint (**a**) interface microstructure; (**b**) Si; (**c**) In; (**d**) Ag; and (**e**) Ti.

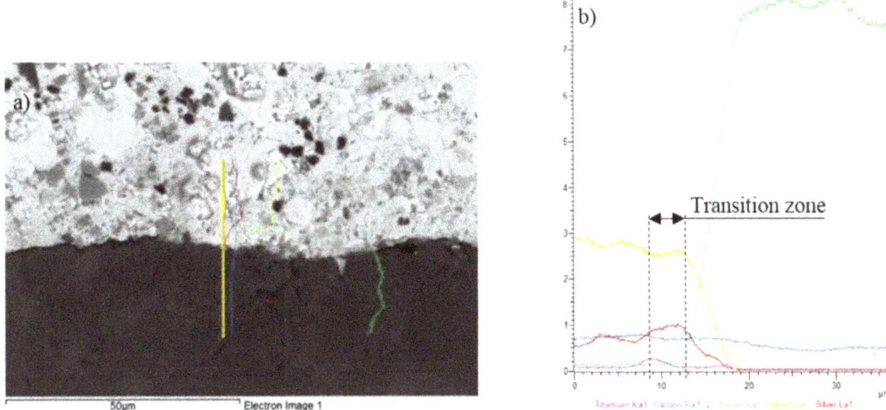

Figure 13. The line EDX analysis of the SiC–In10Ag4Ti joint (**a**) transition zone with a marked line; (**b**) concentration profiles of Ti, C, Si, In, and Ag elements.

3.4. Analysis of the Transition Zone in Cu–In10Ag4Ti Joint

Analysis primarily focused on the transition zone of the joint. Two intermetallic phases, $(CuAg)_6In_5$ and $(AgCu)In_2$, were analyzed in the interface of the Cu–In10Ag4Ti joint; the phases were the result of an interaction between the indium solder and copper substrate. The effect of the active Ti element on bond formation with the copper substrate was negligible.

The $(CuAg)_6In_5$ phase was narrow, with a thickness of approximately 1 μm, this was more similar to copper and had a higher Cu content when compared to Ag. The $(CuAg)_6In_5$ phase, shown in Figure 14, corresponded to the composition at the point of Spectrum 1. The $(AgCu)In_2$ phase was closer to the solder and had a higher silver content (approximately 23 wt %), when compared to copper (approximately 7 wt %); it was wettable by the solder. The $(AgCu)In_2$ phase was relatively heavy with a thickness of up to 13 μm. It corresponded to the composition at the measured point in Spectrum 2, see Table 6.

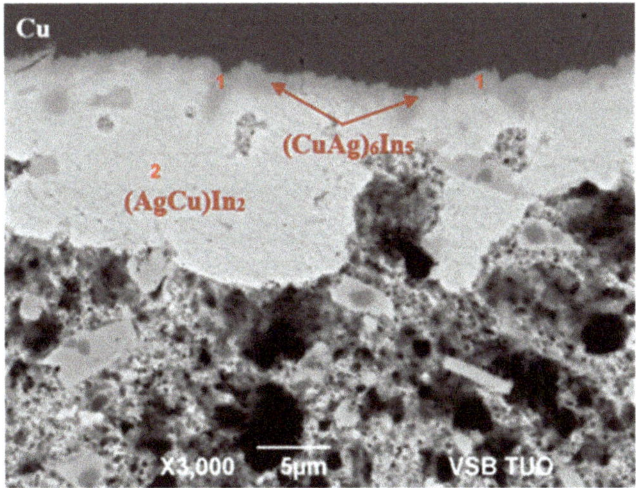

Figure 14. EDX point analysis of the interface of the Cu–In10Ag4Ti joint.

Table 6. EDX point analysis of the interface of the Cu–In10Ag4Ti joint.

Spectrum	In [wt %]	Ag [wt %]	Cu [wt %]	Component
Spectrum 1	58.5	8.2	33.3	phase $(CuAg)_6In_5$
Spectrum 1	57.6	7.3	35.1	phase $(CuAg)_6In_5$
Spectrum 2	69.8	23.2	7.0	phase $(AgCu)In_2$

The results of the EDX point analysis were proven by the course of concentration profiles of In, Ag, and Cu elements in Figure 15. An increase in concentration of Ag can also be observed. This relates to the formation of IMC $(AgCu)In_2$, which contains up to 23 wt % Ag.

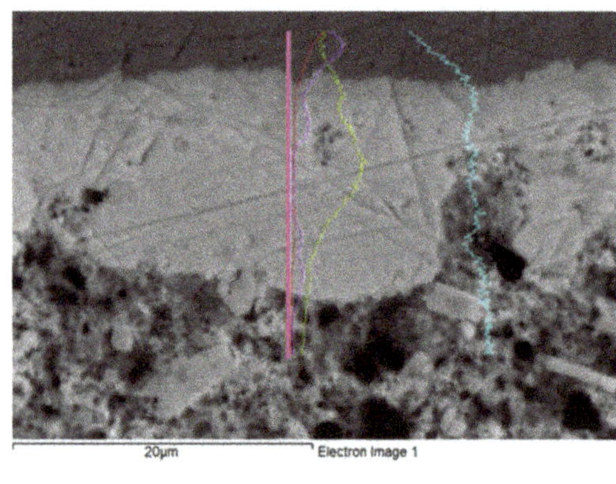

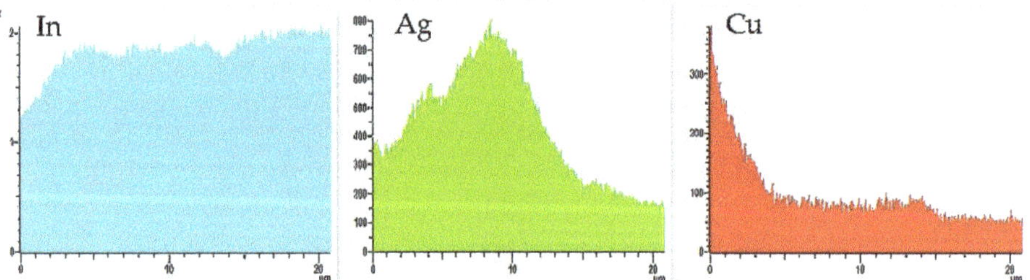

Figure 15. The concentration profiles of Cu, In, and Ag elements in the interface of the Cu–In10Ag4Ti joint in the zone of formation of the new intermetallic phases, $(CuAg)_6In_5$ and $(AgCu)In_2$.

3.5. Shear Strength of Soldered Joints

This study was primarily oriented toward the soldering of SiC ceramics with copper substrate. Owing to the potential application and further use of active In10Ag4Ti solder in industrial practice, the testing of shear strength was also extended to other metals (Cu, Ag, Ni, Al, and stainless steel type AISI 316) and ceramics (Si_3N_4, Al_2O_3, ZrO_2, and AlN).

The ceramics were always tested in combination with a copper substrate. The metals were mutually tested as Cu–Cu, Ag–Ag, etc. Measurements were performed with 3 specimens of each material. The results of the average shear strength testing are documented in Figure 16. Marked deviations represent the minimum and maximum values measured.

The greatest shear strength of the ceramic-metal combinations, 19 MPa, was observed in the AlN–Cu joint. A similar strength of 18 MPa was observed in the ZrO_2–Cu joint. Other material combinations, such as SiC–Cu, Si_3N_4–Cu, and Al_2O_3–Cu, demonstrated comparable average shear strengths ranging between 13 and 13.5 MPa. Of the metals, the greatest average shear strength was observed in the joint of two metallic Ni materials—19 MPa.

However, this metal also exerted the highest scatter of measurements, between 14 and 24 MPa. Metals Al, Ag, Cu, and AISI 316 displayed average shear strengths between 15 and 16 MPa.

From these results it can be concluded that the strength of ceramic-metal joints, in the case of In10Ag4Ti solder, is comparable to that of metal-metal joints. This is caused by the excellent wettability of indium on ceramic materials in conjunction with ultrasonic activation.

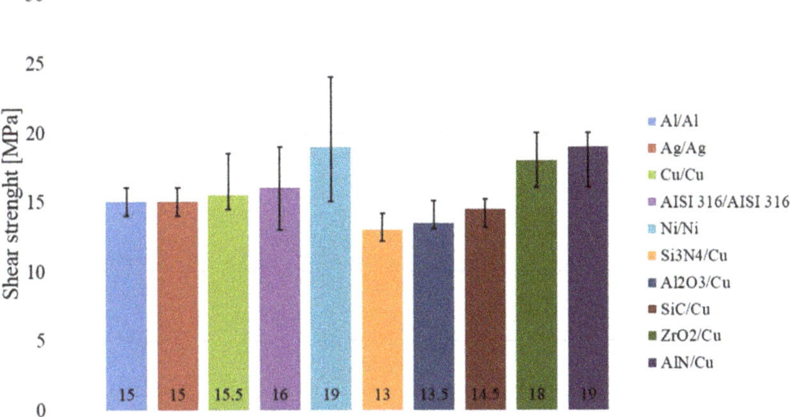

Figure 16. Shear strength of soldered joints with In10Ag4Ti solder.

3.6. Analysis of Fractured Surfaces

The fractured surfaces of joints were analyzed for more exact identification of the bond formation mechanism. Figure 17a,b shows the fractured surface at the interface of a SiC–In10Ag4Ti–Cu joint.

The fractured surface, on the side containing SiC ceramics, remained completely covered with solder. A ductile fracture occurred within the solder. An analysis of the planar distribution of Si, C, In, Ag, and Ti elements was carried out on the fractured surface, as is documented in Figure 18b–f. Regarding the planar distribution of Si, which represents the SiC ceramics in Figure 18b, local spots may be observed, caused by ripping out of solder from the substrate surface.

The character of Ti distribution on the fractured surface, Figure 18f, suggests that Ti is partially bound with the SiC ceramics and may thus contribute locally to bond formation. Therefore, an XRD analysis of the fractured surface in the interface of the SiC–In10Ag4Ti joint was performed.

Nevertheless, it was found that Ti from the solder locally reacted with the surface of the ceramic SiC material at the formation of new phases. Thus, two minority phases, namely the titanium silicide, SiTi, and titanium carbide, C_5Ti_8, were identified, which proves the interaction of titanium with the surface of the SiC ceramics. The record of XRD analysis is documented in Figure 19. Besides the SiTi and C_5Ti_8 phases, other minority phases, such as (In), (Ti), $AgIn_2$, and Ti_3In_4, were also proved by EDX analysis.

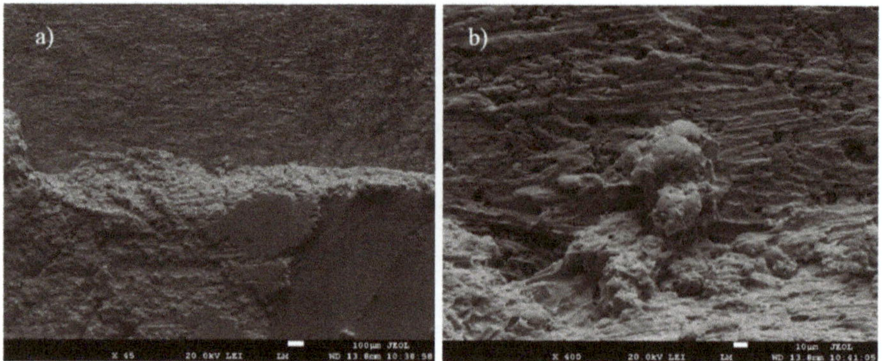

Figure 17. Fractured surface of soldered joint of SiC–In10Ag4Ti–Cu (**a**) 45 times magnification, (**b**) 400 times magnification.

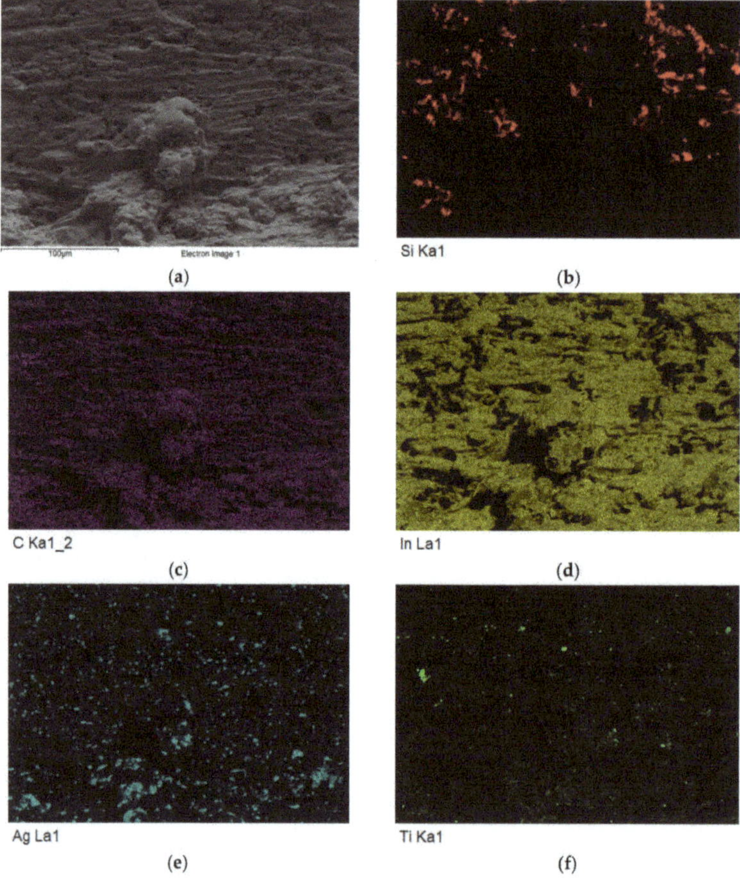

Figure 18. Fractured surface of the soldered joint of SiC–In10Ag4Ti–Cu and the planar distribution of individual elements (**a**) structure of fracture; (**b**) Si; (**c**) C; (**d**) In; (**e**) Ag; and (**f**) Ti.

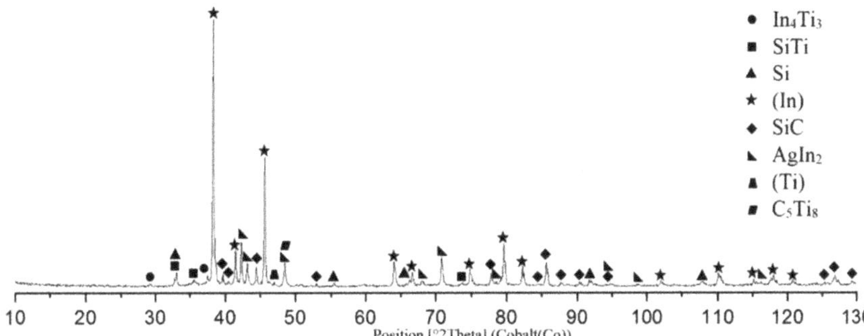

Figure 19. XRD analysis of the interface of SiC–In10Ag4Ti joint.

4. Conclusions

The aim of our research was to characterize the In–Ag–Ti soldering alloy type. We also studied whether the designed composition of the experimental soldering alloy is suitable for soldering SiC ceramics and copper with ultrasound assistance. The following results were achieved;

- DSC analysis was employed to determine the melting point of solder. The DSC curve shows four basic peaks. The first peak, with a maximum temperature of 146.9 °C, corresponds to the temperature of eutectic transformation in the binary system of Ag–In. 65% of the solder volume was molten at this peak. The second and third peak, with maximum temperatures of 178 °C and 215 °C, respectively, represent the peritectic reactions in the Ag–In system, 19% of the solder volume was molten. The fourth and final peak, with a maximum temperature of 245.6 °C, represented the termination of melting of the components in the Ag–In system.

- The microstructure of the In–Ag–Ti type solder was composed of a solder matrix formed by fine eutectics—(In) + Ti$_3$In$_4$ phase. The solder matrix contains uniformly distributed intermetallic phases of silver, mainly AgIn$_2$, and non-uniformly distributed phases of titanium solid solution—α-Ti.

- The SiC-solder bond is formed due to the interaction of indium with the surface of ceramic SiC material: During the soldering process, the indium particles are distributed to the interface with SiC ceramics under the effect of ultrasonic activation, where they combine with the silicon of the SiC ceramics. The indium-SiC bond has an adhesion character, barring the formation of new contact phases. However, XRD analysis proved a local interaction of titanium with the ceramic materials during the formation of new minority phases of titanium silicide (SiTi) and titanium carbide (C$_5$Ti$_8$).

- Two new intermetallic phases, namely (CuAg)$_6$In$_5$ and (AgCu)In$_2$, were identified in the interface of the Cu–In10Ag4Ti joint, this was the result of the interaction between indium solder and copper substrate. The effect of an active Ti element on bond formation with the copper substrate was negligible.

- The measurement of shear strength was performed across a wide scope of metallic and ceramic materials. The average shear strength of a combined joint of SiC–Cu, fabricated with In10Ag4Ti solder, was 14.5 MPa. The results of measurements show that the strength of ceramic–metal joints in the case of In10Ag4Ti solder is comparable to that of metal-metal joints. This is because of the excellent wettability of indium on the ceramic materials at ultrasonic activation.

Acknowledgments: This paper was prepared with the support of the VEGA 1/0089/17 project: Research of new alloys for direct soldering of metallic and ceramic materials. The authors thank Ing. Marián Drienovský for DSC analysis and Martin Kusý for XRD analysis.

Author Contributions: Roman Koleňák and Igor Kostolný conceived, designed the experiments and wrote the paper. Ján Urminský and Jaromír Drápala performed the experiments. Martin Sahul contributed materials tools.

Conflicts of Interest: The authors declare no conflicts of interest.

References

1. Chen, X.; Xie, R.; Lai, Z.; Liu, L.; Zou, G.; Yan, J. Ultrasonic-Assisted Brazing of Al–Ti Dissimilar Alloy by a Filler Metal with a Large Semi-solid Temperature Range. *Mater. Des.* **2016**, *95*, 296–305. [CrossRef]
2. Watanabe, T.; Sakuyama, H.; Yanagisawa, A. Ultrasonic Welding between Mild Steel Sheet and Al-Mg Alloy Sheet. *J. Mater. Process. Technol.* **2009**, *209*, 5475–5480. [CrossRef]
3. Wang, Q.; Zhu, L.; Chen, X. Si Particulate-reinforced Zn-Al Based Composites Joints of Hypereutectic Al-50Si Alloys by Ultrasonic-assisted Soldering. *Mater. Des.* **2016**, *107*, 41–46. [CrossRef]
4. Nagaoka, T.; Morisada, Y.; Fukusumi, M.; Takemoto, T. Ultrasonic-assisted Soldering of 5056 Aluminum Alloy Using Quasi-melting Zn-Sn Alloy. *Metall. Mater. Trans. B* **2010**, *41*, 864–871. [CrossRef]
5. Lai, Z.; Chen, X.; Pan, C. Joining Mg Alloys with Zn Interlayer by Novel Ultrasonic-Assisted transient Liquid Phase Bonding Method in Air. *Mater. Lett.* **2016**, *166*, 219–222. [CrossRef]
6. Lai, Z.; Xie, R.; Pan, C. Ultrasound-assisted Transient Liquid Phase Bonding of Magnesium Alloy Using Brass Interlayer in Air. *J. Mater. Sci. Technol.* **2017**, *33*, 567–572. [CrossRef]
7. Guo, W.; Leng, X.; Luan, T. Ultrasonic-promoted Rapid TLP Bonding of Fine-grained 7034 High Strength Aluminum Alloys. *Ultrason. Sonochem.* **2017**, *36*, 354–361. [CrossRef] [PubMed]
8. Qian, W.; Chen, X.; Lin, Z. Rapid Ultrasound-induced Transient-liquid-phase Bonding of Al-50Si Alloys with Zn Interlayer in Air for Electrical Packaging Application. *Ultrason. Sonochem.* **2017**, *34*, 947–952.
9. Zhao, H.Y.; Liu, J.H.; Li, Z.L. Non-interfacial Growth of Cu3Sn in Cu/Sn/Cu Joints During Ultrasonic-assisted Transient Liquid Phase Soldering Process. *Mater. Lett.* **2016**, *186*, 283–288. [CrossRef]
10. Kirschmann, R.K.; Sokolowski, W.M.; Kolawa, E.A. A Scattering-Mediated Acoustic Mismatch Model for the Prediction of Thermal Boundary Resistance. *J. Heat Transf.* **2000**, *123*, 105–112.
11. Swenson, C.A. Properties of Indium and Thalium at Low Temperatures. *Phys. Rev. J. Arch.* **1955**, *100*, 1067.
12. Glazer, J. Metallurgy of Low Temperature Pb-free solders for electronic assembly. *Int. Mater. Rev.* **1995**, *40*, 65–93. [CrossRef]
13. Plotner, M.; Donat, B.; Benke, A. Deformation properties of indium-based solders at 294 and 77 K. *Cryogenics* **1991**, *31*, 159–162. [CrossRef]
14. Chang, R.W.; McCluskey, F.P. Constitutive Relations of Indium in Extreme-Temperature Electronic Packaging Based on Anand Model. *J. Electron. Mater.* **2009**, *38*, 1855–1859. [CrossRef]
15. Cheong, J.; Goyal, A.; Tadigadapa, S.; Rahn, C. Reliable Bonding Using Indium-Based Solders. In *Proceedings of the SPIE 5343, Reliability, Testing, and Characterization of MEMS/MOEMS III*; SPIE; Bellingham, WA, USA, 2003.
16. Choi, W.K.; Premachandran, C.S.; Chiew, O.S.; Ling, X.; Ebin, L.; Khairyanto, A.; Ratmin, B.; Sheng, K.C.W.; Thaw, P.P.; Lau, J.H. Development of Novel Intermetallic Joints Using Thin Film Indium Based Solder by Low Temperature Bonding Technology for 3D IC Stacking. In Proceedings of the 59th Electronic Components and Technology Conference (ECTC 2009), San Diego, CA, USA, 26–29 May 2009.
17. Cheng, X.; Liu, C.; Silberschmidt, V.V. Numerical Analysis of Thermo-Mechanical Behavior of Indium Micro-joint at cryogenic temperature. *Comput. Mater. Sci.* **2012**, *52*, 274–281. [CrossRef]
18. Zhu, L.; Wang, Q.; Shi, L.; Zhang, X.; Yang, T.; Yan, J.; Zhou, X.; Chen, S. Ultrarapid Formation of Multi-Phase Reinforced Joints of Hypereutectic Al-Si Alloys via an Ultrasound-Induced Liquid Phase Method Using Sn-51In Interlayer. *Mater. Sci. Eng. A* **2018**, *711*, 94–98. [CrossRef]
19. Wang, S.S.; Tseng, Y.H.; Chuang, T.H. Intermetalic Compounds Formed during the Interfacial Reactions between Liquid In-49Sn Solder and Ni substrate. *J. Electron. Mater.* **2006**, *35*, 165–169. [CrossRef]
20. Baren, M.R. *Binary Alloy Phase Diagrams*; Massalski, T.B., Ed.; ASM International: Metal Park, OH, USA, 1990; p. 47.
21. Baraballi, O.M.; Kova, N. *Struktura i Svojstva Metalov i Splavov*; Kyjev Naukova Dumka: Kyjev, Ukraine, 1986.

22. Chang, S.Y.; Tsao, L.C.; Chiang, M.J.; Chuang, T.H.; Tung, C.N.; Pan, G.H. Active Soldering of Indium Tin Oxide (ITO) with Cu in Air Using an $Sn_{3.5}Ag_4Ti(Ce, Ga)$ Filler. *J. Mater. Eng. Perform.* **2003**, *12*, 383–390. [CrossRef]
23. Koleňák, R.; Šebo, P.; Provazník, M.; Koleňáková, M.; Ulrich, K. Shear strength and wettability of active $Sn_{3.5}Ag_4Ti(Ce, Ga)$ solder on Al_2O_3 ceramics. *Mater. Des.* **2011**, *32*, 3997–4003. [CrossRef]

 *metals*

Article

The Role of a MDP/VBATDT-Primer Composition on Resin Bonding to Zirconia

Anuj Aggarwal [1] and Grace M. De Souza [2,*]

[1] Faculty of Dentistry, University of Toronto, 124 Edward Street, Toronto, ON M5G1G6, Canada; dranuj@yahoo.com

[2] Department of Clinical Sciences, Faculty of Dentistry, University of Toronto, 124 Edward Street, Toronto, ON M5G1G6, Canada

* Correspondence: grace.desouza@dentistry.utoronto.ca; Tel.: +1-416-8648206

Received: 26 February 2018; Accepted: 4 April 2018; Published: 7 April 2018

Abstract: Yttria-tetragonal zirconia polycrystal (Y-TZP) is a difficult substrate to bond to due to the absence of a glass phase and the material's chemical inertness. This study evaluated the effect of two monomers for metal, MDP (10-methacryloyloxydecyl dihydrogen phosphate) and VBATDT (6-(4-vinylbenzyl-n-propyl)amino-1,3,5-trizaine-2,4-dithiol) on bond strength to Y-TZP. Seven combinations with different concentrations of MDP and VBATDT-monomers (0.0, 0.1, 0.5, or 1.0 wt %) in acetone solution were developed and applied to the surface of Y-TZP slabs, which were bonded to composite resin substrates using a resin cement under standard loading. Non-primed samples were used as controls. Bonded specimens were cut for microtensile testing and tested after either 48 h or 180 days in water storage at room temperature. All samples from control group (no primer) and MV5 group (0% MDP/0.5% VBATDT) debonded spontaneously. Two-way ANOVA showed that the primer had a significant effect ($p < 0.001$) on bond strength to zirconia, whilst storage time did not ($p = 0.203$). Tukey HSD (honest significant difference) test indicated that groups with at least 0.5% of each monomer resulted in higher initial bond strength values. Although chemical bonding to zirconia is credited to MDP, a correct balance between MDP and VBATDT may imply in better bond strength results. The minimum concentration of each monomer should not be lower than 0.5 wt %.

Keywords: high-crystalline content zirconia; active monomers; oxide layer

1. Introduction

Three mol % yttria-tetragonal zirconia polycrystalline (Y-TZP) is a ceramic used in a variety of applications, such as thermal barrier protection in aero and industrial gas turbines, oxygen sensors and fuel cell membranes [1]. Y-TZP is also the strongest ceramic available for application in dentistry and orthopedics, which makes it highly indicated to replace missing posterior teeth [2,3]. Furthermore, the combination of zirconia and porcelain veneer coverage results in a highly aesthetic dental restoration [4]. Nonetheless, achieving good adhesion between zirconia-based prosthesis and resin cements is a challenge given the low bond strength values between these two substrates which are a consequence of the crystalline content of Y-TZP [5,6]. Hydrofluoric acid, for example, dissolves the glass phase of silica-based ceramics increasing surface area, but it does not modify the surface morphology of Y-TZP [7]. Airborne particle abrasion is an alternative to increase the inner surface roughness of Y-TZP copings, in an attempt to improve micromechanical interlocking and, consequently, the bond strength to resin cements [8,9]. However, it has been shown that sandblasting the zirconia surface for 5 s with particles of ~50 μm creates surface damage and reduces cyclic fatigue resistance in approximately 30%, by introducing flaws that would propagate under cyclic loading and, therefore, compromise the longevity of all-ceramic restorations [10]. A decrease in reliability of 110 μm particle-abraded zirconia specimens has also been demonstrated [11]. Nonetheless, according to Scherrer et al. [12], airborne particle abrasion using 30 μm alumina particles not only avoids undesired damage to the surface, but also improves the fatigue behavior of some materials [12]. However, sandblasting, itself, does not promote high bond strength to different luting systems [5,13] and the composition of the resin-based luting system is more critical than the size of the particles employed for surface abrasion [14].

The chemical inertness of zirconia also represents a challenge in the luting process, due to its low surface energy and wettability [15]. Additionally, its nonpolar surface has high corrosion resistance and does not chemically interact with potentially adhesive materials [16]. Considering that Y-TZP is essentially zirconium oxide [17], the application of primers developed for metal bonding may improve the bond strength between zirconia and luting systems, without causing further mechanical damage to the structure of the material [18]. 10-Methacryloyloxydecyl dihydrogen phosphate (MDP) is a monomer with high affinity to base metals [19]. The hydroxyl groups of the oxide layer on Y-TZP seem to react with the phosphate ester monomer of the MDP, leading to strong chemical reactions at the interface between the two materials [20]. The better bond strength may be due to either Van der Walls forces or hydrogen bonds [21]. MDP-based materials have been used in association with airborne particle abrasion [8,9,22] or without any mechanical modification of the surface [19,23,24], and results indicate that the bond strength between zirconia and resin cement is higher when MDP-based materials are employed, irrespective of the mechanical treatment [20,25]. However, the six-month stability of the MDP-mediated bonding and non-sandblasted zirconia is questionable [24].

Interestingly, in a previous study developed by our research group, only the application of a commercial primer containing MDP and 6-(4-vinylbenzyl-n-propyl)amino-1,3,5-triazine-2,4-dithione (VBATDT) was capable of promoting high and stable bond strength to Y-TZP intaglio's surface as opposed to a primer containing only MDP [19]. Similar results were obtained when that primer was compared to other commercial primers making use of different monomers [23]. The successful primer is a solution of MDP and VBATDT dispersed in acetone and the ratio between the two monomers is not disclosed by the manufacturer. Considering that this primer was developed for bonding to metal infrastructures [26], it can be hypothesized that the formulation is not optimized for Y-TZP-based substrates, and that variations in the ratio between the two monomers may clarify the importance of each of them on bonding to Y-TZP.

Therefore, this study evaluated the effect of both, MDP and VBATDT, on bonding and stability of the bonding between Y-TZP and resin-based cement. This was assessed by developing experimental solutions with different ratios of both molecules and evaluation of the bond strength at different storage times. The null hypotheses were that bond strength is not affected by the composition of the primer and that storage time has no effect on bond strength.

2. Materials and Methods

Materials used in this study are listed in Table 1. Fully-sintered zirconia cylinders (97% zirconium dioxide stabilized with 3% yttria-lava frame, 3 M ESPE, St. Paul, MN, USA) with 19 mm diameter and 100 mm height were cut to obtain 20 slices with 4 mm thickness. The disc-shaped slices had both faces ground up to 600 grit carbide silicon paper (Buehler Canada, Whitby, ON, Canada) under water cooling pressure. An impression of one of the slices was taken and 6 mm-thick composite resin substrates (Clearfil Majesty Esthetic, Kuraray America Inc., Huston, TX, USA) were incrementally built, with each 1 mm layer being light-activated for 40 s (850 mW/cm^2, VIP Jr., Bisco Inc., Schaumburg, IL, USA). Light-output was confirmed throughout the study using a laboratory-graded spectroradiometer (CheckMARC, BlueLight Analytics, Halifax, NS, Canada). The composite resin substrates were aged in deionized water for 30 days to provide hydration and avoid later hygroscopic expansion. After the storage period, both sides of each composite resin sample were finished using the same method described for Y-TZP samples, so that flat surfaces of standard roughness could be produced. Y-TZP and composite resin substrates were cut in half with a diamond blade under water cooling, to generate 40 samples of each material. Samples were ultrasonically cleaned in distilled water for 10 min and stored in distilled water at room temperature.

Table 1. Names, formulation, and additional information about the materials employed.

Commercial Materials		
Material *Manufacturer*	Classification	Composition
Lava Frame *3M ESPE, St. Paul, MN, USA*	High-crystalline content zirconia	97 mol % zirconium dioxide and 3 mol % yttrium dioxide
Clearfil Majesty Esthetic *Kuraray America, Inc., New York, NY, USA*	Light-cure nanohybrid resin composite, shade A2	BisGMA, hydrophobic aromatic dimethacrylate, hydrophobic aliphatic methacrylate, silanated barium glass filler, pre-polymerized organic filler dl-Camphorquinone
Clearfil Esthetic Cement EX *Kuraray America, Inc., New York, NY, USA*	Self-etch dual-cure resin-based cement	BisGMA, TEGDMA, hydrophobic aromatic dimethacrylate, silanated silica filler, silanated barium glass filler, colloidal silica
Experimental primers		
MV1		0.5 wt % MDP and 0.0 wt % VTATDT in acetone
MV2		0.5 wt % MDP and 0.1 wt % VBATDT in acetone
MV3		0.5 wt % MDP and 0.5 wt % VBATDT in acetone
MV4		0.5 wt % MDP and 1.0 wt % VBATDT in acetone
MV5		0.0 wt % MDP and 0.5 wt % VBATDT in acetone
MV6		0.1 wt % MDP and 0.5 wt % VBATDT in acetone
MV7		1.0 wt % MDP and 0.5 wt % VBATDT in acetone

BisGMA: bisphenol A diglycidylmethacrylate; TEGDMA: triethylene glycol dimethacrylate; MDP: 10-Methacryloyloxydecyl dihydrogen phosphate; VBATDT: 6-(4-vinylbenzyl-n-propyl)amino-1,3,5- triazine-2,4-dithione.

Zirconia and composite resin blocks were randomly assigned to one of 8 test groups (n = 5), seven experimental and one control group, according to the treatment to be applied. Seven combinations of MDP (10-methacryloyloxydecyl dihydrogen phosphate) and/or VBATDT (6-(4-vinylbenzyl-n-propyl)amaino-1,3,5-trizaine-2,4-dithiol) primers were developed and the final composition may be seen in Table 1. Monomer ratios were calculated by weight percent (wt %) and acetone was used as a solvent. One group with no primer application was designated as control.

The Y-TZP surface was air-dried and two layers of the primer were applied with a microbrush. After 60 s, solvent was removed with a gentle air blast. Composite resin surfaces were prepared following the directions of the resin cement manufacturer. Dual cure cement (Clearfil Esthetic Cement EX, Kuraray America Inc., New York, NY, USA) was placed on the treated zirconia surface and the composite substrate was placed over the cement and held under a 600 gf load. The excess cement was removed with a spatula and light-activation was performed perpendicular to the adhesive interface at four different locations around the sample (40 s each at 850 mW/cm^2, VIP Jr., Bisco Inc., Schaumburg, IL, USA). Cemented samples were stored in deionized water at room temperature ($\sim$22 °C).

After 24 h storage, slabs were obtained by cutting the blocks perpendicularly to the adhesive interface with a diamond blade under water cooling. A second cut, perpendicular to the first one, was carried out resulting in beams of approximately 1 mm^2 cross-sectional area and 10 mm length. Specimens were then stored in deionized water at room temperature ($\sim$22 °C) and tested after two aging periods: 48 h and 180 days. For the longer storage groups (180 days), water was replaced on a weekly basis.

Five beams were randomly selected from each sample for the microtensile testing. The remaining beams were kept in deionized water. Each specimen was individually attached to a special jig with cyanoacrylate glue (Krazy Glue, Elmers Products, High Point, NC, USA). Testing was performed in a universal testing machine (Model 5565, Instron Corp., Norwood, MA, USA) at a crosshead speed of 0.5 mm/min and the load at failure was recorded. The dimensions of each beam at the bonded interface were measured with a digital caliper to determine the cross-sectional area and nominal bond strength (in MPa) was calculated. After 180 days, five beams were randomly selected again from each sample and the bond strength test was repeated following the same protocol. Both portions of each tested beam were evaluated under stereomicroscope (60× magnification, Olympus SZ61,

Olympus America Inc., Center Valley, PA, USA) and the mode of failure was classified as type 1 (adhesive between ceramic and cement), type 2 (adhesive between composite and cement), or type 3 (mixed failure). One entire sample (both zirconia and composite resin portions) from each group and each mode of failure was selected for electron scanning microscopy (SEM-JEOL JSM-66 10 LV, Tokyo, Japan) after sputter coating with carbon for 10 s.

For statistical purposes, each cemented sample was considered as the experimental unit ($n = 5$) and beams were considered repetitions within the same sample. A two-way ANOVA was conducted to evaluate the effect of treatment and storage time on bond strength. Tukey HSD (honest significant difference) was applied to compare the experimental groups ($p = 0.05$). If present, it was determined beforehand that any pre-test failure would be treated as 0 MPa.

For the analysis of the atomic composition of the primers used, flat surfaces of Y-TZP ($4 \times 4 \times 1$ mm^3) were prepared as previously described and treated with two layers of the corresponding primer, which was left undisturbed for one minute, after which it was vigorously air-dried. Non-sputter-coated control and primer-treated samples were analyzed under energy dispersive X-ray spectroscopy (EDS) analysis in an electron scanning microscope (JEOL, Tokyo, Japan) fitted to an EDS detector (silicon drift detector with INCA data acquisition, Oxford Instruments, Abingdon, UK).

3. Results

3.1. Microtensile Bond Strength Test

All the samples from group MV5 and control group debonded prior to the microtensile test, and their bond strength values were included as 0 MPa (zero MPa) for statistical purposes. No pre-test failures were observed for any of the other groups, and approximately 12–16 beams were obtained from each sample. The two-way ANOVA was performed for the remaining groups. Primer had a significant effect on bond strength results ($p < 0.001$) and there was no effect of storage time ($p = 0.203$). The interaction primer versus time ($p = 0.190$) was not significant.

Descriptive statistics (means and standard deviations) of the bond strength data and results of the Tukey HSD test ($p = 0.05$) are shown in Table 2. The highest mean bond strength was presented by samples treated with MV3 (13.7 $\pm$ 5.0 MPa), which was similar to MV4 (12.2 $\pm$ 3.8 MPa) and MV7 (11.9 $\pm$ 4.0 MPa) treated samples. Groups with the lowest percentage of any of the components (MV1, MV2 and MV6) presented the lowest bond strength results (Table 2).

Table 2. Means (standard deviation–SD) of microtensile bond strength and Tukey HSD results for primer irrespective of aging time.

Treatment	Mean* (SD)
MV3	13.7 (5.0) [A**]
MV4	12.2 (3.8) [AB]
MV7	11.9 (4.0) [AB]
MV1	11.0 (3.5) [B]
MV2	10.3 (3.4) [B]
MV6	9.3 (3.8) [B]
MV5	0.0 (0.0) [C]
Control	0.0 (0.0) [C]

* Bond strength values expressed in MPa; ** Similar letters within the same column indicate statistically similar bond strength results at 5% significance level.

3.2. Mode of Failure

Incidence of mode of failure for all experimental groups at both storage times is shown in Figure 1. Analysis of mode of failure indicated that there was a decrease in type 1 (between ceramic and cement) failure after aging for all groups evaluated. MV3-treated groups presented higher incidence of type 2 failure (between composite and cement) at both storage times. Lower bond strength values were associated with higher incidence of type 1 failure (groups MV1, MV2, and MV6). Pre-test failed samples (control and MV5-treated) showed 100% incidence of type 1 failure.

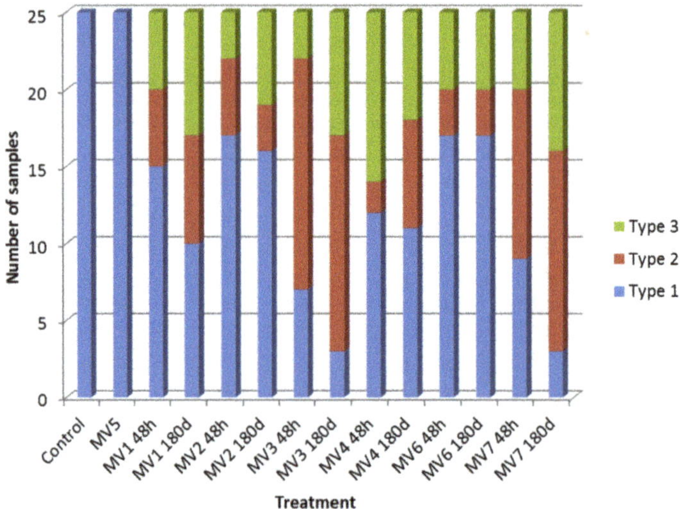

Figure 1. Incidence of mode of failure for each experimental group at different storage times.

3.3. Energy Dispersive Spectroscopy

Analysis of the atomic composition of the primers applied on the Y-TZP surface indicated that the molecules were not evenly distributed. Control sample, as expected, only indicated the presence of zirconium, oxygen, and carbon (yttrium was not identified as a possible element to be encountered) (Figure 2A). MV1 resulted in a film with some areas having higher concentration of phosphorous and structures that appeared like "droplets" of higher carbon concentration, which was similar to the MV2-treated surface. Application of MV3 resulted in a homogeneous film spread on the surface, with even distribution of sulfur-rich and phosphorous-rich areas. MV4 resulted in a thin film characterized by higher concentration of zirconium (Figure 2B) and droplets with higher concentration of carbon and sulfur (Figure 2C). This was similar to MV7-treated surface, which showed a thin film (gray area) with a distribution of dark spots rich in carbon, phosphorous, and sulfur (Figure 2D). The MV5-treated surface showed a thin film throughout the surface where high concentration of zirconium was observed, and seldom areas with low peaks of sulfur could also be encountered. For the MV6-treated surface, there was a more subtle incidence of darker spots with slightly higher concentration of sulfur and carbon in a thin film indicated by higher peaks of zirconium.

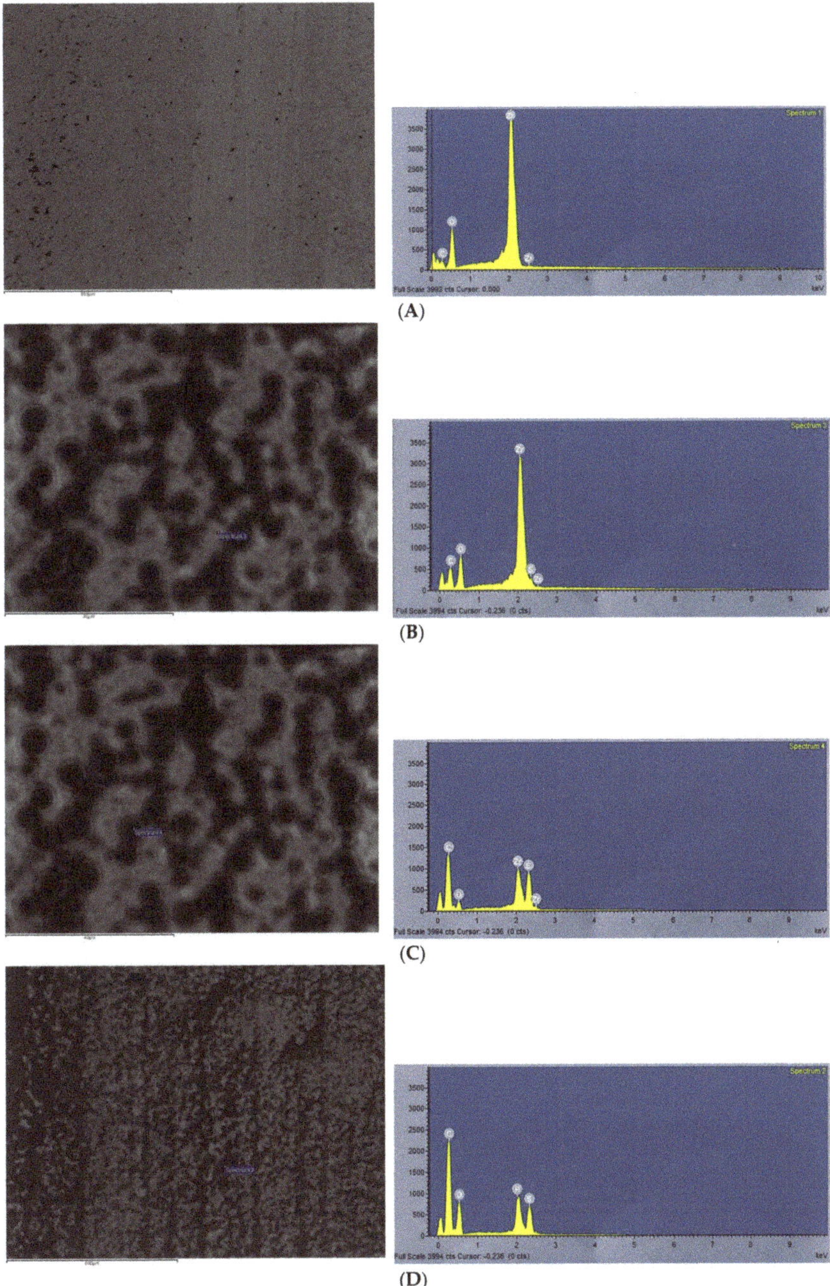

Figure 2. SEM of primer-treated Y-TZP samples and spectra presented by energy dispersive X-ray spectroscopy. (**A**) Control sample indicating high peaks of zirconium, oxygen, and carbon; (**B**) MV4-treated surface, with a thin film coverage evidenced by high peaks of zirconium; Darker spots seen on (**C**) (spectrum 4) show a higher concentration of sulfur and carbon covering the zirconium surface. (**D**) MV7-treated surface, where high peaks of carbon, phosphorous, and sulfur are evidenced in the darker areas (spectrum 2), compatible with high concentration of both monomers (carbon and phosphorous from MDP and carbon and sulfur from VBATDT).

4. Discussion

The present study evaluated the effect of different ratios of two monomers—MDP and VBATDT—on bond strength between Y-TZP and resin-based luting system. To avoid any unexpected effect of the cement's chemistry on bond strength, the resin-based cement selected did not contain either MDP or any other functional monomer in its composition. It has already been demonstrated that MDP-containing cements result in higher and more stable bond strength than the one used in the current investigation (Clearfil Esthetic Cement) [27]. Indeed, depending on the chemistry of the resin-based cement, it can outperform the chemical impact of the primer treatment, eliminating possible effects of the primer on bonding to Y-TZP [23,28]. Additionally, Y-TZP samples were not air-abraded with alumina particles, which would have had an effect on surface roughness [29], possibly increasing micromechanical interlocking by the penetration of the resin cement into the microretentions created [14,30]. Therefore, it is possible to say that the bond strength results reported in the present study are strictly related to the chemical interaction between Y-TZP oxide layer and primer, and between primer and resin cement.

Analysis of the microtensile bond strength results indicated that primer had a significant effect ($p < 0.001$) on bond strength between Y-TZP and resin cement. Therefore, the first null hypothesis was rejected. The treatment with MV3 resulted in superior bond strength results when compared to MV1, MV2, and MV6. The effectiveness of the bonding between MV3 and Y-TZP was also confirmed by the high incidence of type 2 mode of failure, indicating that the weak link on those samples was between resin cement and composite resin substrate. MV3 primer had a similar ratio (0.5 wt %) between the two monomers of interest–MDP and VBATDT, and the bond strength results of MV3 at 48 h were significantly higher than those presented by groups where VBATDT content was absent or lower (MV1 and MV2–0.0 wt % and 0.1 wt %, respectively), indicating that VBATDT may be a critical molecule for bonding to zirconia substrates, which has not yet been reported in the literature. A previous study published by our research group indeed showed that a MDP only-based primer resulted in significantly lower bond strength than a MDP/VBATDT-based primer [19]. Many other studies have already reported successful outcomes when MDP/VBATDT-primer is used to bond to Y-TZP [18,19,28,31–33]. One might argue that the successful combination between MDP and VBATDT has already been demonstrated by the extensive published research evaluating the efficacy of Alloy Primer (commercial brand) on bonding to zirconia [26]. However, the ratio between the two monomers has not been disclosed by the manufacturer (Kuraray America Inc., Huston, TX, USA), making it difficult to assess which one of the monomers plays the most important role on Y-TZP surface adhesion. Indeed, the importance of the synergistic effect of both, MDP and VBATDT, on initial bond strength between Y-YZP and resin cement, had not been considered before.

The similar importance of both monomers on the chemical interaction with Y-TZP is evidenced when MV3 results are compared to MV4 and MV7. The higher concentration of either MDP (MV7) or VBATDT (MV4) was not capable of improving bond strength results (Table 2). MDP has two functional groups, one is a di-valent phosphoryl group which is possibly adsorbed onto the Y-TZP surface, and the other is a methacryloyl group that copolymerizes with resin monomers either in the adhesive or in the resin cement composition [27]. VBATDT is a thione-thiol tautomer developed to improve the bonding between noble metal alloys and methacrylate-based resins. The coupling mechanism occurs by transforming thione in thiol and formation of the bond on the metal surface, and copolymerization of the vinyl groups with the resin cement [34]. Although further studies are necessary to precisely describe the mechanism of action of the VBATDT-molecule on the surface of Yttria-stabilized zirconium dioxide, we hypothesize that the thione-thiol group bonds to the largely available oxide layer of zirconia, somewhat contributing to the bond strength values. FTIR analysis has been able to demonstrate the crystalline VBATDT embedded in the amorphous MDP solution, indicating good miscibility between them [34] and leading to the hypothesis that one helps the other on bonding to metal substrates, increasing bond strength values [31]. Based on our findings, it is possible to speculate that the same is true for the surface of zirconium dioxide (ZrO_2), that the miscibility of both monomers created a thin and uniform layer that resulted in superior bonding due to the homogeneous distribution of stresses through the adhesive interface. The bonding was also chemically stable. The interaction between both monomers is evidenced by the EDS images of the Y-TZP surface treated with different primers. Figure 2B (MV4–spectrum 3) shows high peaks of carbon and zirconium and is related to the MDP amorphous phase. Figure 2C (MV4–spectrum 4) shows higher peaks of carbon and sulfur, which is present in the VBATDT molecule. The zirconium peak observed in this spectrum is much lower, indicating that zirconium has been shadowed by the crystalline structure of VBATDT. The homogeneous distribution of both monomers in the solution is characterized on Figure 2D (MV7), where VBATDT is dispersed throughout the MDP amorphous phase. The analysis of the results also shows that the MV4 and MV7-treated samples presented bond strength values that were similar to MV1, MV2, and MV6-treated samples. This result may reflect the dissimilar ratio between MDP and VBATDT monomers in MV4 and MV7 compositions, whereby higher and balanced concentration of both monomers might have led to improved bond strength. The investigation of the effect of higher and balanced concentrations of MDP and VBATDT up to a saturation level is of great interest to clarify whether or not the chemical bonding between resin cement and Y-TZP can be maximized.

The lowest bond strength results were obtained when the concentration of one of the monomers was reduced, either MDP (MV6) or VBATDT (MV2). This further indicates that the association of both monomers seem to be the key to succeed when bonding to Y-TZP. When MDP was absent in the formulation (MV5), all of the samples debonded prematurely. Sample preparation for microtensile testing is very critical, especially when a very hard and tough substrate, like fully-sintered Y-TZP, needs to be cut through the adhesive interface. The vibration of the diamond blade may cause a high incidence of pre-test failures. Although this was not experienced with any other group, samples treated with MV5 could not withstand the challenges of sample preparation. A bond strength of at least 5 MPa is necessary for the samples to survive microtensile sample preparation [35]. However, bond strength values as low as 4.3 MPa were reported in a previous study using the same technique [24], indicating that the bonding promoted by MV5 primer was probably inferior to 4.3 MPa. The primer application per se was expected to improve bond strength between Y-TZP and resin cement due to the improved wettability [30,36]. Kim et al. (2011) observed that the contact angle between zirconia and resin cement may significantly impact bond strength and, the lower the contact angle, the higher the shear bond strength [5]. Nonetheless, the surface wettability and the small contact angle between zirconia and cement after MV5 application was not enough to promote sufficient adhesion to survive sample preparation and pre-test failure was as high as that of control samples, without any primer application. Results also indicate that at least 0.1 wt % of MDP (MV6) is required in the formulation to avoid spontaneous failure.

When evaluating the effect of artificial aging on microtensile bond strength after primer application, results indicated that storage time had no significant effect. Therefore, this research fails to reject the second null hypothesis. In this study, beams with a cross-sectional area of approximately 1 mm 2 were stored in water at room temperature for either 48 h or 180 days. Given the small cross-section of the adhesive interface, some hydrolytic degradation of the bonding was expected after 180 days water storage. However, the mode of failure presented by the experimental groups (Figure 1) do not indicate the degradation of the zirconia-cement adhesive interface, but the degradation of the bonding in other sites of the adhesive interface. In a previous study, only samples treated with MDP/VBATDT commercial primer and bonded with a methacrylated phosphoric ester–based cement showed stable bond strength after aging [19]. Nonetheless, in another study the same adhesive strategy resulted in significant decrease of bond strength after six months [28]. Many factors may be related to the stability of the bonding in the present study. The acidic functional monomer present in the MDP formulation has been considered relatively stable to hydrolysis due to its long carbonyl chain [14,20]. The homogenous coverage of the surface with the MDP/VBATDT solution and the chemical affinity between both molecules and the zirconia oxide layer may have reduced water penetration at the zirconia-primer interface. Additionally, an effective bonding between the methacryloyl groups in the MDP molecules, the vinyl groups in the VBATDT molecules and the methacrylate-based resin cement may have occurred, minimizing water penetration at the primer/resin cement interface.

The bond strength values for the primer-treated specimens that withstood sample preparation ranged from 9.1 to 15.1 MPa. As mentioned above, cutting is a critical step for the adhesive interface of zirconia-bonded specimens, due to the resistance offered by the fully sintered Y-TZP substrate and the possible vibration of the blade during the extensive cutting procedures. Therefore, it is possible to assume that the same chemical treatment would offer better bond strength values in a regular scenario. In dentistry, for example, it has been stated that the minimum bond strength values for acceptable clinical bonding should stay within 10 and 13 MPa [24,37]. Hence, the findings of the present study indicate that it is possible to generate Y-TZP-based restorations with clinically-acceptable bond strength without the need for alumina blasting the surface, as long as the primer treatment combines MDP and VBATDT molecules in a concentration of at least 0.5 wt % MDP and 0.1 wt % VBATDT. Therefore, to expand the results of the current study, the interaction between the chemical treatments and cements with dissimilar composition should be further investigated.

5. Conclusions

The findings of the present study demonstrate that both monomers, MDP and VBATDT, are critical for bonding to non-air particle abraded Y-TZP. The balanced highest concentration of both monomers (0.5 wt %) resulted in the highest initial bond strength results and aging did not affect the overall bond strength values.

Acknowledgments: The authors acknowledge Kuraray America Inc. for the preparation of the experimental primers and other materials' donation. We also would like to express our gratitude to Paulo F. Cesar for proofreading the article. Costs with chemical and mechanical analyses were supported by Grace M. De Souza's University of Toronto start-up fund.

Author Contributions: Grace M. De Souza conceived and designed the experiments; Anuj Aggarwal performed the experiments; Grace M. De Souza and Anuj Aggarwal analyzed the data and wrote the manuscript.

Conflicts of Interest: The authors declare no conflict of interest.

References

1. Jiang, S.P.; Love, J.G.; Zhang, J.P.; Hoang, M.; Ramprakash, Y.; Hughes, A.E.; Badwal, S.P.S. The electrochemical performance of LSM/zirconia-yttria interface as a function of a-site non-stoichiometry and cathodic current treatment. *Solid State Ionics* **1999**, *121*, 1–10. [CrossRef]
2. Christel, P.; Meunier, A.; Heller, M.; Torre, J.P.; Peille, C.N. Mechanical properties and short-term in-vivo evaluation of yttrium-oxide-partially-stabilized zirconia. *J. Biomed. Mater. Res.* **1989**, *23*, 45–61. [CrossRef] [PubMed]
3. Piconi, C.; Maccauro, G. Zirconia as a ceramic biomaterial. *Biomaterials* **1999**, *20*, 1–25. [CrossRef]
4. Devigus, A.; Lombardi, G. Shading vita In-ceram YZ substructures: Influence on value and chroma, part II. *Int. J. Comput. Dent.* **2004**, *7*, 379–388. [PubMed]
5. Kim, M.J.; Kim, Y.K.; Kim, K.H.; Kwon, T.Y. Shear bond strengths of various luting cements to zirconia ceramic: Surface chemical aspects. *J. Dent.* **2011**, *39*, 795–803. [CrossRef] [PubMed]
6. Behr, M.; Proff, P.; Kolbeck, C.; Langrieger, S.; Kunze, J.; Handel, G.; Rosentritt, M. The bond strength of the resin-to-zirconia interface using different bonding concepts. *J. Mech. Behav. Biomed. Mater.* **2011**, *4*, 2–8. [CrossRef] [PubMed]
7. Borges, G.A.; Sophr, A.M.; de Goes, M.F.; Sobrinho, L.C.; Chan, D.C. Effect of etching and airborne particle abrasion on the microstructure of different dental ceramics. *J. Prosthet. Dent.* **2003**, *89*, 479–488. [CrossRef]
8. Blatz, M.B.; Oppes, S.; Chiche, G.; Holst, S.; Sadan, A. Influence of cementation technique on fracture strength and leakage of alumina all-ceramic crowns after cyclic loading. *Quintessence Int.* **2008**, *39*, 23–32. [PubMed]
9. Pereira, L.D.; Campos, F.; Dal Piva, A.M.D.; Gondim, L.D.; Souza, R.; Ozcan, M. Can application of universal primers alone be a substitute for airborne-particle abrasion to improve adhesion of resin cement to zirconia? *J. Adhes. Dent.* **2015**, *17*, 169–174.

10. Zhang, Y.; Lawn, B.R.; Rekow, E.D.; Thompson, V.P. Effect of sandblasting on the long-term performance of dental ceramics. *J. Biomed. Mater. Res. B* **2004**, *71*, 381–386. [CrossRef] [PubMed]

11. Kosmac, T.; Oblak, C.; Jevnikar, P.; Funduk, N.; Marion, L. Strength and reliability of surface treated Y-TZP dental ceramics. *J. Biomed. Mater. Res.* **2000**, *53*, 304–313. [CrossRef]

12. Scherrer, S.S.; Cattani-Lorente, M.; Vittecoq, E.; de Mestral, F.; Griggs, J.A.; Wiskott, H.W. Fatigue behavior in water of Y-TZP zirconia ceramics after abrasion with 30 mum silica-coated alumina particles. *Dent. Mater.* **2011**, *27*, e28–e42. [CrossRef] [PubMed]

13. Casucci, A.; Goracci, C.; Chieffi, N.; Monticelli, F.; Giovannetti, A.; Juloski, J.; Ferrari, M. Microtensile bond strength evaluation of self-adhesive resin cement to zirconia ceramic after different pre-treatments. *Am. J. Dent.* **2012**, *25*, 269–275. [PubMed]

14. Gomes, A.L.; Castillo-Oyague, R.; Lynch, C.D.; Montero, J.; Albaladejo, A. Influence of sandblasting granulometry and resin cement composition on microtensile bond strength to zirconia ceramic for dental prosthetic frameworks. *J. Dent.* **2013**, *41*, 31–41. [CrossRef] [PubMed]

15. Matinlinna, J.P.; Lassila, L.V.; Vallittu, P.K. Pilot evaluation of resin composite cement adhesion to zirconia using a novel silane system. *Acta Odontol. Scand.* **2007**, *65*, 44–51. [CrossRef] [PubMed]

16. Lohbauer, U.; Zipperle, M.; Rischka, K.; Petschelt, A.; Muller, F.A. Hydroxylation of dental zirconia surfaces: Characterization and bonding potential. *J. Biomed. Mater. Res. B* **2008**, *87*, 461–467. [CrossRef] [PubMed]

17. Yun, J.Y.; Ha, S.R.; Lee, J.B.; Kim, S.H. Effect of sandblasting and various metal primers on the shear bond strength of resin cement to Y-TZP ceramic. *Dent. Mater.* **2010**, *26*, 650–658. [CrossRef] [PubMed]

18. Yagawa, S.; Komine, F.; Fushiki, R.; Kubochi, K.; Kimura, F.; Matsumura, H. Effect of priming agents on shear bond strengths of resin-based luting agents to a translucent zirconia material. *J. Prosthodont. Res.* **2017**, *17*, 1883–1958. [CrossRef] [PubMed]

19. De Souza, G.M.; Silva, N.R.; Paulillo, L.A.; De Goes, M.F.; Rekow, E.D.; Thompson, V.P. Bond strength to high-crystalline content zirconia after different surface treatments. *J. Biomed. Mater. Res. B* **2010**, *93*, 318–323. [CrossRef] [PubMed]

20. De Oyague, R.C.; Monticelli, F.; Toledano, M.; Osorio, E.; Ferrari, M.; Osorio, R. Influence of surface treatments and resin cement selection on bonding to densely-sintered zirconium-oxide ceramic. *Dent. Mater.* **2009**, *25*, 172–179. [CrossRef] [PubMed]

21. Sciasci, P.; Abi-Rached, F.O.; Adabo, G.L.; Baldissara, P.; Fonseca, R.G. Effect of surface treatments on the shear bond strength of luting cements to Y-TZP ceramic. *J. Prosthet. Dent.* **2015**, *113*, 212–219. [CrossRef] [PubMed]

22. Kern, M.; Wegner, S.M. Bonding to zirconia ceramic: Adhesion methods and their durability. *Dent. Mater.* **1998**, *14*, 64–71. [CrossRef]

23. Dias de Souza, G.M.; Thompson, V.P.; Braga, R.R. Effect of metal primers on microtensile bond strength between zirconia and resin cements. *J. Prosthet. Dent.* **2011**, *105*, 296–303. [CrossRef]

24. De Souza, G.; Hennig, D.; Aggarwal, A.; Tam, L.E. The use of MDP-based materials for bonding to zirconia. *J. Prosthet. Dent.* **2014**, *112*, 895–902. [CrossRef] [PubMed]

25. Kim, J.H.; Chae, S.; Lee, Y.; Han, G.J.; Cho, B.H. Comparison of shear test methods for evaluating the bond strength of resin cement to zirconia ceramic. *Acta Odontol. Scand.* **2014**, *72*, 745–752. [CrossRef] [PubMed]

26. Akazawa, N.; Koizumi, H.; Nogawa, H.; Nakayama, D.; Kodaira, A.; Matsumura, H. Effect of mechanochemical surface preparation on bonding to zirconia of a trin-butylborane initiated resin. *Dent. Mater. J.* **2017**, *36*, 19–26. [CrossRef] [PubMed]

27. Koizumi, H.; Nakayama, D.; Komine, F.; Blatz, M.B.; Matsumura, H. Bonding of resin-based luting cements to zirconia with and without the use of ceramic priming agents. *J. Adhes. Dent.* **2012**, *14*, 385–392. [PubMed]

28. Da Silva, E.M.; Miragaya, L.; Sabrosa, C.E.; Maia, L.C. Stability of the bond between two resin cements and an yttria-stabilized zirconia ceramic after six months of aging in water. *J. Prosthet. Dent.* **2014**, *112*, 568–575. [CrossRef] [PubMed]

29. Shin, Y.J.; Shin, Y.; Yi, Y.A.; Kim, J.; Lee, I.B.; Cho, B.H.; Son, H.H.; Seo, D.G. Evaluation of the shear bond strength of resin cement to Y-TZP ceramic after different surface treatments. *Scanning* **2014**, *36*, 479–486. [CrossRef] [PubMed]

30. Yi, Y.A.; Ahn, J.S.; Park, Y.J.; Jun, S.H.; Lee, I.B.; Cho, B.H.; Son, H.H.; Seo, D.G. The effect of sandblasting and different primers on shear bond strength between yttria-tetragonal zirconia polycrystal ceramic and a self-adhesive resin cement. *Oper. Dent.* **2015**, *40*, 63–71. [CrossRef] [PubMed]

31. Sanohkan, S.; Kukiattrakoon, B.; Larpboonphol, N.; Sae-Yib, T.; Jampa, T.; Manoppan, S. The effect of various primers on shear bond strength of zirconia ceramic and resin composite. *J. Conserv. Dent.* **2013**, *16*, 499–502. [PubMed]
32. Maeda, F.A.; Bello-Silva, M.S.; de Paula Eduardo, C.; Miranda Junior, W.G.; Cesar, P.F. Association of different primers and resin cements for adhesive bonding to zirconia ceramics. *J. Adhes. Dent.* **2014**, *16*, 261–265. [PubMed]
33. Stefani, A.; Brito, R.B., Jr.; Kina, S.; Andrade, O.S.; Ambrosano, G.M.; Carvalho, A.A.; Giannini, M. Bond Strength of Resin Cements to Zirconia Ceramic Using Adhesive Primers. *J. Prosthodont.* **2016**, *25*, 380–385. [CrossRef] [PubMed]
34. Silikas, N.; Wincott, P.L.; Vaughan, D.; Watts, D.C.; Eliades, G. Surface characterization of precious alloys treated with thione metal primers. *Dent. Mater.* **2007**, *23*, 665–673. [CrossRef] [PubMed]
35. Pashley, D.H.; Sano, H.; Ciucchi, B.; Yoshiyama, M.; Carvalho, R.M. Adhesion testing of dentin bonding agents: a review. *Dent. Mater.* **1995**, *11*, 117–125. [CrossRef]
36. Hummel, M.; Kern, M. Durability of the resin bond strength to the alumina ceramic Procera. *Dent. Mater.* **2004**, *20*, 498–508. [CrossRef] [PubMed]
37. Luthy, H.; Loeffel, O.; Hammerle, C.H. Effect of thermocycling on bond strength of luting cements to zirconia ceramic. *Dent. Mater.* **2006**, *22*, 195–200. [CrossRef] [PubMed]

Article

Microstructural Characteristics and M23C6 Precipitate Behavior of the Course-Grained Heat-Affected Zone of T23 Steel without Post-Weld Heat Treatment

Seong-Hyeong Lee [1], Hye-Sung Na [2], Kyong-Woon Lee [3], Youngson Choe [4] and Chung Yun Kang [2,*]

1   Department of Hybrid Materials & Machining Technology, Graduate School of Convergence Science, Pusan National University, Busan 46241, Korea; reach486@naver.com
2   Department of Material Science and Engineering, Pusan National University, Busan 46241, Korea; joyclubman@daum.net
3   Corporate R&D Institute, Doosan Heavy Industries, Changwon 642-792, Korea; kyongwoon.lee@doosan.com
4   Department of Chemical and Biomolecular Engineering, Pusan National University, Busan 46241, Korea; choe@pusan.ac.kr
*   Correspondence: kangcy@pusan.ac.kr; Tel.: +82-10-8329-8429

Received: 10 December 2017; Accepted: 8 March 2018; Published: 9 March 2018

Abstract: The microstructural characteristics of a simulated heat-affected zone (HAZ) in SA213-T23 (2.25Cr-1.6W steel) used for boiler tubes employed in thermal power plants were investigated using nital, alkaline sodium picrate, and Murakami's etchants. In order to investigate the microstructure formation process of the HAZ in the welding process, simulated HAZ specimens were fabricated at intervals of 100 °C for peak temperatures between 950 and 1350 °C, and the microstructural features and precipitate behavior at various peak temperatures were observed. The alkaline-sodium-picrate-etched microstructures exhibited a black dot or band, which was not observed in the natal-etched microstructure. As the temperature increased from 950 to 1350 °C, the black dot and band became wider and thicker. Experimental analyses using an electron probe micro-analyzer, electron backscatter diffraction, and transmission electron microscopy revealed the appearance of austenite in the black dot region at a peak temperature of 950 °C; its amount increased up to a peak temperature of 1050 °C and thereafter decreased as the peak temperature further increased. The amount of M23C6 decreased with an increase in peak temperature. Based on these results, we investigated the behaviors of austenite and M23C6 as functions of the peak temperature.

Keywords: SA213-T23; Cr-Mo steel; heat-affected zone; M23C6; carbide dissolution; tint etching

1. Introduction

Ultra-supercritical (USC) power plants are being operated under increasing pressure and high-temperature conditions to reduce fuel costs, environmental pollutants, and greenhouse gas emissions. SA213-T22 (2.25Cr-1Mo steel), a conventional boiler tube material used in water-wall tubes in power plants, should be thickened for high-temperature and high-pressure applications [1,2]. However, if the tube becomes thicker than the ones commercially used, the construction costs will increase owing to the change in boiler design and the increase in the number of welding passes [3–6]. Furthermore, there is an increased risk of thermal stress and welding defects during the operation of the power plant.

Therefore, SA213-T23 (2.25Cr-1.6W steel), which has superior high-temperature creep properties compared to the existing material SA213-T22 (2.25Cr-1Mo steel), was developed. In order to enhance the high-temperature creep strength, T23 was developed by adding tungsten (1.6%) instead of reducing the molybdenum (0.2%) content compared to T22; additionally, vanadium, niobium, and boron were added [7–11].

However, in the case of T23, reheat cracking has been reported during post-weld heat treatment (PWHT) to mitigate the residual stress in the weld metal [12–20]. Low alloy Cr-Mo steel is known to have a high risk of reheat cracking [21–23], which is typically due to the intergranular cracks developing in the weld metal or heat-affected zone (HAZ) during PWHT or high-temperature service [18,19]. Reheat cracking is also known as stress-relaxation cracking because the cracks develop during the process of relieving the residual stresses [24]. The reheat cracking phenomenon in low-alloy Cr-Mo steels occurs when the plastic deformation when stress relaxation, rather than being strengthened transgranular, is concentrated along the weakened grain boundary [22–24]. The transgranular strengthening can be attributed to precipitation strengthening by fine and uniformly distributed precipitates [18–20]. The grain boundary is weakened due to the formations of a depleted zone [15–17] and coarse precipitates [18–20] and due to the segregation of tramp elements [22–27].

T23 is also reported to be susceptible to reheat cracking because the transgranular is strengthened and the grain boundary is weakened. The transgranular strengthening of T23 is attributed to the homogeneous transgranular precipitation of a fine metal carbide containing the alloying elements V and Nb [15–17]. The grain boundary weakening is attributed to the precipitation of an incoherent intergranular carbide M3C at the boundaries [8,9], the grain boundary segregation of the Al and P elements [15,16], and the Cr- and W-depleted zones formed by the M3C and M23C6 carbides of Fe, Cr, and W in the grain boundaries [16,17]. Therefore, it can be observed that the reheat crack of SA213-T23 is closely related to the nature of the precipitate.

Studies to identify the causes of the reheat cracks in T23 have been based only on an analysis of the microstructure of the coarse-grained HAZ (CGHAZ) after PWHT. However, it is necessary to analyze the microstructural features of CGHAZ before PWHT in order to understand the root cause of crack initiation in this material. In particular, it is necessary to characterize the types and distribution of the grain boundaries and intergranular precipitates that can cause reheat cracking, and analyze precipitate behavior and phase transformation of the matrix during the formation of the HAZ via the welding thermal cycle.

Therefore, in order to investigate the microstructure formation process of the HAZ in the welding process, simulated HAZ specimens were fabricated at intervals of 100 °C for peak temperatures between 950 and 1350 °C, and the microstructural features and precipitate behavior at various peak temperatures were observed. The microstructures were analyzed using various etching solutions such as nital, alkaline sodium picrate, and Murakami's etchant and observed under an optical microscope in order to develop a method to easily identify the behavior of the precipitates and the change in grain size. Further verifications were performed using characterization methods such as transmission electron microscopy (TEM), electron probe micro-analyzer (EPMA), and electron backscatter diffraction (EBSD).

2. Materials and Methods

Table 1 presents the chemical composition of the SA213-T23 (2.25Cr-1.6W) material used in this study. Chemical composition was analyzed using an optical emission spectrometer (LAB LAVM 10, SPECTRO Analytical Instruments GmbH, Kleve, Germany).

Table 1. Chemical composition of the T23 steel.

Material	Chemical Composition (atom %)													
	C	Si	Mn	Ni	Cr	Mo	V	Nb	Al	Ti	B	W	N	Fe
SA213-T23	0.07	0.24	0.50	0.11	2.09	0.15	0.22	0.028	0.018	0.042	0.0017	1.66	0.0095	Bal

In actual welding, it is difficult to evaluate the microstructure characteristics because the HAZ is a narrow region owing to various thermal cycles. Therefore, the simulated HAZ was fabricated by using Gleeble 3500 (Dynamic Systems Inc., New York, NY, USA) thermal cycle simulator and the microstructure was observed. The thermal cycle of the HAZ was determined using the Sysweld software 9.5 (ESI, Paris, France) for the simulation of welding analysis.

Figure 1 shows the thermal cycle curve of the HAZ according to the distance along the fusion line when gas tungsten arc welding is applied to the tube material with a thickness of 6 mm with the heat input of 16 kJ/cm. Using Gleeble 3500 thermal cycle simulator, the HAZ was simulated at intervals of 100 °C from the peak temperature of 950 °C (temperature above the AC3 point) to the peak temperature of 1350 °C (CGHAZ temperature) using the thermal cycle curve calculated using Sysweld software. The validation of the thermal profile has been verified with a thermocouple.

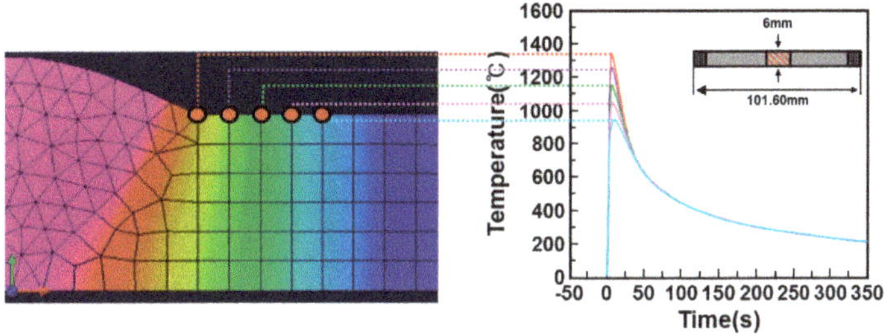

Figure 1. Schematic illustration of simulated various heat-affected zone (HAZ) thermal cycle and geometry of sample.

The microstructure of the simulated HAZ was observed by etching with nital etchant (2 mL HNO_3, 50 mL methanol) and alkaline sodium picrate etchant (2 g picric acid, 25 g NaOH, 100 mL Boiled water). Murakami etchant (10 g $K_3Fe(CN)_4$, 10 g NaOH, 100 mL water) was used to observe the distribution of the precipitates. Prior γ etchant (54 g NaOH, 5.4 g picric acid, 2–3 drops glycerol, 200 mL boiled water + 2 mL HNO_3, 50 mL methanol) was used to observe grain size.

The phase analysis of the simulated HAZ at various peak temperatures was performed using an electron probe micro-analyzer (EPMA) (JXA-8530F, JEOL, Tokyo, Japan) and electron backscatter diffraction (EBSD) (SUPRA40VP, Carl Zeiss, Oberkochen, Germany), and the equilibrium diagram was constructed using Thermo-Calc TCF.6 software to analyze the experimental results. For the precipitation analysis, precipitates were extracted via the carbon replica method, and image observation, energy-dispersive X-ray spectroscopy (EDS), and diffraction pattern analysis were performed using a field-emission transmission electron microscope (FE-TEM) (TALOS F200X, FEI, Hillsboro, OR, USA). Using the image analysis program Image-Pro Plus (Ver.4.5.0.29, MEDIA CYBERNETICS, Rockville, MD, USA), the precipitate size distribution at various peak temperatures was measured using Murakami-etched optical microscope photographs.

3. Results

3.1. Microstructure Evolution of the Base Metal

Figure 2 shows the microstructure of the base metal (2.25Cr-1.6W steel) using various etching methods. Figure 2a,d show the optical and scanning electron microscope (SEM) microstructures obtained using nital etching. Figure 2b,e show the optical and SEM microstructures obtained using alkaline sodium picrate etching. Figure 2c shows the optical microstructure obtained using Murakami etchant. In order to compare the differences between nital and alkaline sodium picrate etching, the microstructure was observed after nital etching, and it was subsequently finely polished. Afterward, alkaline sodium picrate etching was performed to observe the same position as that observed in nital etching, and the differences in the microstructures owing to etching were compared. The nital-etched microstructure in Figure 2a,d shows grain boundaries appearing as prior austenite grain boundaries, and precipitates in the prior austenite grain boundaries and fine precipitates in transgranular are observed. In the alkaline-sodium-picrate-etched microstructure in Figure 2b,e, the grain boundaries appear as austenite grain boundaries as in the case of nital etching. The precipitates are visible in the prior austenite grain boundaries, but the precipitates in transgranular are not visible. However, the matrix is observed to be monochromatic in nital etching in Figure 2a, whereas the matrix is observed to be in two colors—ivory and brown—in alkaline sodium picrate etching in Figure 2b. In alkaline sodium picrate, the average hardness of the ivory color phase was 207.4 $Hv_{0.01}$ and the average hardness of the brown color phase was 191.7 $Hv_{0.01}$. The base material of the 2.25Cr-1.6W steel (SA213-T23) is generally known to consist of tempered martensite, bainite, and M23C6 carbide in the grain boundary [28–30]. Therefore, the ivory color phase of relatively high hardness is considered tempered martensite, and the brown color phase of relatively low hardness is considered bainite. In the case of nital etching, the optical microstructure in Figure 2a exhibits a monochromatic color, whereas the SEM microstructure in Figure 2d exhibits a height difference between the two colored parts in Figure 2b. This is because tempered martensite has a higher hardness than bainite, so bainite is reduced more during polishing. Thus, it is considered that the bainite part was reduced more than the tempered martensite part in nital etching. Murakami etching illustrated in Figure 2e is an etching technique for carbide etching, and it is possible to observe the distribution of precipitates in the optical microscope before SEM observation. Thus, the appearance of the precipitates in the grain boundaries and the transgranular can be observed.

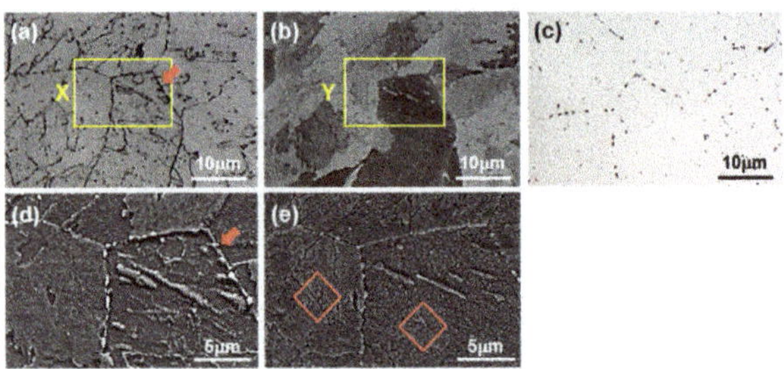

Figure 2. Optical and SEM microstructure of the base metal. (**a**,**d**) nital etchant, (**b**,**e**) alkaline sodium picrate etchant, and (**c**) Murakami etchant. (**d**,**e**) SEM micrograph at the region denoted as X, Y in (**a**,**b**).

Figure 3 shows the result of composition analysis using EPMA for the phase analysis of the base metal. The precipitates where the alloying elements of C, W, and Cr are concentrated in a point shape according to the prior austenite grain boundary can be observed.

The results of TEM analysis of the precipitates using carbon replica method are shown in Figure 4 to analyze the precipitate in the transgranular and grain boundary. In the TEM photographs, Figure 4a,b, a coarse precipitate of rectangular or rhombic shape is observed at the prior austenite grain boundary, and fine precipitates are observed in the transgranular. From the pattern analysis, it can be concluded that the coarse precipitate of Y_1 is the M23C6 carbide of [011] FCC, and the circular fine precipitate within the grain of Y_2 is the MC carbide of [001] FCC. From the EPMA results in Figure 3, the M of M23C6 carbide is the Fe, Cr, and W alloy elements. As MC is approximately 50 nm or less, it is considered that V, W, and Nb were not detected in the EPMA analysis.

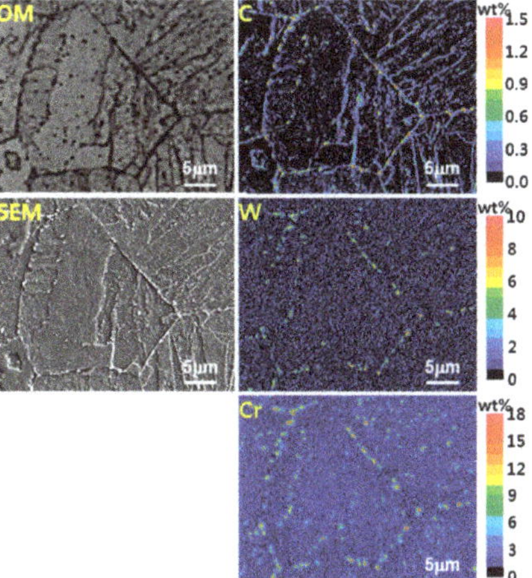

Figure 3. Electron probe micro-analyzer (EPMA) mapping results showing the distribution of elements in the base metal.

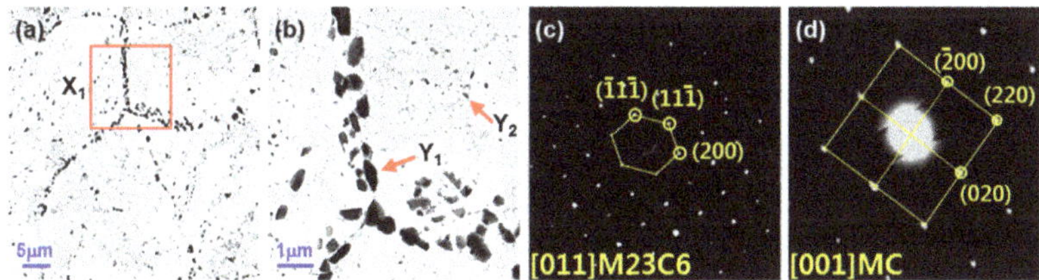

Figure 4. (**a**) TEM micrograph of the base metal at the region denoted as X_1. (**b**) An enlarged portion of (**a**). (**c,d**) Diffraction patterns at the regions denoted as Y_1 and Y_2.

3.2. Microstructure Evolution of the Simulated HAZ at Various Peak Temperatures

Figure 5 shows the optical microstructure of the HAZ at various peak temperatures (950–1350 °C) using nital, alkaline sodium picrate, Murakami, and prior γ etching. Similar to Figure 2, the observation positions of the microstructure with nital and alkaline etching were observed again. First, a change in microstructure was observed with nital etching as the peak temperature increased. It can be observed that, as the peak temperature increased from 950 to 1350 °C, the former austenite grain size (marked with an arrow) gradually increased. Grain size increase was clearly observed in the prior-γ-etched microstructure. Further, as the peak temperature increased from 1150 to 1350 °C, a lath shape in the matrix was observed. The microstructure in the case of alkaline sodium picrate etching shows that, as the temperature increased from 950 to 1350 °C, the prior austenite grain size gradually increased, similar to nital etching. Notably, a black dot or band (marked with an arrow) other than the prior austenite grain boundary was observed. As the temperature increased from 950 to 1350 °C, it became thicker and changed from a black dotted line to a black band. This phenomenon was observed only in alkaline sodium picrate etching, but not nital etching. With Murakami etching, it was observed that, as the peak temperature increased, the point-shaped precipitates in the prior austenite grain boundary decreased. Precipitates were observed up to 1150 °C, but hardly observed at temperatures above 1250 °C. Figure 2 shows that the grain boundary precipitates appearing in Murakami etching were M23C6. The reason for the disappearance of the M23C6 carbide with the increase in temperature from 950 to 1350 °C appears to be related to the dissolution of precipitates.

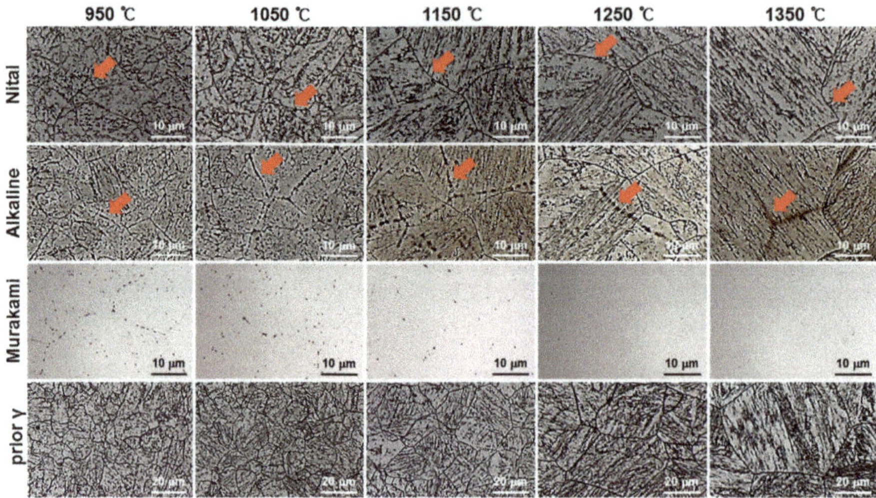

Figure 5. Optical microstructure with various etchants (nital, alkaline sodium picrate, Murakami, and prior γ) and various peak temperatures.

Figure 6 shows the SEM photograph of nital etching and alkaline sodium picrate etching at various peak temperatures (950–1350 °C). The observed position was the same as in Figure 5, and the black band that appeared in alkaline sodium picrate etching was mainly examined. The SEM photographs (marked with an arrow) of nital etching at the peak temperature of 950 °C exhibited white spherical precipitates. Precipitates were not observed in the prior austenite grain boundary, and the precipitates were present in a lined form in the transgranular. The precipitate size increased at the peak temperature of 1050 °C, but precipitates gradually dimmed at the peak temperature of 1150 °C and disappeared above the peak temperature of 1250 °C. The SEM photographs (marked with an arrow) of alkaline sodium picrate etching at the peak temperature of 950 °C exhibited white spherical precipitates and holes around the precipitates. At the peak temperature of 1050 °C, the hole size became larger and the precipitate size at the center of the hole became smaller. At the peak temperature of 1150 °C, the precipitate disappeared and only a hole was observed. Further, at the peak temperature of 1250 °C, a black band region did not appear in the SEM photograph as in the case of nital etching. In alkaline sodium picrate etching at 1350 °C, a black band was apparent in the optical microstructure, shown in Figure 5, but black bands are not easily observed in the SEM microstructure in Figure 6. These results indicate that the precipitate was dissolved and the black band appeared in the region where the precipitate was present. The black band was observed only in alkaline sodium picrate etching, but not in nital and Murakami etching.

Figure 7 shows the result of EPMA composition mapping to observe the changes of the phase composition at various peak temperatures. The black dot (band) was the region where C, W, and Cr elements were concentrated. As the peak temperature increased, the C, W, and Cr elements gradually diffused. It can be observed that the alloy composition was similar to M23C6, which is the intergranular precipitate of the base material (Figure 3). This is because the M23C6 precipitates composed of C, W, and Cr elements dissolved in the matrix as the peak temperature increased, but the C, W, and Cr elements did not dissolve completely in the matrix owing to insufficient dissolution time of the welding heat cycle conditions. W and Cr are considered to be partially dissolved.

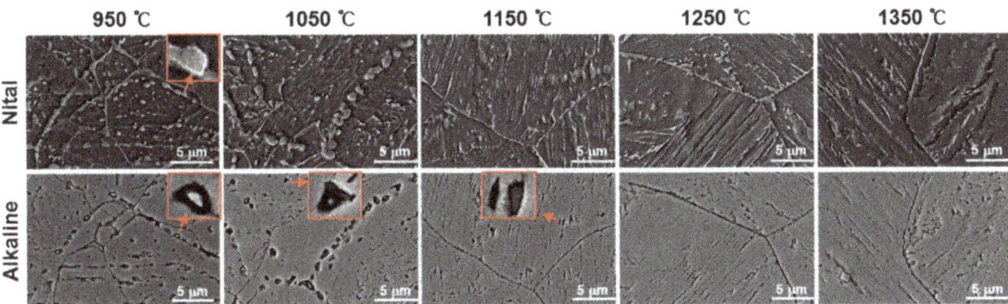

Figure 6. SEM microstructure with various etchants (nital, alkaline sodium picrate) and various peak temperatures.

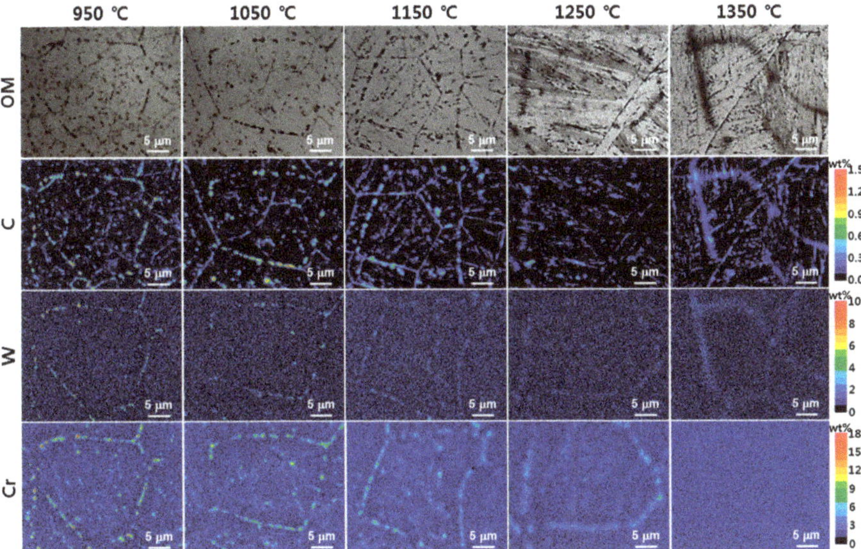

Figure 7. EPMA mapping results showing distribution of element at various peak temperatures.

In order to analyze the phase identification of the partially dissolved black dot or band, EBSD analysis was conducted at each peak temperature, and the results are shown in Figure 8. Alkaline etching was performed on the EBSD specimens to observe the same position as that corresponding to the EBSD results.

In the EBSD phase map at the peak temperature of 950 °C, a black dot region existed as M23C6 and gamma or as iron. M23C6 and gamma existed alone or together. In the EBSD phase map, iron (gamma) is an austenite, and iron (alpha) of the matrix could be ferrite (BCC), bainite (BCC), or martensite (BCT), but it is considered to be bainite or martensite based on the CCT diagram of T23 and hardness (270 Hv). Therefore, based on the SEM photographs (the alkaline-sodium-picrate-etched microstructure) at the peak temperature of 950 °C (Figure 6), the precipitate was M23C6, and the hole around the M23C6 or the lone hole was austenite or martensite. Similarly, at the peak temperature of 1050 °C, M23C6 and iron (gamma) were present alone or together in the black dot region, or iron (alpha) existed. However, compared to the peak temperature of 950 °C, the amount of M23C6 decreased and that of iron (gamma) increased. At the peak temperature of 1150 °C, the amount of M23C6 and austenite decreased in the black dot region. At the peak temperatures of 1250 and 1350 °C, the black band region was almost iron (alpha) without M23C6 and gamma. The black dots had different directions in the inverse pole figure. It can be observed that a microstructure of lath shape appeared beyond the peak temperature of 1150 °C.

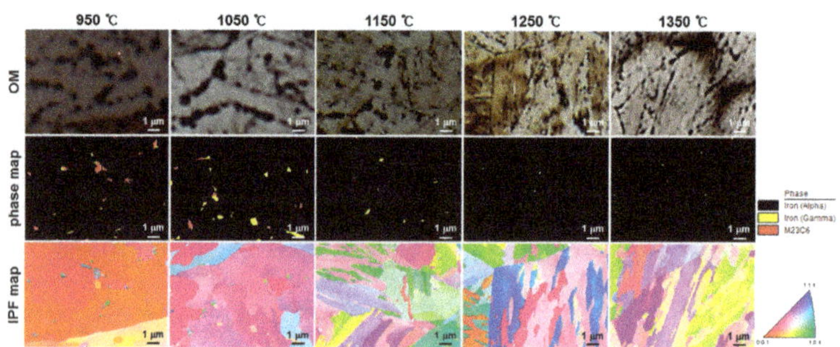

Figure 8. EBSD results at various peak temperatures.

In the EBSD analysis, austenite appeared at the peak temperature of 950 °C, and the amount increased at the peak temperature of 1050 °C and thereafter decreased as the peak temperature increased. M23C6 decreased with the increase in peak temperature. At the peak temperatures of 950, 1050, and 1150 °C, M23C6 and iron (gamma) were present alone or together in the black dot region, or iron (alpha) existed. However, at the peak temperatures of 1250 and 1350 °C, the black band was almost iron (alpha) without either of them.

Figure 9 shows the result of TEM analysis of a thin film at the peak temperatures of 1050 °C in order to analyze the microstructure of the phase identification of austenite and M23C6 in EBSD. Rectangular precipitates can be observed in the BF image of Figure 9a, and an elliptical phase can be observed around the rectangular precipitate. This is similar to the shape of the precipitate and the hole around the precipitate in the SEM photograph (alkaline-sodium-picrate-etched microstructure) at the peak temperature of 950 °C of Figure 6. In order to analyze the phase of the rectangular precipitate and elliptical phase, the diffraction pattern of the X_1 region was analyzed. In the diffraction pattern analysis, the precipitate was M23C6 and the phase around the precipitate was austenite. The diffraction pattern analysis of the X_2 region in the matrix exhibited the BCC structure. As the lath shape was not observed, the matrix was regarded as plate-like bainite. In the EDS line analysis of the Y_1 region in Figure 9d, Cr, C, and W were concentrated in M23C6 as shown in the EPMA analysis, and these elements gradually decreased from M23C6 to austenite and the matrix. It is believed that the M23C6 alloying elements such as C, W, and Cr were diffused as M23C6 dissolved in the matrix.

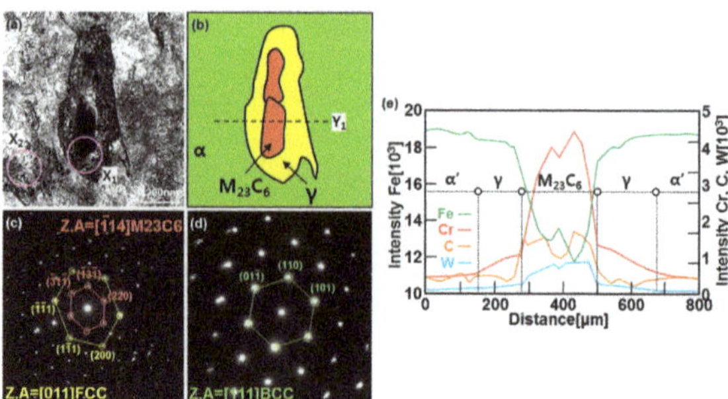

Figure 9. (**a**) TEM micrograph and (**b**) schematic illustration of 1050 °C. (**c**,**d**) Diffraction patterns at the regions denoted as X_1 and X_2. (**e**) EDX line-scan result of Y_1 across particle.

4. Discussion

4.1. Correlation between Peak Temperature and Grain Size

Figure 5 shows that the precipitation gradually decreased and the prior austenite grain size increased with the increase in peak temperature (950–1350 °C). In the results of Figures 2–8, the M23C6 precipitates of base metal were observed to dissolve with the increase in peak temperature. In order to investigate the dissolution of M23C6 as the peak temperature increased, the equilibrium diagram of the phase volume fraction was calculated using Thermo-Calc TCF.6 software (Thermo-Calc Software, Stockholm, Sweden). Figure 10 shows the result of Thermo-Calc TCF.6 calculation for T23 steel using an Fe-based alloy database. The phases shown in Figure 10a are austenite and ferrite phases, and the precipitates are M6C, M23C6, MC (V-W rich), and MC (Nb rich) phases. T23 (2.25Cr-1.6W) is a steel with excellent hardenability. It transforms into bainite (BCC) or martensite (BCT) without transformation into ferrite (BCC) when cooling in the austenite region [8]. The M6C precipitate is a high-temperature stable phase and is not formed during tempering at the time of fabrication. It is known that M23C6 phase transforms into the M6C phase during the long-term operation (400–600 °C) of a power plant [31].

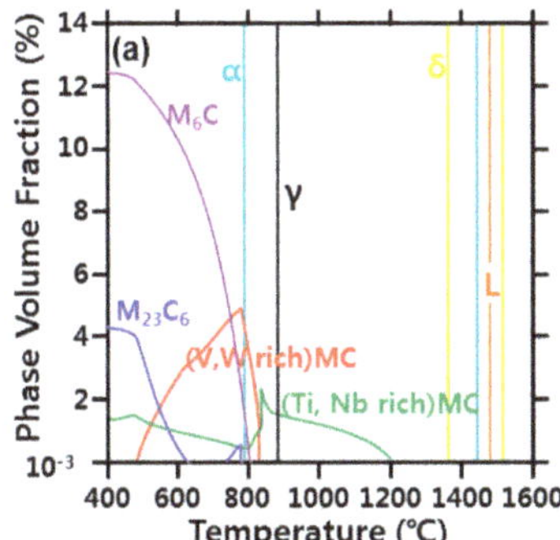

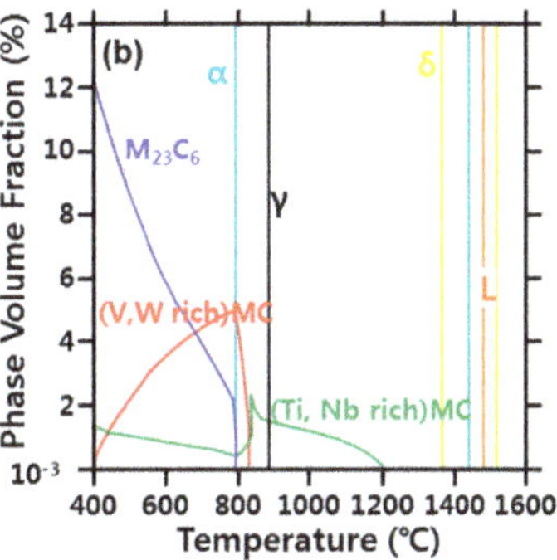

Figure 10. Phase diagram of the SA213-T23 (2.25Cr-1.6W) material calculated using Thermo-calc TCF.6: (**a**) with the precipitation of M6C carbide and (**b**) without the precipitation of M6C carbide.

Therefore, an equilibrium diagram without the M6C phase was constructed, and the result is shown in Figure 10b. In Figure 10b, it can be observed that M23C6 was a stable region at a temperature lower than 800 °C. The behavior of M23C6 in the thermal cycle at the peak temperature of 950 °C was predicted, and when heated above 800 °C, the M23C6 phase transformed into an austenite matrix. It can also be deduced that, when cooled from 950 to 800 °C, it was dissolved in the austenite matrix. The reason for the presence of carbide at the peak temperature of 950 °C is that the coarse carbides (particle size: 0.91 μm) among the precipitates present in the base metal were partially dissolved because they did not have sufficient time for full dissolution at the peak temperature of 950 °C.

As the peak temperature increased from 950 to 1350 °C, the M23C6 dissolution accelerated because the holding time was longer than that above 800 °C and the diffusion of carbon increased with temperature. Therefore, precipitation decreased as the peak temperature increased from 950 to 1350 °C.

According to Zener's theory of the relationship between precipitate and grain growth, the grain boundary migration is suppressed when the precipitate is coarse particle, high volume fraction, and uniform distribution [32–34]. In order to investigate the correlation among size, fraction, and distribution of the precipitates and grain size, the precipitate size was measured quantitatively with the increase in peak temperature using a Murakami etching photograph. Figure 11 shows the distribution of precipitate size according to the increase in peak temperature (950–1350 °C). In the size distribution of carbide particles shown in Figure 11, as the peak temperature increased, the amount of precipitate decreased and the particle size became finer. Therefore, the grain size increased with the increase in peak temperature because the driving force of the grain boundary migration was increased as the peak temperature increased, and the pinning effect was reduced owing to the dissolution of the precipitates, which interfered with the grain boundary migration.

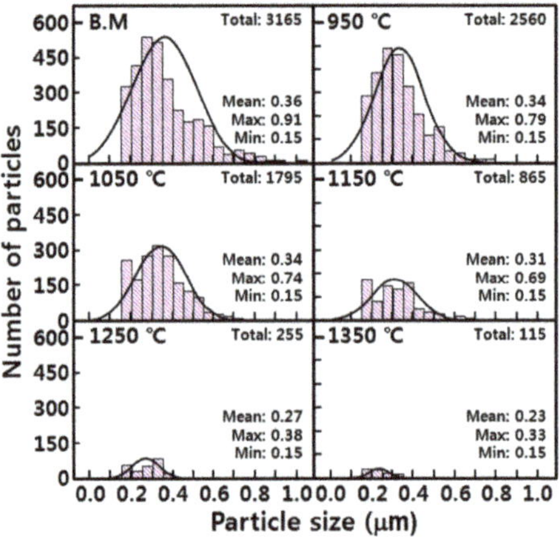

Figure 11. Carbides particles size distribution: the base metal and at 950, 1050, 1150, 1250, and 1350 °C.

4.2. Correlation between Carbide Dissolution and Austenite Formation

In order to investigate the reason for the appearance of the black dot or band region as austenite (FCC) at the peak temperature of 1050 °C but as a matrix (BCC) at the peak temperature of 1350 °C, the following analysis was performed.

Figure 12 shows the result of EPMA line analysis by enlarging the black dot or band regions at the peak temperatures of 1050 and 1350 °C. In the preceding analysis, the alloying elements of the black dot at 1050 °C identified as austenite were as follows: C: 0.6–1.0%; W: 3–4%; Cr: 7–12%. Further, the alloying elements of the black band at 1350 °C identified as BCC were as follows: C: 0.2–0.3%; W: 2–3%; Cr: 2–3%. It can be observed that the black dot region at the peak temperature of 1050 °C had a higher alloying element of C, W, and Cr than the black band region at the peak temperature of 1350 °C. This is because, as illustrated in Figure 10, the dissolution of M23C6 was further accelerated and the diffusion of the alloying elements was higher at 1050 °C than at 1350 °C.

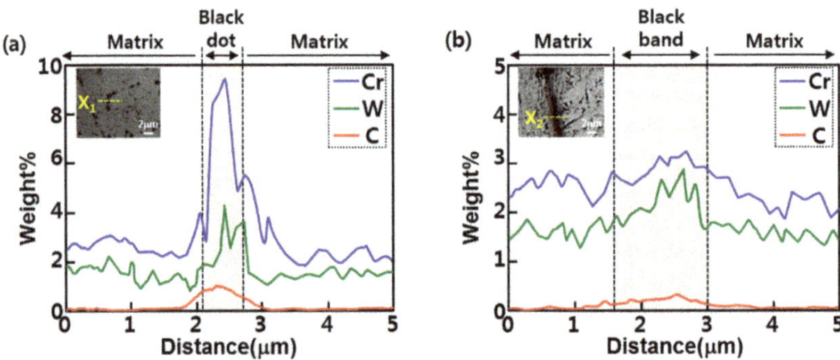

Figure 12. EPMA line results showing distribution of elements at (**a**) 1050 °C and (**b**) 1350 °C.

The stabilization region of the phase depending on the alloy composition is shown in Figure 13 in the Schaeffler diagram modified according to H. Schneider [35]. In the Schaeffler diagram modified according to H. Schneider, the alloy component (C: 0.6–1.0%; W: 3–4%; Cr: 7–12%) of the black dot at the peak temperature of 1050 °C was either the austenite or the austenite + martensite region. Moreover, the alloy component (C: 0.12–0.29%; Cr: 2.3–3.2%; W: 1.8–2.9%) of the black band at the peak temperature of 1350 °C was the martensite region.

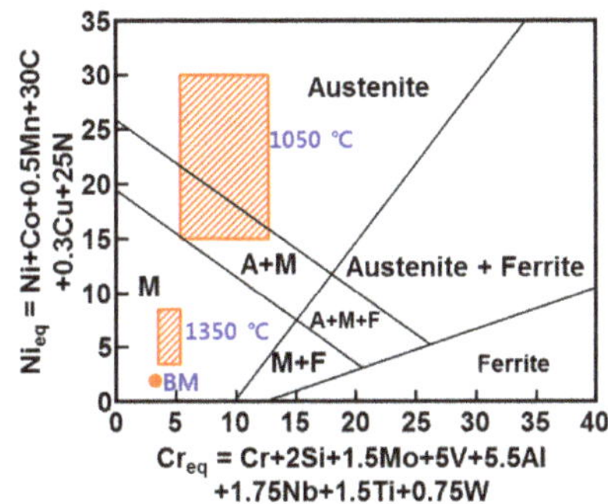

Figure 13. Schaeffler diagram modified according to H. Schneider.

Therefore, austenite appeared in the black dot region at the peak temperature of 1050 °C because the concentration of C, W, and Cr was high owing to the low diffusion of the M23C6 alloying element. However, the black band region at the peak temperature of 1350 °C had a lower concentration of C, W, and Cr than the black dot region at the peak temperature of 1050 °C owing to the high diffusion of the M23C6 alloying element in the former as compared to the latter. Consequently, the black band at the peak temperature of 1350 °C had a greater concentration of C, W, and Cr compared to the matrix, but C, W, and Cr were not sufficiently concentrated for austenite to appear.

The reason for the decrease of M23C6 and the formation of austenite and its subsequent decrease with the increase in the peak temperature is shown in Figure 14 with a schematic diagram. The alloying elements of M23C6 were C, W, and Cr, and the dissolution of M23C6 increased as the peak temperature increased. In the equilibrium diagram, M23C6 should be completely dissolved at over 800 °C, but it was almost dissolved in the peak temperature range of 1250–1350 °C by the thermal cycle of rapid heating and quenching. Subsequently, the C, W, and Cr alloying elements of M23C6 did not completely diffuse to the matrix and created concentrated regions. As the peak temperature increased, the diffusion of C, W, and Cr increased and the concentration of C, W, and Cr decreased. Therefore, when the concentrations of C, W, and Cr were high, austenite was stable. However, as the peak temperature increased, the alloying component region where the austenite was stable moved to the alloy component region where the martensite was stable.

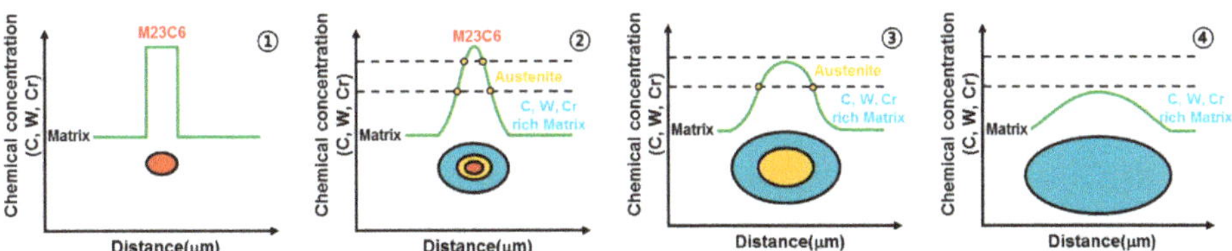

Figure 14. Schematic illustration of carbide and austenite behavior.

5. Conclusions

The microstructural features and precipitate behaviors of CGHAZ (before PWHT) in SA213-T23 (2.25Cr-1.6W) material were analyzed. The findings are summarized as follows:

(1) When the base metal of T23 was etched with alkaline sodium picrate, the ivory color obtained could be identified as tempered martensite and the brown color obtained was identified as bainite. In the case of Murakami etching, carbide was observed, and the grain boundary precipitate was M23C6 and the transgranular precipitate was MC.

(2) In order to investigate the microstructure formation process of CGHAZ, the microstructure of the simulated HAZ at various peak temperatures (950–1350 °C) was observed by etching separately with alkaline sodium picrate and nital. Alkaline-sodium-picrate-etched microstructures exhibited a black dot or band, which was not observed in the nital-etched microstructure. The black dot or band had a transgranular distribution, regardless of the nature of the prior austenite grain boundary. As the temperature increased from 950 to 1350 °C, the black dot became wider and thicker and turned into black bands.

(3) Rectangular M23C6 precipitate was present at the center of the black dot with the size of a few micrometers at the peak temperatures of 950 and 1050 °C, and austenite was present around the precipitates. As the peak temperature increased (1150–1350 °C), the M23C6 and austenite decreased, and the martensite phase, which is the same as the matrix, increased in the black band region.

(4) In the EPMA analysis, the M23C6 (C-, W-, and Cr-rich phases) precipitate of the prior austenite grain boundary in the base metal gradually dissolved with the increase in peak temperature (950–1350 °C). Therefore, it was observed that C, W, and Cr, which are the main alloying elements of M23C6, gradually diffused into the matrix. Therefore, the M23C6 with a size less than 1 μm in the base metal changed to a black dot that was coarser than the M23C6 in the base material when the peak temperature increased to 950 and 1050 °C. As the peak temperature increased from 1050 to 1350 °C, the M23C6 carbide changed into a thick black band because M23C6 (C, W-, and Cr-rich phases) gradually dissolved and diffused into the matrix. According to the amount of M23C6 diffusion, the Schaeffler diagram showed that the alloying component of the black dot (C: 0.6–1.0%; W: 3–4%; Cr: 7–12%) at the peak temperature of 1050 °C was the austenite region and that the alloy component of the black band (C: 0.2–0.3%; W: 2–3%; Cr: 2–3%) at the peak temperature of 1350 °C was the martensite region.

(5) In the calculation using Thermo-Calc, the M23C6 carbide (C-, W-, and Cr-rich phases) was stable at 800 °C or less. M23C6 dissolved above 800 °C, and the size and volume fraction of the M23C6 carbide decreased rapidly as the peak temperature increased (950–1350 °C). Therefore, it was concluded that the austenite grain size increased because the precipitates that interfered with grain boundary migration decreased.

As a preliminary study of reheat cracking, the main purpose of this study is to show that C, W, and Cr segregations exist in the CGHAZ due to the carbide of the base metal. The segregation zone of CGHAZ was found to precipitate M7C3 and M23C6 during PWHT, which confirmed that it affected reheat cracking. Therefore, while this first paper investigates the mechanism by which segregation is generated in the CGHAZ, a second paper will present the effect of segregation of the CGHAZ on reheat cracking.

Acknowledgments: This research was supported by the Global Frontier Program through the Global Frontier Hybrid Interface Materials (GFHIM) of the National Research Foundation of Korea (NRF) funded by the Ministry of Science, ICT & Future Planning (No. 2013M3A6B1078869) and the Brain Korea 21 Plus project.

Author Contributions: Seong-Hyeong Lee, Hye-Sung Na, Kyong-Woon Lee, Youngson Choe, and Chung-Yun Kang conceived and designed the experiments; Seong-Hyeong Lee and Hye-Sung Na performed the experiments; Seong-Hyeong Lee and Chung-Yun Kang analyzed the data; Seong-Hyeong Lee wrote this paper.

Conflicts of Interest: The founding sponsors had no role in the design of the study; in the collection, analyses, or interpretation of data; in the writing of the manuscript; or in the decision to publish the results.

References

1. Klueh, R.; Nelson, A. Ferritic/martensitic steels for next-generation reactors. *J. Nucl. Mater.* **2007**, *371*, 37–52. [CrossRef]
2. Bendick, W.; Gabrel, J.; Hahn, B.; Vandenberghe, B. New low alloy heat resistant ferritic steels T/P23 and T/P24 for power plant application. *Int. J. Press. Vessels Pip.* **2007**, *84*, 13–20. [CrossRef]
3. Viswanathan, R.; Bakker, W. *Materials for Ultra Supercritical Fossil Power Plants*; EPRI Report No. TR-114750; Elsevier: Amsterdam, The Netherlands, 2000.
4. Dhooge, A.; Vekeman, J. New generation 21/4Cr steels T/P 23 and T/P 24 weldability and high temperature properties. *Weld. World* **2005**, *49*, 75–93. [CrossRef]
5. Evans, G. The effect of heat input on the microstructure and properties of C-Mn all-weld-metal deposits. *Weld. J.* **1982**, *61*, 125–132.
6. Dixon, B.; Hakansson, K. Effects of welding parameters on weld zone toughness and hardness in 690 MPa steel. *Weld. J.* **1995**, *74*, 122S–132S.
7. Haarmann, K.; Kottmann, G.; Vaillant, J. *The T23/24 Book*; Vallourec & Mannesmann Tubes: Houston, TX, USA, 2000.
8. Nawrocki, J.; Dupont, J.; Robino, C.V.; Marder, A. *The Stress-Relief Cracking Susceptibility of a New Ferritic Steel-Part I: Single-Pass Heat-Affected Zone Simulations*; Sandia National Labs.: Albuquerque, NM, USA; Livermore, CA, USA, 1999.
9. Nawrocki, J.; Dupont, J.; Robino, C.; Puskar, J.; Marder, A. The mechanism of stress-relief cracking in a ferritic alloy steel. *Weld. J. N. Y.* **2003**, *82*, 25S–35S.
10. Masuyama, F.; Yokoyama, T.; Sawaragi, Y.; Iseda, A. *Service Exposure and Reliability Improvement: Nuclear, Fossil, and Petrochemical Plants*; The American Society of Mechanical Engineers (ASME): New York, NY, USA, 1994.
11. Jiménez, J.A.; Carsí, M.; Ruano, O.A. Effect of rhenium on the microstructure and mechanical behavior of Fe–2.25 Cr–1.6 W–0.25 V–0.1 C bainitic steels. *J. Mater. Sci. Technol.* **2017**, *33*, 1487–1493. [CrossRef]
12. Di Gianfrancesco, A. *Materials for Ultra-Supercritical and Advanced Ultra-Supercritical Power Plants*; Woodhead Publishing: Cambridge, UK, 2016.
13. Wang, B.; Xue, X.; Feng, Z.Y.; An, F.Q.; Zheng, R.L.; Bo, Z.; Yong, Q.J.; Xiao, L. Research on water wall tubes and welded joints of 1000 MW USC tower boiler. In Proceedings of the 7th International Conference on Advances in Materials Technology for Fossil Power Plants, Waikoloa, HI, USA, 22–25 October 2013.
14. Viswanathan, R. Advances in materials technology for fossil power plants. In Proceedings of the Fifth International Conference, Marco Island, FL, USA, 3–5 October 2007.
15. Heo, N.; Chang, J.; Kim, S.-J. Elevated temperature intergranular cracking in heat-resistant steels. *Mater. Sci. Eng. A* **2013**, *559*, 665–677. [CrossRef]
16. Park, K.; Kim, S.; Chang, J.; Lee, C. Post-weld heat treatment cracking susceptibility of T23 weld metals for fossil fuel applications. *Mater. Des.* **2012**, *34*, 699–706. [CrossRef]
17. Chang, J.; Kim, B.; Heo, N. Stress relief cracking on the weld of T/P 23 steel. *Procedia Eng.* **2011**, *10*, 734–739. [CrossRef]
18. Heo, N.; Chang, J.; Yoo, K.; Lee, J.; Kim, J. The mechanism of elevated temperature intergranular cracking in heat-resistant alloys. *Mater. Sci. Eng. A* **2011**, *528*, 2678–2685. [CrossRef]
19. Chang, J.; Heo, N.; Lee, C. Intergranular cracking susceptibility of 2.25 Cr1. 3W and 9Cr1MoVNb weld metals at elevated temperatures. *Met. Mater. Int.* **2010**, *16*, 981–985. [CrossRef]
20. Chang, J.; Heo, N.; Lee, C. Effects of mo addition on intergranular cracking behavior of 2.25 CrW (P23) weld metal at elevated temperatures. *Met. Mater. Int.* **2011**, *17*, 131–135. [CrossRef]
21. Bang, H.-S.; Kim, J.-M. Proposed specimen for reheating cracking susceptibility and mechanical behaviour assessment in 2.25 Cr–1Mo steel pressure vessel weld joint. *Sci. Technol. Weld. Join.* **2001**. [CrossRef]
22. TAMAKI koreaki. Reheat cracking of Cr-Mo steels. *J. Jpn. Weld. Soc.* **2003**, *72*, 193–200.
23. Lundin, C.; Kelley, S.; Menon, R.; Kruse, B. Stress rupture behavior of postweld heat treated 2. 25 Cr-1 Mo steel weld metal. *Weld. Res. Counc. Bull.* **1986**, *315*, 1–66.

24. Hippsley, C.; Knott, J.; Edwards, B. A study of stress relief cracking in 214 Cr 1 mo steel—II. The effects of multi-component segregation. *Acta Metall.* **1982**, *30*, 641–654. [CrossRef]

25. Shin, J.; McMahon, C. Mechanisms of stress relief cracking in a ferritic steel. *Acta Metall.* **1984**, *32*, 1535–1552. [CrossRef]

26. Hippsley, C.; Knott, J.; Edwards, B. A study of stress relief cracking in 214Cr 1 Mo steel—I. The effects of P segregation. *Acta Metall.* **1980**, *28*, 869–885. [CrossRef]

27. Balaguer, J.; Wang, Z.; Nippes, E. Stress-relief cracking of a copper-containing HSLA steel. *Weld. J.* **1989**, *68*, 121–131.

28. Vaillant, J.; Vandenberghe, B.; Hahn, B.; Heuser, H.; Jochum, C. T/P23, 24, 911 and 92: New grades for advanced coal-fired power plants—properties and experience. *Int. J. Press. Vessels Pip.* **2008**, *85*, 38–46. [CrossRef]

29. Zieliński, A.; Golański, G.; Sroka, M.; Skupień, P. Microstructure and mechanical properties of the T23 steel after long-term ageing at elevated temperature. *Mater. High Temp.* **2016**, *33*, 154–163. [CrossRef]

30. Orr, J.; Robertson, D.G. Low alloy steels–the foundation of the power generation industry. In Proceedings of the 2nd International ECCC Conference, Creep & Fracture in High Temperature Components-Design & Life Assessment, Zurich, Switzerland, 21–23 April 2009; pp. 585–598.

31. Wang, X.; Li, Y.; Li, H.; Lin, S.; Ren, Y. Effect of long-term aging on the microstructure and mechanical properties of T23 steel weld metal without post-weld heat treatment. *J. Mater. Process. Technol.* **2018**, *252*, 618–627. [CrossRef]

32. Manohar, P.A.; Ferry, M.; Chandra, T. Five decades of the zener equation. *ISIJ Int.* **1998**, *38*, 913–924. [CrossRef]

33. Wang, Y.; Ding, M.; Zheng, Y.; Liu, S.; Wang, W.; Zhang, Z. Finite-element thermal analysis and grain growth behavior of HAZ on argon tungsten-Arc welding of 443 stainless steel. *Metals* **2016**, *6*, 77. [CrossRef]

34. Du, Z.; Xiao, S.; Liu, J.; Lv, S.; Xu, L.; Kong, F.; Chen, Y. Hot deformation behavior of Ti-3.5 Al-5Mo-6V-3Cr-2Sn-0.5 Fe alloy in $\alpha + \beta$ field. *Metals* **2015**, *5*, 216–227. [CrossRef]

35. Onoro, J. Martensite microstructure of 9–12% Cr steels weld metals. *J. Mater. Process. Technol.* **2006**, *180*, 137–142. [CrossRef]

Article

The Application of 40Ti-35Ni-25Nb Filler Foil in Brazing Commercially Pure Titanium

Shan-Bo Wang [1], Chuan-Sheng Kao [1], Leu-Wen Tsay [2] and Ren-Kae Shiue [1,*

[1] Department of Materials Science and Engineering, National Taiwan University, Taipei 106, Taiwan; j9280108@gmail.com (S.-B.W.); d03527005@ntu.edu.tw (C.-S.K.)

[2] Institute of Materials Engineering, National Taiwan Ocean University, Keelung 202, Taiwan; b0186@mail.ntou.edu.tw

* Correspondence: rkshiue@ntu.edu.tw; Tel.: +886-2-33664533

Received: 7 February 2018; Accepted: 26 February 2018; Published: 1 March 2018

Abstract: The clad ternary 40Ti-35Ni-25Nb (wt %) foil has been applied in brazing commercially pure titanium (CP-Ti). The wavelength dispersive spectroscope (WDS) was utilized for quantitative chemical analyses of various phases/structures, and electron back scattered diffraction (EBSD) was used for crystallographic analyses in the brazed joint. The microstructure of brazed joint relies on the Nb and Ni distributions across the joint. For the β-Ti alloyed with high Nb and low Ni contents, the brazed zone (BZ), consisting of the stabilized β-Ti at room temperature. In contrast, eutectoid decomposition of the β-Ti into Ti_2Ni and α-Ti is widely observed in the transition zone (TZ) of the joint. Although average shear strengths of joints brazed at different temperatures are approximately the same level, their standard deviations decreased with increasing the brazing temperature. The presence of inherent brittle Ti_2Ni intermetallics results in higher standard deviation in shear test. Because the Ni content is lowered in TZ at a higher brazing temperature, the amount of eutectoid is decreased in TZ. The fracture location is changed from TZ into BZ mixed with α and β-Ti.

Keywords: vacuum brazing; titanium; eutectoid; clad filler foil; microstructure

---

1. Introduction

Titanium (Ti) and its alloys are characterized with high specific strength, good corrosion resistance, and excellent biocompatibility [1]. They are currently applied in aerospace, petroleum, and bio industries [2]. For instance, pure Ti has received great attention in medical applications. Improvement of its mechanical properties by selective laser melting has been performed in order to enhance the biomechanical compatibility of Ti implants [3]. High-strength and ductile β-Ti was successfully proposed for structural application [4]. Selective laser melting was applied to manufacture the fully dense Ti/TiB composite for medical application [5]. The importance of Ti and its alloys are increasing in recent years.

Commercially pure titanium (CP-Ti) is unalloyed in purity from 99.0 to 99.5 in wt % [2,6]. There are four grades of CP-Ti according to the contents of impurities, such as iron and interstitial elements, hydrogen, carbon, nitrogen, and oxygen. Increasing the amount of impurities in CP-Ti results in increasing its strength, but its ductility is deteriorated [2,7]. Grade 2 CP-Ti is widely used due to its comprised strength and ductility.

Brazing of Ti and its alloys can be an important issue in application of such materials [8]. Most titanium alloys can be successfully brazed by Ti-(Zr)-Cu-Ni braze alloys [8–10]. Traditional Ti-based brazing fillers are alloyed with Cu and Ni as the melting point depressants, and they are featured with moderate brazing temperature and excellent bonding strength. However, the introduction of Cu into the braze alloy results in great concern, since the Cu is considered as a toxic ingredient for most bio applications [11]. The Nb metal is considered as a non-toxic element, and it shows good biocompatibility [12]. Although the Nb has a high melting point of 2469 °C, the melting temperature of Ti-based braze alloy is slightly increased for Nb addition below 15 at % [13]. Therefore, the Cu in Ti-Cu-Ni braze alloy can be replaced by Nb in the Ti-Ni-Nb braze alloy for biomedical application. The clad 40Ti-35Ni-25Nb (wt %) foil is a promising filler metal in brazing many titanium alloys. Based on the related binary alloy phase diagrams, the β-Ti is soluble with Nb, and the Ni addition into the Ti-based braze alloy is served as a melting point depressant [13]. However, the Ni content in the braze alloy is prone to react with the Ti-based substrate, and forms brittle intermetallics to deteriorate the bonding strength of the brazed joint [14,15].

Traditional analysis tools, such as scanning/transmission electron microscope (SEM/TEM), electron probe microanalyzer (EPMA) and high-power X-ray diffractometer (XRD), are not satisfactory in analysis of a brazed joint involved with phase transformation. XRD structural analysis is not suitable for miniature phase and/or structure identification of the brazed joint. On the other hand, SEM/EPMA observations present quantitative chemical compositions of the specific phase(s), but lack structural data to identify them. For example, the transformation between α and β phases in the titanium alloy could not be accurately identified from SEM/EPMA analyses. On the other hand, the width of a brazed joint is usually below 100 μm, so slicing different brazed zones in order to make TEM examination are quite difficult. The electron backscatter diffraction (EBSD) technique has made a great achievement in recent years. The combination of morphology, element mapping, and crystallographic data makes it possible to analyze such a brazed joint undergoing phase transformation [16,17]. The purpose of this research is to perform phase identifications of 40Ti-35Ni-25Nb brazed CP-Ti joints using EPMA/WDS and SEM/EBSD. Microstructural evolution and shear strengths of vacuum brazed joints are also evaluated in this study.

2. Materials and Experimental Procedures

CP-Ti templates with the dimension of 15 mm × 7 mm × 4.2 mm were used in the experiment. Brazing surfaces were ground by SiC papers up to grit 800 and subsequently ultrasonically cleaned in ethanol solution prior to vacuum brazing. The clad 40Ti-35Ni-25Nb (wt %) foil was used as the brazing filler metal with a thickness of 50 μm. Vacuum brazing of CP-Ti substrate was performed with a heating rate of 0.33 °C/s under a vacuum of 5×10^{-3} Pa, and brazed at 1000, 1100, 1200 °C for 600 s, respectively. All of the vacuum brazed specimens were preheated at 900 °C for 1800 s in order to achieve temperature equilibrium of brazed joints.

The brazed joint was cut by a low-speed diamond saw and experienced a standard metallographic procedure before inspection. A JEOL 8600SX electron probe microanalyzer (EPMA, JEOL Ltd., Tokyo, Japan) equipped with the wavelength dispersive spectroscope (WDS) was utilized for quantitative chemical analyses of various phases/structures in the brazed joint. Its operation voltage was 15 kV, and the minimum spot size was 1 μm. An FEI Quanta 650 field emission scanning electron microscope (FESEM, FEI Corp., Hillsboro, OR, USA) equipped with the Oxford Nordlys Max 3 electron back scattered diffraction (EBSD) was used for crystallographic analyses in order to identify various phases/structures in the brazed joint. Its operation voltage was set at 20 or 25 kV. For the specimen that was prepared for EBSD analysis was ground by SiC paper, polished by colloidal silica, and finally modified by using a Fischione SEM Mill 1060 (E.A. Fischione Instruments, Inc., Export, PA, USA).

Bonding strengths of brazed joints were evaluated by shear tests, and tests were carried out on three specimens for each brazing condition. Symmetrical double lap joints, CP-Ti/40Ti-35Ni-25Nb/CP-Ti, were applied in shear tests [14,18]. Figure 1 displayed the schematic diagram of shear test specimen, and the shear test specimen was enclosed in a graphite fixture [18]. Two bold black lines, 3.5-mm wide, in the middle of the graph indicated the brazing filler foil. Shear tests were conducted using a Shimadzu AG-10 universal testing machine (Shimadzu Corp., Kyoto, Japan) with a constant crosshead speed of 0.0167 mm/s. Cross-sections of joints after brazing were cut with a low-speed diamond saw. Failure analysis of the joint after shear test was examined by a JEOL JSM 6510 scanning electron microscope (SEM, JEOL Ltd., Tokyo, Japan), with the operation voltage of 15 kV.

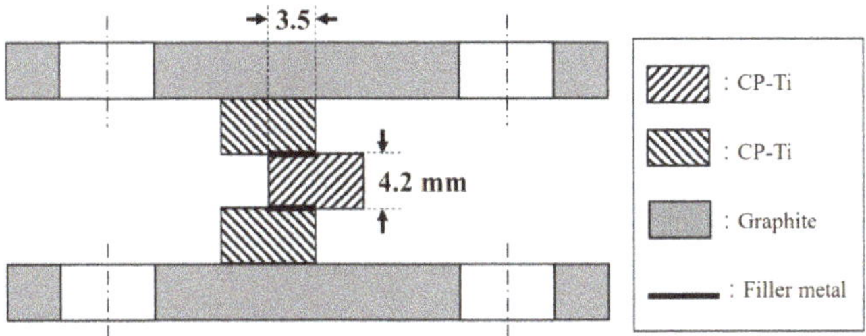

Figure 1. Schematic diagram of the shear test specimen [18].

3. Results and Discussion

Figure 2 displays the microstructural evolution of CP-Ti/Ti-35Ni-25Nb/CP-Ti joints at different brazing temperatures. Microstructures of central brazed zones (BZs) at 1000 and 1100 °C are similar. Increasing the brazing temperature causes an increase in the width of BZ. The microstructure of BZ at 1200 °C is different from those brazed at 1000 and 1100 °C. BZ at 1200 °C contains two separated acicular phases. The microstructure of CP-Ti substrate is not included in Figure 2, and there is no acicular α-Ti in CP-Ti substrate after brazing. At least two phases with acicular feature are observed in the transition zone (TZ) between the BZ and the CP-Ti substrate. One is white and the other is black in the backscattered electron images (BEIs) of Figure 2. The acicular shape of α-Ti in the TZ is originated from the transformation of β-Ti upon cooling cycle of brazing. It is deduced that the white phase contains high atomic number element, such as Nb, in BEIs, and will be discussed later.

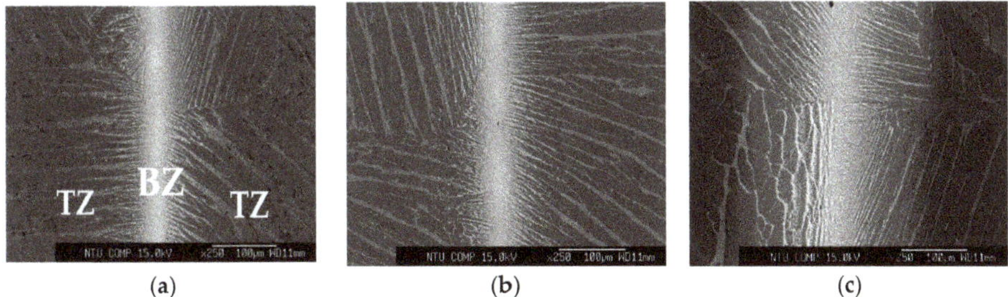

Figure 2. Electron probe microanalyzer (EPMA) backscattered electron images (BEIs) of CP-Ti/40Ti-35Ni-25Nb/CP-Ti joints brazed at (**a**) 1000, (**b**) 1100 and (**c**), 1200 °C for 600 s.

Figure 3 shows the EPMA BEIs of the joint brazed at 1000 °C for 600 s, and WDS chemical analysis results are displayed in Table 1. The brazed joint contains several distinct phases and structure. According to the EPMA chemical analysis, the white phase in the BZ is β-Ti, as marked by A in Figure 3a. The acicular white phase in TZ is identified as retained β-Ti as marked by B in Figure 3b. Both A and B have similar chemical composition but different morphologies. The black acicular phase in TZ is α-Ti alloyed with minor Nb and Ni, as marked by D in the figure. It is noted that a eutectoid of Ti₂Ni and α-Ti is observed in the TZ next to CP-Ti substrate, as marked by E in Figure 3b. Microstructures of TZ is changed from retained β-Ti plus acicular α-Ti close to the BZ into eutectoid plus acicular α-Ti next to CP-Ti substrate.

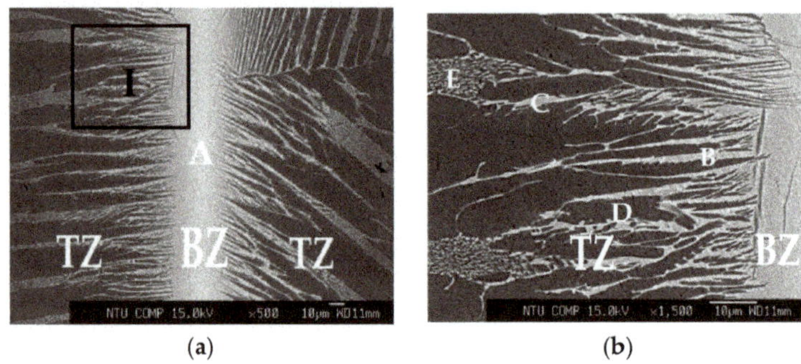

Figure 3. EPMA BEIs of CP-Ti/40Ti-35Ni-25Nb/CP-Ti joint brazed at 1000 °C for 600 s: (**a**) cross section overview, (**b**) higher magnification of location I in (**a**).

Table 1. EPMA quantitative chemical analysis results in Figure 3.

at %	A	B	C	D	E
Nb	10.0	8.4	0.9	1.7	0.3
Ni	4.3	4.3	29.0	0.1	8.5
Ti	85.7	87.3	70.1	98.2	91.2
Phase	β-Ti	retained β-Ti	Ti$_2$Ni	α-Ti	eutectoid

Figure 4 shows SEM/EBSD crystallographic analysis of the TZ between the BZ and CP-Ti substrate brazed at 1000 °C for 600 s. According to Figure 4b, there are mixtures of retained β-Ti and Ti$_2$Ni along lath boundaries of acicular α-Ti. It is worth mentioning that retained β-Ti and Ti$_2$Ni have similar color in BEIs (Figures 3b and 4a). They can be easily distinguished by SEM/EBSD crystallographic analysis. According to Figure 4, the TZ comprises of Ti$_2$Ni/α-Ti eutectoid, retained β-Ti, and Ti$_2$Ni along lath boundaries of acicular α-Ti matrix. Because Nb-Ni-Ti ternary alloy phase diagram in the Ti-rich corner is still not available, related binary alloy phase diagrams are cited in order to unveil microstructures of brazed joints [13,19].

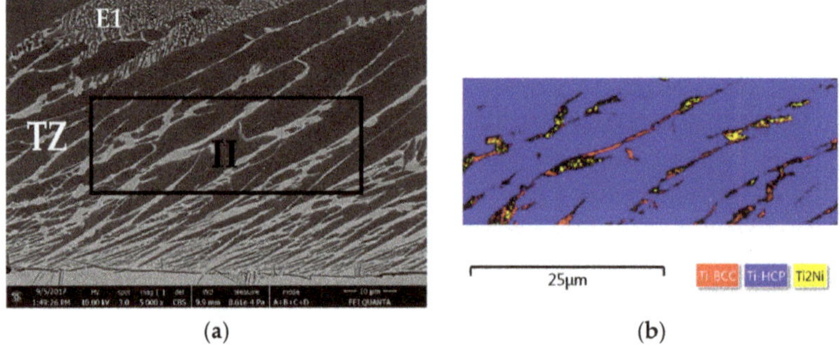

Figure 4. Scanning electron microscope (SEM)/electron back scattered diffraction (EBSD) analysis results of the transition zone (TZ) between the brazed zone (BZ) and CP-Ti substrate brazed at 1000 °C for 600 s: (**a**) SEM/BEI, (**b**) EBSD map of location II in (**a**).

According to binary alloy phase diagrams of Ti-Nb and Ti-Ni, the Ni could combine with Ti to form Ti2Ni, but it did not react with Nb in the experiment [13]. The α-Ti dissolves Nb up to 2.5 at %, and the Nb can be completely dissolved into β-Ti, i.e., (β-Ti, Nb). The chemical composition of

40Ti-35Ni-25Nb filler foil in at % is 49.1 Ti, 35.1 Ni, and 15.8 Nb. The melting point of Ti2Ni is 984 °C, so it is formed upon cooling cycle of brazing if the brazing temperature exceeds 1000 °C. Increasing the brazing temperature enhances both dissolution of the CP-Ti substrate into the braze melt and diffusion of Nb and Ni from the braze melt into CP-Ti substrate. The Ni is depleted from BZ and dissolved into the β-Ti up to 10 at % during brazing [13]. Consequently, the coarse solidified Ti2Ni is absent from the joint brazed above 1000 °C. The β-Ti is stabilized by alloying of Nb. If the content of Nb is high enough to stabilize the β-Ti, retained β-Ti is obtained after brazing. In contrast, if the β-Ti alloyed with high Ni concentration does not consist of enough Nb content, then the β-Ti will decompose into fine eutectoid of α-Ti and Ti2Ni, as demonstrated by label E in Figure 3b and E1 in Figure 4a. Phase diagrams show strong support with experimental observations.

Figure 5 displays EPMA BEIs of CP-Ti/40Ti-35Ni-25Nb/CP-Ti joint brazed at 1100 °C for 600 s. EPMA quantitative chemical analysis results of Figure 5 are listed in Table 2. In BZ, the β-Ti alloyed with 11.5 Nb and 1.8 Ni in at % as marked by F in Figure 5b. High Nb concentration stabilizes the β-Ti to room temperature. The TZ close to BZ primarily comprises of retained β-Ti marked by G in Figure 5b. The acicular α-Ti marked by I in Figure 5b is alloyed with low Nb and Ni concentrations. In contrast, the retained β-Ti is stabilized by much higher Nb and Ni concentrations, as marked by G in Figure 5b.

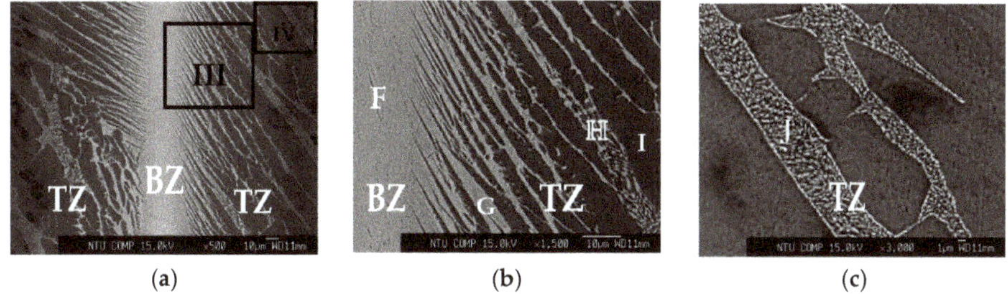

Figure 5. EPMA BEIs of CP-Ti/40Ti-35Ni-25Nb/CP-Ti joint brazed at 1100 °C for 600 s: (**a**) cross section overview, (**b**) higher magnification of location III in (**a**), (**c**) higher magnification of location IV in (**a**).

Table 2. EPMA quantitative chemical analysis results in Figure 6.

at %	F	G	H	I	J
Nb	11.5	8.3	0.2	1.2	0.1
Ni	1.8	5.4	15.0	0.1	9.3
Ti	86.7	86.3	84.8	98.7	90.6
Phase	β-Ti	retained β-Ti	eutectoid	α-Ti	eutectoid

Although both Nb and Ni can stabilize β-Ti, they take part in different roles [2]. Nb-Ti belongs to β isomorphous, and Ni-Ti is classified as β eutectoid. The transport mechanism of Nb and Ni from the BZ into CP-Ti substrate primarily relies on solid-state diffusion of Nb and Ni in β-Ti during brazing. The diffusion activation energies of Nb and Ni in β-Ti are 39.3 and 29.6 kcal/mol, respectively [20]. Therefore, Ni has a much higher diffusivity in β-Ti than Nb. According to the binary phase diagrams of Nb-Ti and Ni-Ti, the β-Ti can be alloyed with Ni up to 10 at %, but it cannot be stabilized to room temperature without alloying with Nb [13]. The Ni is alloyed in the retained β-Ti with 8.3 at % Nb, as illustrated at location G of Figure 5b. It is also noted that the eutectoid of Ti$_2$Ni and α-Ti is found in the TZ close to the CP-Ti substrate marked by H and J in Figure 5b,c. The chemical compositions in at % of location J are 90.6 Ti, 9.3 Ni, and 0.1 Nb. Slow diffusion of Nb in the β-Ti results in forming low Nb and high Ni contents of β-Ti, and it decomposes into fine eutectoid of Ti$_2$Ni and α-Ti, as shown in Figure 5c.

Figure 6 displays EPMA BEIs of CP-Ti/40Ti-35Ni-25Nb/CP-Ti joint brazed at 1200 °C for 600 s. EPMA quantitative chemical analyses of Figure 6 are listed in Table 3. The microstructure of BZ at 1200 °C is quite different from those at 1000 and 1100 °C. The BZ is not a single phase anymore, and the β-Ti is decomposed into α-Ti and retained β-Ti alloyed with 10.5 at % Nb and 4.6 at % Ni (marked by K) in Figure 6b. High brazing temperature, e.g., 1200 °C, results in depletion diffusion of Nb from the BZ to TZ. Therefore, the β-Ti in BZ is decomposed into acicular α-Ti and is retained β-Ti via eutectoid reaction. The microstructure of TZ contains retained β-Ti and eutectoid of Ti_2Ni and α-Ti along lath boundaries of acicular α-Ti matrix. The chemical composition of eutectoid, as marked by M in Figure 6c is 91.3 Ti, 0.3 Nb, and 8.4 Ni in at %. The β-Ti alloyed with high Ni and low Nb concentrations promotes eutectoid transformation into Ti_2Ni and α-Ti upon the cooling cycle of brazing.

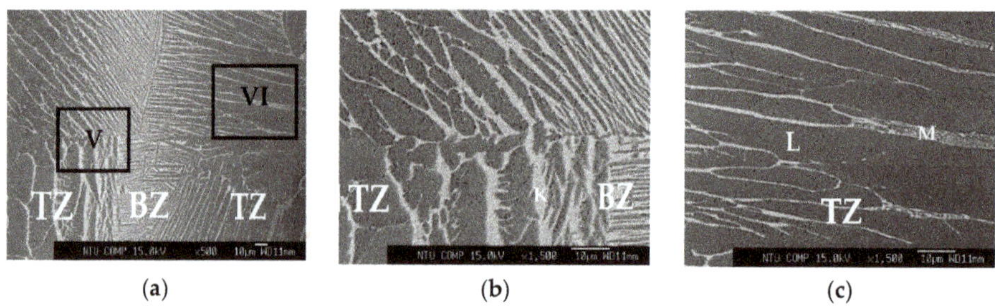

Figure 6. EPMA BEIs of CP-Ti/40Ti-35Ni-25Nb/CP-Ti joint brazed at 1200 °C for 600 s: (**a**) cross section overview, (**b**) higher magnification of location V in (**a**), (**c**) higher magnification of location VI in (**a**).

Table 3. EPMA quantitative chemical analysis results in Figure 6.

at %	K	L	M
Nb	10.5	2.3	0.3
Ni	4.6	0.2	8.4
Ti	84.9	97.5	91.3
Phase	retained β-Ti	α-Ti	eutectoid

Figure 7 shows SEM/EBSD analysis results of the joint brazed at 1200 °C for 600 s. According to Figure 7a,b, the BZ consists of acicular α and β-Ti. It is consistent with the EPMA analysis results (Figure 6 and Table 3). In the TZ, the eutectoid of Ti_2Ni precipitates and α-Ti is demonstrated in EBSD phase distribution map of location VIII in Figure 7c. It agrees with the EPMA chemical results of Figure 6c and Table 3. However, there are a few red retained β-Ti streaks in the eutectoid, as illustrated in Figure 7d. Streak-like Ti_2Ni and retained β-Ti are identified along the lath boundaries of acicular α-Ti matrix due to insufficient Nb alloyed in the β-Ti during brazing. Consequently, the stability of β-Ti is strongly related to its Nb content.

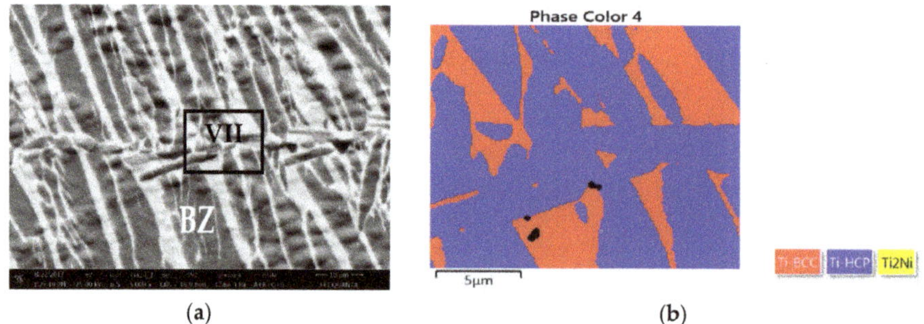

Figure 7. *Cont.*

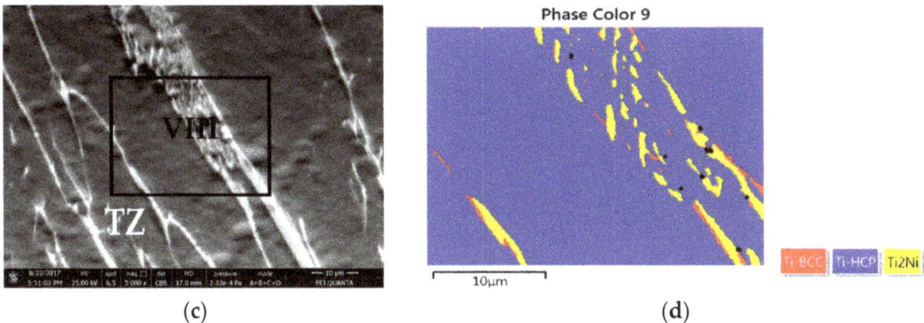

(c)   (d)

Figure 7. SEM/EBSD analysis results of the joint brazed at 1200 °C for 600 s: (**a**) BEI of the BZ, (**b**) EBSD phase distribution map of location VII in (**a**), (**c**) BEI of the TZ, (**d**) EBSD phase distribution map of location VIII in (**c**).

Figure 8 displays the average shear strengths with standard deviations of CP-Ti/Ti-35Ni-25Nb/CP-Ti joints that are brazed at different temperatures. Both specimens brazed at 1000 and 1100 °C exhibit similar average shear strengths between 341 and 351 MPa. The average shear strength is slightly increased to 402 MPa for the specimen brazed at 1200 °C. In previous study, average shear strengths of dissimilar brazed high-strength Ti-6Al-4V and Ti-15-3 joints using Ti-Cu-Ni fillers are between 282 and 545 MPa, depending on the thermal history of brazing cycles [21]. The maximum average shear strength of brazed CP-Ti joint using 40Ti-35Ni-25Nb foil is 402 MPa and it is acceptable in brazing CP-Ti.

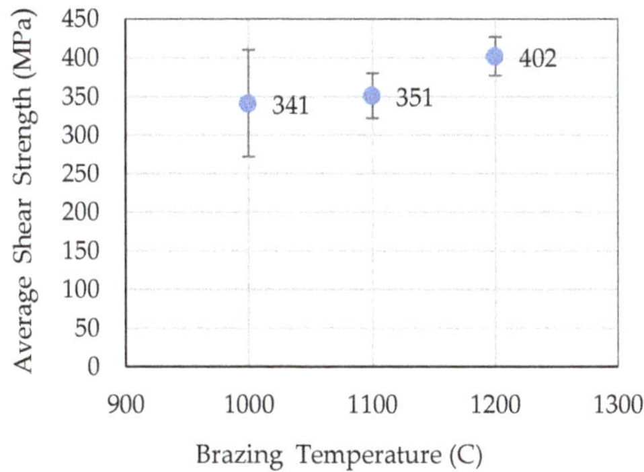

Figure 8. Average shear strengths of CP-Ti/40Ti-35Ni-25Nb/CP-Ti brazed joints.

Figure 9 shows BEI cross-sections and SEI fractographs of joints brazed at 1000, 1100, and 1200 °C for 600 s, respectively. For specimens that were brazed at 1000 and 1100 °C, cracks initiate from Ti$_2$Ni and eutectoid of TZ, and the cracking of acicular lath causes quasi-cleavage fracture being widely observed from SEI fractographs, as illustrated in Figure 9a,b. The fracture location changes from eutectoid and/or Ti$_2$Ni in TZ to α/β-Ti in BZ of the joint brazed at 1200 °C. Higher brazing temperature enhances Ni depletion from BZ into CP-Ti substrate, so a wider joint is obtained. Volume fraction of Ti$_2$Ni in TZ is decreased due to a lower Ni concentration in TZ. The SEI fractograph reveals isothermal solidified α/β-Ti in BZ, as illustrated in Figure 9c.

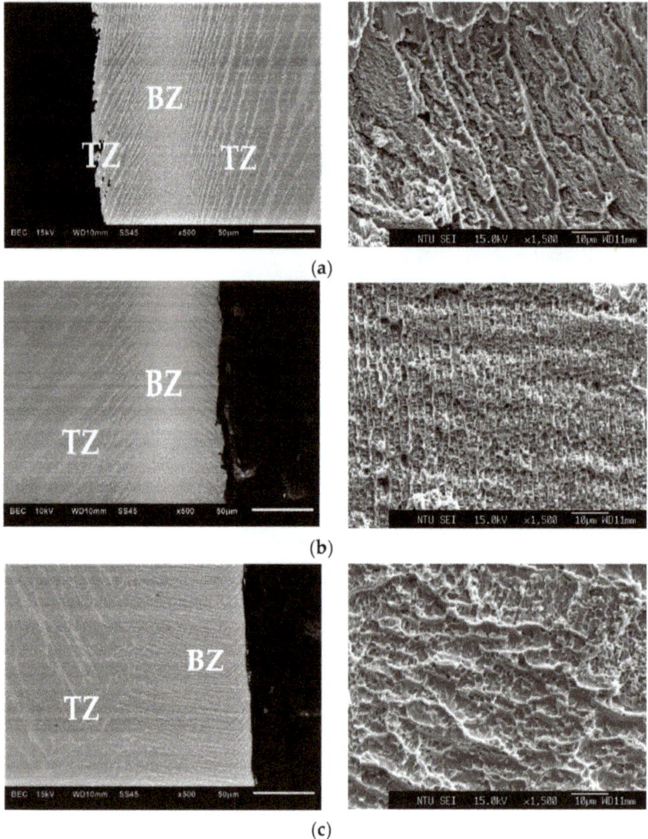

Figure 9. SEM BEI cross-sections and EPMA fractographs of joints brazed at (**a**) 1000, (**b**) 1100 and (**c**) 1200 °C for 600 s.

Although the average shear strengths of joints brazed at different temperatures are approximately the same level, standard deviations of average shear strengths are quite different, as illustrated in Figure 8. The standard deviation of the joint brazed at 1000 °C demonstrates the highest value of 69 MPa. Increasing the brazing temperature to 1100 or 1200 °C results in decreasing the standard deviation to 29 or 25 MPa, respectively. The presence of inherent brittle Ti_2Ni intermetallics results in higher standard deviation in the shear test result. It is preferred that a higher brazing temperature, such as 1200 °C, contributes to decrease the amount of Ti_2Ni in the joint. The reliability of the brazed joint is improved in practical engineering application.

Heat treatment of the joint after brazing could be helpful to decrease the amount of brittle Ti_2Ni in TZ, and a tough joint could be obtained after brazing. However, the dissolution of Ti_2Ni into β-Ti is much more prominent than that into α-Ti. The temperature of heat treatment must exceed β transus temperature of the titanium alloy due to the high solubility of Ni in the β-Ti. It is worth mentioning that the β transus temperature of the titanium alloy is strongly related to Nb and Ni concentrations in the Ti alloy. It had better be clarified before heat treating the brazed joint. Additionally, the β-Ti is completely soluble with Nb, and dissolves Ni up to 10 at % [13]. The application of 40Ti-35Ni-25Nb filler foil in brazing β-Ti alloy may obtain a joint free of the Ti-Ni intermetallic compound. It deserves to study in the future.

4. Conclusions

Vacuum brazing of CP-Ti using the clad 40Ti-35Ni-25Nb (wt %) filler foil was performed at 1000, 1100, and 1200 °C for 600 s. Microstructures of the brazed joints are strongly related to Nb and Ni depletion from the brazed joint into CP-Ti substrate. Although both Nb and Ni stabilize β-Ti, they take part in different actions. The Nb belongs to β isomorphous, and the Ni is categorized as β eutectoid. The diffusion of Ni in β-Ti is much faster than that of Nb in β-Ti. The β-Ti is stabilized by alloying Nb over 10 at % in the BZ of joints brazed at 1000 and 1100 °C. Increasing the brazing temperature to 1200 °C results in the depletion of the Nb from BZ into CP-Ti substrate. The mixture of acicular α and β-Ti is identified in the BZ. In TZ of the brazed joint, retained β-Ti, eutectoid of Ti$_2$Ni/α-Ti are formed along acicular α-Ti lath boundaries. For the β-Ti alloyed with high Ni and low Nb contents, it decomposes into Ti$_2$Ni/α-Ti eutectoid and/or lath boundary Ti$_2$Ni intermetallics. Although the average shear strengths of joints brazed at different temperatures are approximately the same level, standard deviations of average shear strengths are quite different. The standard deviation of the joint brazed at 1000 °C demonstrates the highest value of 69 MPa. Increasing the brazing temperature to 1100 or 1200 °C results in decreasing the standard deviation to 29 or 25 MPa, respectively. The presence of inherent brittle Ti$_2$Ni intermetallics results in higher standard deviation in the shear test result. For joints brazed at 1000 and 1100 °C, quasi-cleavage fracture in TZ is widely observed from the fractured surface. The fracture location is changed from eutectoid and/or Ti$_2$Ni in TZ into α/β-Ti in BZ of the joint brazed at 1200 °C. A higher brazing temperature enhances Ni depletion from BZ into CP-Ti substrate, so a wider joint is obtained. Volume fraction of Ti$_2$Ni in TZ is decreased due to lower Ni concentration in TZ. The fractured surface consists of isothermal solidified α/β-Ti in BZ. It is preferred that a higher brazing temperature, such as 1200 °C, contributes to a decrease the amount of Ti$_2$Ni in BZ and TR of the joint. The application of clad 40Ti-35Ni-25Nb filler foil demonstrates the potential in brazing CP-Ti for industrial use.

Acknowledgments: The authors gratefully acknowledge the financial support of this investigation by the Ministry of Science and Technology of Taiwan (Contract No. MOST 106-2221-E-002-174-MY3). Thanks to H.C.L. and C.Y.K. of Instrumentation Center, National Taiwan University for EPMA experiments.

Author Contributions: R.-K.S. and L.-W.T. designed the experiment and explained the data. S.-B.W. performed the experiment. C.-S.K. performed the EBSD analysis.

Conflicts of Interest: The authors declare no conflict of interest.

References

1. Davis, J.R. *ASM Handbook Volume 2 Properties and Selection: Nonferrous Alloys and Special-Purpose Materials*; ASM International: Cleveland, OH, USA, 1990; pp. 661–669.
2. Smith, W.F. *Structure and Properties of Engineering Alloys*; McGraw-Hill Book, Co.: New York, NY, USA, 1993; pp. 433–484.
3. Attar, H.; Calin, M.; Zhang, L.C.; Scudino, S.; Eckert, J. Manufacture by selective laser melting and mechanical behavior of commercially pure titanium. *Mater. Sci. Eng. A* **2014**, *593*, 170–177. [CrossRef]
4. Attar, H.; Bonisch, M.; Calin, M.; Zhang, L.C.; Scudino, S.; Eckert, J. Selective laser melting of in situ titanium–titanium boride composites: Processing, microstructure and mechanical properties. *Acta Mater.* **2014**, *76*, 13–22. [CrossRef]
5. Okulov, I.V.; Wendrock, H.; Volegov, A.S.; Attar, H.; Kühn, U.; Skrotzki, W. High strength beta titanium alloys: New design approach. *Mater. Sci. Eng. A* **2015**, *628*, 297–302. [CrossRef]
6. Roger, R.; Collings, E.W.; Welsch, G. *Materials Properties Handbook: Titanium Alloys*; ASM International: Cleveland, OH, USA, 1993; pp. 1–3.
7. Walter, J.L.; Jackson, M.R.; Sims, C.T. *Titanium and Its Alloys: Principles of Alloying Titanium*; ASM International: Cleveland, OH, USA, 1988; pp. 23–33.
8. Yeh, T.Y.; Shiue, R.K.; Chang, C.S. Microstructural evolution of brazed CP-Ti using the clad Ti-20Zr-20Cu-20Ni foil. *Metall. Mater. Trans. A* **2013**, *44*, 9–14. [CrossRef]

9.   Yeh, T.Y.; Shiue, R.K.; Chang, C.S. Microstructural observation of brazed Ti-15-3 alloy using the clad Ti-20Zr-20Cu-20Ni foil. *ISIJ Int.* **2013**, *53*, 726–728. [CrossRef]

10. Humpston, G.; Jacobson, D.M. *Principles of Soldering and Brazing*; ASM International: Cleveland, OH, USA, 1993; pp. 31–69.

11. Okulov, I.V.; Volegov, A.S.; Attar, H.; Bonisch, M.; Ehtemam-Haghighi, S.; Calin, M. Composition optimization of low modulus and high-strength TiNb-based alloys for biomedical application. *J. Mech. Behav. Biomed. Mater.* **2017**, *65*, 866–871. [CrossRef] [PubMed]

12. Ehtemam-Haghighi, S.; Prashanth, K.G.; Attar, H.; Chaubey, A.K.; Cao, G.H.; Zhang, L.C. Evaluation of mechanical and wear properties of Ti-xNb-7Fe alloys designed for biomedical applications. *Mater. Des.* **2016**, *111*, 592–599. [CrossRef]

13. Massalski, T.B. *Binary Alloy Phase Diagrams*; ASM International: Cleveland, OH, USA, 1990.

14. Shiue, R.K.; Wu, S.K.; Chen, Y.T. Strong bonding of infrared brazed α_2-Ti$_3$Al and Ti-6Al-4V using Ti-Cu-Ni fillers. *Intermetallics* **2010**, *18*, 107–114. [CrossRef]

15. Zou, Z.H.; Zeng, F.H.; Wu, H.B.; Liu, J.; Li, Y.; Gu, Y.; Yuan, T.C.; Zhang, F.Q. The Joint strength and fracture mechanisms of TC4/TC4 and TA0/TA0 brazed with Ti-25Cu-15Ni braze alloy. *J. Mater. Eng. Perform.* **2017**, *26*, 2079–2085. [CrossRef]

16. Gey, N.; Humbert, M. Specific analysis of EBSD data to study the texture inheritance due to the $\beta \rightarrow \alpha$ phase transformation. *J. Mater. Sci.* **2003**, *38*, 1289–1294. [CrossRef]

17. Kurpaska, L.; Jozwik, I.; Jagieiski, J. Study of sub-oxide phases at metal-oxide interface in oxidized pure zirconium and Zr-1.0% Nb alloy by using SEM/FIB/EBSD techniques. *J. Nucl. Mater.* **2016**, *476*, 56–62. [CrossRef]

18. Lin, C.Z.; Shiue, R.K. Vacuum brazing niobium using the clad 50Ti-35Ni-15Nb foil. *Int. J. Refract. Met. Hard Mater.* **2018**, *71*, 206–210. [CrossRef]

19. Villars, P.; Prince, A.; Okamoto, H. *Handbook of Ternary Alloy Phase Diagrams Volume 10*; ASM International: Cleveland, OH, USA, 1997; pp. 12805–12808.

20. Lide, D. *CRC Handbook of Chemistry and Physics*, 74th ed.; CRC Press: Ann Arbor, MI, USA, 1993; pp. 12–151.

21. Chang, C.T.; Du, Y.C.; Shiue, R.K.; Chang, C.S. Infrared brazing of high-strength titanium alloys by Ti-15Cu-15Ni and Ti-15Cu-25Ni filler foils. *Mater. Sci. Eng. A* **2006**, *420*, 155–164. [CrossRef]

Article

Investigation on the Effect of Tool Pin Profiles on Mechanical and Microstructural Properties of Friction Stir Butt and Scarf Welded Aluminium Alloy 6063

Pankul Goel [1], Arshad Noor Siddiquee [2], Noor Zaman Khan [2,*], Mohd Azmal Hussain [3], Zahid A. Khan [2], Mustufa Haider Abidi [4,*] and Abdulrahman Al-Ahmari [4]

[1] Department of Mechanical Engineering, IMS Engineering College, National Highway 24, Ghaziabad 201009, India; pankul_goel@rediffmail.com
[2] Department of Mechanical Engineering, Jamia Millia Islamia, New Delhi 110025, India; arshadnsiddiqui@gmail.com (A.N.S.); zakhanusm@yahoo.com (Z.A.K.)
[3] Department of Mechanical Engineering, Noida Institute of Engineering and Technology, Greater Noida, Uttar Pradesh 201306, India; ajmal0201@gmail.com
[4] Princess Fatima Alnijris's Research Chair for Advanced Manufacturing Technology, Advanced Manufacturing Institute, King Saud University, Riyadh-11421, Saudi Arabia; alahmari@ksu.edu.sa
* Correspondence: noor_0315@yahoo.com (N.Z.K.); mabidi@ksu.edu.sa (M.H.A.); Tel.: +91-11-26981259 (N.Z.K); +966-11-4698773 (M.H.A.)

Received: 14 December 2017; Accepted: 17 January 2018; Published: 19 January 2018

Abstract: In the present study, friction stir welding (FSW) of butt and scarf joints of Al 6063-T6 were investigated. Five different tool pin profiles (cylindrical, tapered cylindrical, square, triangular, and hexagonal) were applied for performing welding. Scarf joint, being a new joint configuration, was used and effect of pin profiles was investigated on this type of joint configuration. The effect of pin profiles on microstructure, micro-hardness, impact and tensile properties of friction stir welded Al 6063-T6 was investigated. Scanning electron and optical microscopy were employed to characterize the different zones of welded joints. A thorough discussion on correlation between mechanical properties and microstructure has been made. In addition, the formation of various defects during the FSW was discussed with the help of fractography of the fractured surfaces.

Keywords: friction stir welding; scarf joint; butt joint; tool pin profiles; mechanical properties; microstructure; metallurgy

---

1. Introduction

Aluminum alloy has been widely acknowledged in fabrication of lightweight structures especially for aviation, automobile and the entire transportation sector as it has high strength to weight ratio, corrosion resistance and good formability. Conventionally, for joining certain classes of aluminum alloys (e.g., age hardenable alloys), a new hot shear welding technique known as Friction Stir Welding (FSW), proved a great success [1–3]. It is evident from the research available that flaws like porosity and hot cracking are not found in FSW [4]. Moreover, the dendritic structure that is a characteristic feature of the fusion weld microstructure is not present in the FSW, thus any harm to the mechanical properties due its presence is just not possible [5]. There are a number of stages involved in the sequential progress of FSW process that is the pre-heating, initial deformation, extrusion, forging and metallurgical phases during heat rejection as shown in Figure 1. This process is energy efficient and environmentally friendly too [6]. Basically, the welding process operates by governing the amount of frictional heat generated between the rotating tool and the workpiece being welded, through a set of process parameters like tool rotation speed, plunge depth, welding speed, etc., in such a way so as to thermally condition the abutting joint surfaces in the severe plastically deformed region. Schematic of FSW process is presented in Figure 2.

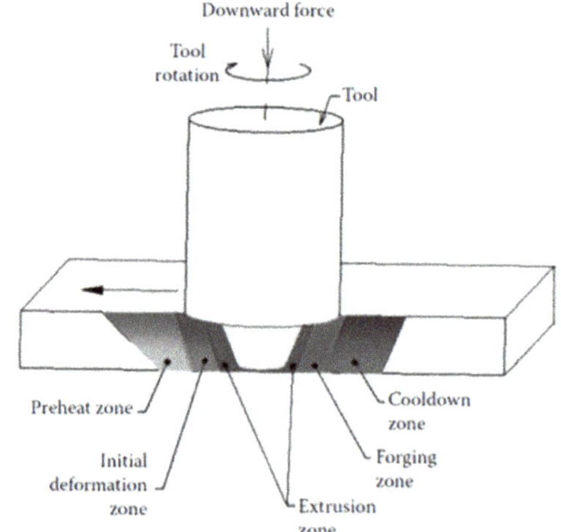

Figure 1. Stages in the Friction Stir Welding (FSW) process [7].

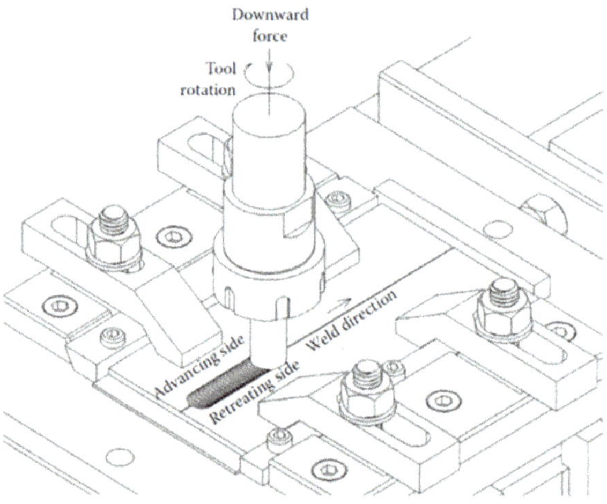

Figure 2. Schematic of FSW process [7].

The rotating cylindrical tool used to perform FSW is comprised of a shoulder (with a specified diameter 'D') and a small pin or probe (of a predefined diameter 'd' and length 'l') attached to the shoulder. The pin or probe can have various geometries, and it is plunged into the faying surface of the materials being welded. The size and designs of shoulder and pin (commonly referred to as tool design) have a significant effect on heat input and material movement. The study conducted by Vijayavel et al. [8] revealed the fluctuations in the tensile strength and hardness of the friction stir processed material because of the variation in the shoulder diameters to pin diameter (D/d) ratio, and this was further linked with associated microstructural changes. Similarly, the research conducted by Khan et al. [9] stated that a D/d ratio of 2.6 gives the maximum tensile strength, whereas a D/d ratio of 2.8 results in the minimum tensile strength and microstructures showed grain refinement in the zone where the material is mixed through stirring action (known as stir zone) due to dynamic recrystallization during FSW. Therefore, the present study used the optimized *D/d* of 2.6 for all the experiments. Another pin/probe dimension i.e., 'pin length' controls the depth of penetration. The design of the probe is such that its length is less than the thickness of the plate and its diameter is typically a little larger than the thickness of the plate [10]. An adequately chosen shoulder diameter provides two important functions: (a) prevents plasticized material, being stirred, from the out of the stir zone in the form of flash and (b) sufficient amount of frictional heat input that is necessary for the softening of material. Apparently, a truncated cone shaped pin akin to shape of flow vortex of the softened material is commonly used. Both the probe and shoulder generate a fair amount of frictional heat and heat of deformation that creates almost hydrostatic condition around the surface in contact with the probe and shoulder. Thus, under the correct conditions of heat and flow material in the front moves from the advancing side (AS) to the retreating side (RS) and consolidates at the back to form a sound joint. The material from the leading edge is exported towards the trailing edge where it is forged into a joint [11]. Roughly, with every complete rotation of the tool, a sufficient amount of energy builds up to squeeze out semi-circular shells from the base material. This is exactly how the weld zone develops or it can be understood as an integrated effect of small extrusions [12]. Moreover, the investigation done by Elangovan et al. [13] shows that the shell extrusion process is responsible for grain refinement and enhancement of material properties. The enhancement of mechanical properties is mainly affected by the changes in the microstructure of the weld region resulting from dynamic recovery [14].

Parameters like tool geometry and joint configuration have a considerable effect on the material flow and temperature evolution, and the same is reflected in the microstructure of joints. Furthermore, as a result of this combination of frictional heat and mechanical intermixing of materials, typical micro-structural zones are evolved after FSW such as (a) the stir zone (SZ), consist of fine and re-crystallized grains, (b) the thermo mechanically affected zone (TMAZ), comprising of plastically deformed grains, and (c) the heat affected zone (HAZ) containing grains similar to base material. The fine grains structure of SZ was a result of severe plastic deformation caused by stirring action of the tool. The region next to SZ is less plastically deformed and is subjected to partial dynamical re-crystallization, and it is named TMAZ. However, no plastic deformation is seen in the HAZ region, and it only experiences a thermal effect [9]. Figure 3 shows the friction-stir welding micro-structural zones.

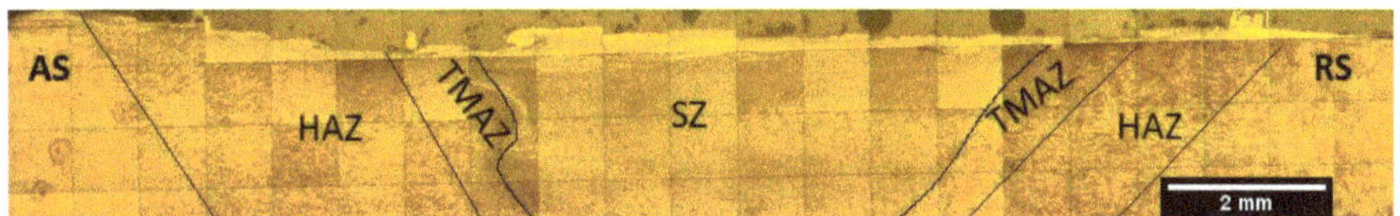

Figure 3. Friction stir welding micro-structural zones.

In case of age-hardenable alloys, the softening temperature is so high that it may lead to coarsening or dissolution of the strengthening precipitate and consequently may adversely affect the joint properties. Available literature reveals that the density, size, and distribution of the strengthening precipitates strongly affect the mechanical properties of the friction-stir welded precipitation-hardenable Al alloys [15–19]. Usually, during the welding thermal cycle [17,19], the strengthening precipitates are dissolved due to the heat generation and thus mechanical properties in the region around the weld may decrease.

Owing to various advantages over fusion welding processes, the FSW process has its own defect such as tunneling defect, kissing bond (KB), joint line remnant, hooking defect, voids, and incomplete root penetration. These defects form due to improper material movement and inadequate heat generation caused by improper selection of process parameters, tool geometry and joint configuration.

Material movement depends on several parameters including tool pin profile. These defects, if not minimized or eliminated, may lead to the degradation of mechanical properties of the joints. Defect formation in butt and lap joint configuration were investigated by many researchers [20–24]. Moreover, some defects such as kissing bond and joint line remnants, are typical to FSW. The main reason for these joints lies in the fact that, in the case of butt joint, the faying surface and the tool axis coincide. The velocities and flow of material at and near the tool axis are negligible and contributes to these defects. To overcome these defects, higher rpm and/or slower traverse speed are suggested. However, these conditions also add more heat, which is detrimental to age-hardenable alloys. Thus, alternate ways such as a change in joint configuration need to be explored so that they produce stronger joints of age-hardenable materials can be better understood. Incidentally, most butt welds by FSW are made in square joint configuration, which makes faying line coincidental to the tool axis. It is worthwhile to mention that shear movement is maximum at the tool periphery and ideally zero at the tool axis. In the case of a regular butt joint, the faying surface and the tool axis coincide and, because of deficient shear and flow, the chances of defect such as joint line remnant and kissing bond are high. The scarf joint has an advantage in which the abutting surface is inclined to the tool axis and can thus assist in alleviating issues that lead to some FSW defects. To the knowledge of the authors, no literature is available for the FSW with scarf joints. This paper has made an attempt to be the torch bearer in investigating the effect of scarf joints. Any further attempt to separately optimize the parameters for scarf configuration may uncover the complete potential of this type of joint. During FSW, the material undergoes extrusion at the leading edge in the AS, which churns around the edge and finally forged behind on the RS at trailing edge, thus consolidating the joint. During the course of this action, the material moves under shear through a set of stick-and-slip actions around the pin and undergoes severe plastic deformation. If the lateral surface of pins has flat faces such as in prismatic/pyramid shape, the material, which otherwise moves purely under shear, experiences support to its flow through pulsating action of the flats. Thus, the pin profile plays a significant role in material movement around the pin. While the effect of pin profiles on square butt joints has been explored by some researchers, its effect on scarf joint is not explored.

2. Experimental Method

In the present work, AA6063-T6 was applied as the base material (BM) and joined by friction stir welding in square butt and scarf joint configurations. Rectangular plates of 200 mm × 45 mm × 4.75 mm were used as welding. A 26° inverted bevel was tested to make up the scarf joint configuration. For the chosen shoulder diameter, the bevel angles were chosen based on comprehensive trail runs. For the scarf joint, the plates with inverted bevels were kept on AS as shown in Figure 4.

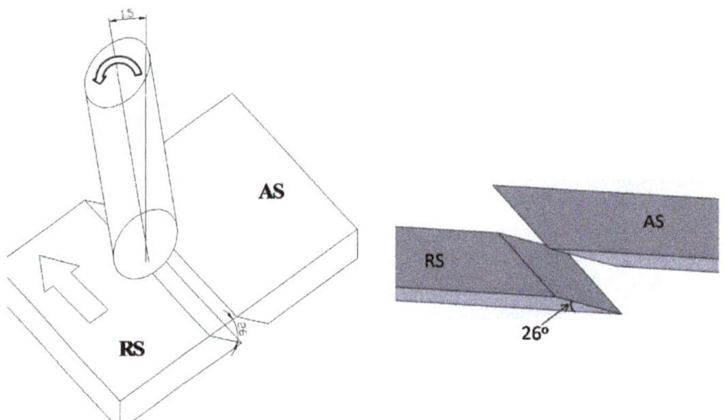

Figure 4. Alignment of plates for scarf joint configurations.

To investigate the effect of pin geometries, hot die steel (H13) tools with five different pin profiles such as cylindrical, tapered cylindrical, triangular, square and hexagonal (as shown in Figure 5) having 20 mm shoulder diameter, 7 mm pin diameter and 4.35 mm pin length were tested.

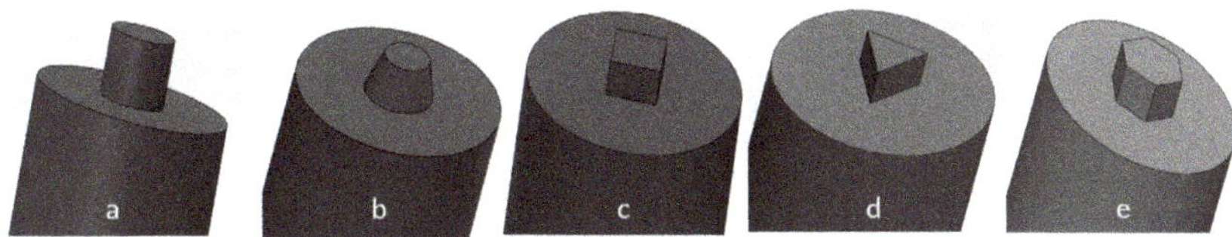

Figure 5. FSW tools with different pin profile (**a**) cylindrical; (**b**) tapered cylindrical; (**c**) triangular; (**d**) square and (**e**) hexagonal.

Chemical composition, mechanical and physical properties of Al 6063-T6 are shown in Tables 1 and 2, respectively. Experiments were conducted on a robust vertical milling machine (Maker: Bharat Fritz Werner (BFW), Bangalore, India) adapted to perform FSW. The intimate contact between the abutting faces has a significant effect on the joint especially defect formation and thus the plates were clamped by using a robust fixture with zero root gaps and adequate lateral contact pressure. Parameters such as tool tilt angle rotational and traverse speed are key to the joint efficiency and they were carefully chosen through a rigorous experimentation. Tool tilt of 1.5° (towards the trailing side of the tool from the vertical axis of rotation), rotational speed of 900 rpm and traverse speed of 50 mm/min was used during experiments. These parameters were chosen by optimization through comprehensive trial experimentation.

Table 1. Chemical composition of AA6063-T6.

Element	Al	Cu	Mg	Mn	Fe	Si	Ti	Cr	Zn	Ni
AA6063-T6	98.75	0.0280	0.489	0.031	0.245	0.426	0.014	0.006	0.0297	0.0029

Table 2. Mechanical and thermal properties of AA6063-T6.

Aluminium Alloy	Ultimate Tensile Strength (UTS) (MPa)	Yield (MPa)	Elongation (%)	Thermal Conductivity	Melting Point
AA6063-T6	220	110	14	200 W/m·K	616 °C

The tensile and impact samples were prepared as per ASTM E-8 and ASTM E-32 specifications, respectively. A tensile test was performed on a computer interface tensometer at a cross head speed of 2 mm/min. Microstructural specimens were prepared by grinding and polishing as per standard procedure and were subsequently polished by applying 1 and 0.3 micron alumina suspension and diamond paste, respectively. The polished samples were etched with Keller's reagent and macrographs of etched specimen are shown in Table 3. Micro-structural analysis was done on an optical microscope, whereas scanning electron microscopy (SEM, Zeiss, Jena, Germany) was applied to analyze fractured tensile samples. Micro-hardness was also traced on the transverse weld section across various weld zones. The Vickers hardness tests were conducted by means of hardness testing machine (Mitutoyo, Utsunomiya, Japan) at 2 N load and 15 s dwell time. It is worthwhile to note that the present work has been performed to investigate, demonstrate and compare the effects of pin profile on two different joint configurations. One of the joint configurations is very scantly reported in the literature; in such a situation, a microstructural analysis and its correlation to the process parameters shall be of greater importance, as it provides sound technical basis on the process mechanics. Thus, rather than a statistical analysis, discussions and analyses based on microstructure property correlation shall be more apt.

The FSW process parameters and tool geometry except tool pin profile were optimized and kept constant during the experiments and are shown in Table 4.

Table 3. Macrograph of friction stir (FS) welded butt and scarf joints showing different weld zones.

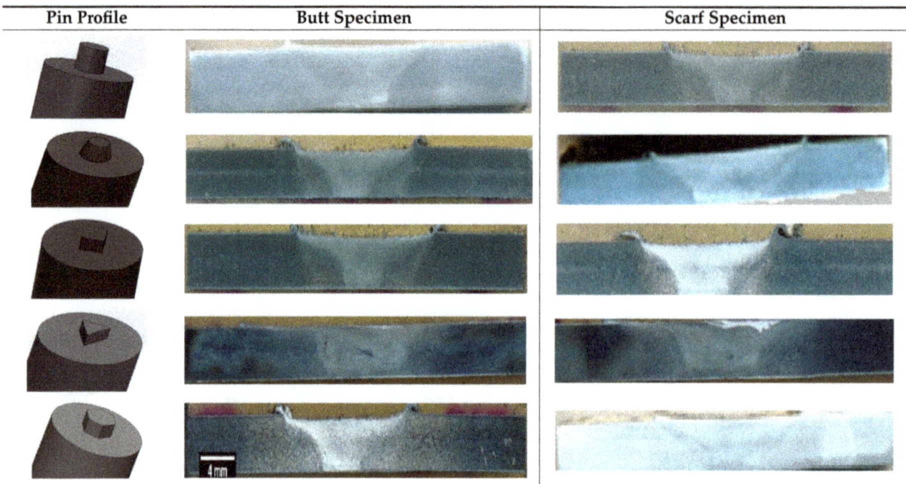

Pin Profile	Butt Specimen	Scarf Specimen

Table 4. Welding parameters that are constant during welding.

Process Parameter	Unit	Value
Tool rotational speed	Rpm	900
Welding speed	mm/min	50
Tool tilt angle	Degree	1.5
Tool shoulder diameter	mm	20
Tool shoulder surface	-	Flat
Pin diameter	mm	7.3
Pin length	mm	4.5

3. Results and Discussion

3.1. Microstructure Evolution of Butt Joint

The base material (Al 6063-T6) microstructure as given in Figure 6 shows that it consists of pancake shaped grain boundaries along with the presence of dark precipitates that are uniformly distributed.

Figure 6. Microstructure of the base material.

In contrast to this, the microstructure of weld is comprised of refined grains (Figures 7 and 8). Micrographs of square butt and scarf configurations are depicted in Figures 7 and 8, respectively. It is evident from Figure 7 that, in square butt configuration, defects like joint line remnant, tunnel and kissing bond (zig-zag line) are produced, which is a commonly reported concern [25–28]. Moreover, such situation prevailed during several pin geometries including hexagonal, conical and triangular sections, but the size of defects is very prominent in welds made with conical pin. As shown in Figure 7e, in the case of triangular pins, the tunnels found on the AS, which indicates that the material was pulled by the leading edge of tool from AS, but the pin could not effectively stir and fill the material behind the tool, and, consequently, a void is formed.

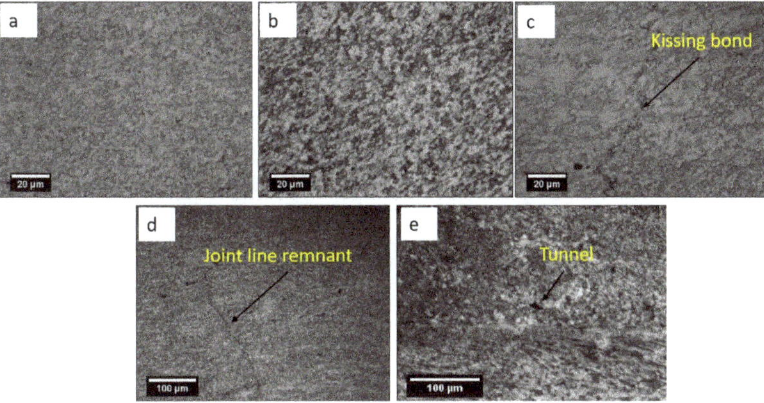

Figure 7. Butt joint micrographs obtained for specimen produced by using tools of (**a**) cylindrical pin; (**b**) tapered cylindrical pin; (**c**) square pin; (**d**) hexagonal pin; (**e**) triangular pin.

Micrograph from the SZ of weld made with conical pin is shown in Figure 7b, which reveals that the joint possesses uniformly distributed grains along with fine particles that may be Mg_2Si [29]. This contributes to enhanced joint strength and the same has been confirmed from the tensile test. The cylindrical pin profile, however, produced joint of maximum tensile strength among the other profiles used. Furthermore, zigzag lines and minute micro-cracks are also observed, which may be due to agglomerated presence of Al_2O_3 particles along the faying line, which could not be merged due these particles [29].

Figure 5c–e depict the micrographs of welds produced using square, hexagonal and triangular pin profiles, respectively. Defects such as kissing bond, tunnel and joint line remnant are visible in these micrographs. Furthermore, as calculated by Elangovan et al. [13], the number of pulses is equal to the product of rotation speed per second and the number of flat faces in the pin. Therefore, at a traverse speed of 50 mm/min (0.8333 mm/s) and rotation speed of 900 rpm (60 rps), the number of pulse produced per seconds were 90, 60, 45 for hexagonal, square and triangular pin profiles, respectively. Pins with flat faces like hexagonal, square and triangular produce pulsating stirring action and assist with material movement, provided the size and number of pulsations are correct. Thus, for effective pulsating action that can aid in effective stirring and homogenous redistribution of the strengthening particles, proper selection of pin-shoulder diameter is also required. However, in the present study, the employed pin-shoulder diameter shows that, in some cases, defects are present even despite pulsating action (like a tunnel defect with triangular pin profile). This may be because of the fact that the material extruded by a pulse has to be pressed by the shoulder, but, a material by a single pulsation being either too much or too less may result in defects. However, in this study, the defects that appear to occur were the results of insufficient availability of material at the site of joint due to too less material being eroded in a single pulse.

However, there is no such pulsating action in the case of cylindrical and tapered pin profiles, and so the extrusion of material by these pin profiles is in the form of continuous fine layers. Thus, the intermixing and deposition of this continuous supply of material in a limited amount are successfully accomplished by the tool shoulder (20 mm), and it is evident from the results that the strength obtained for these profiles is better than the rest of the profiles. Furthermore, study of micrographs reveals that variation in grain size and distribution of strengthening particles Mg_2Si (minute black spots in the micrographs are actually Mg_2Si [13]) are clearly visible.

Moreover, in the case of a triangular tool, the fall in tensile and impact strength of joints are due to the presence of the tunnel defect as seen in Figure 8. However, the appearance of the top surface of the joint may look fine, but the tunneling defect may still be present, as it exists below the surface.

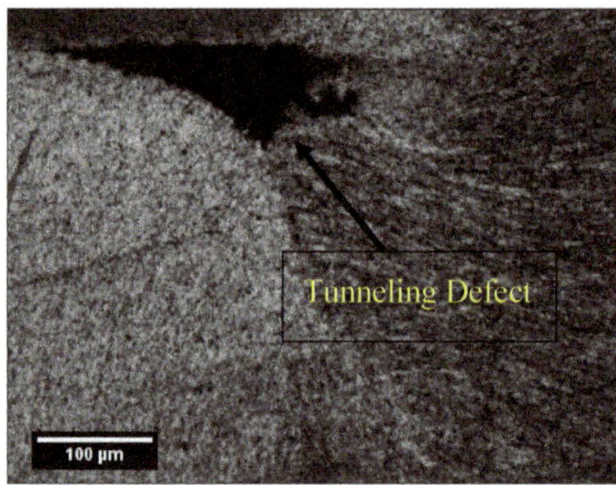

Figure 8. Tunnel defect in the welded joint.

3.2. Microstructure Evolution of Scarf Joint

It was quite clear from available literature that the majority of the FSW work was done over butt joint and lap joint configuration [30], whereas, in the present work, an attempt has been made to make a joint between new configurations, which was actually a combination of lap and butt joints, and it is named scarf joint configuration. From the micrograph of the SZ of scarf joints, the influence of pin profiles on the material flow and defect formation is analyzed. Moreover, the microstructure for scarf joints (Figure 9) shows that, while defects that are more serious like voids, tunnel, and cavities are either absent or significantly reduced in size. Only defects like kissing-bond, hooking (which is typically found in lap joints [31–33]), and zigzag lines were found in specimens of the weld produced by different pin profiles.

As shown in Figure 9a–e, in almost all the micrographs, the initial joint line of two bevel plates was clearly visible, and this was one of the reasons for the decrease in strength of the scarf joint. These surfaces, however, experienced deformations and might deviate from the original straight and flat contact interfaces. The wider kissing bond appears to be the main reason for the reduced strength; otherwise, other bulk defects such as voids, tunnels or cavities are reduced in the case of scarf joints. It is evident from Figure 9b that the SZ is comprised of elongated grains and transition from SZ to TMAZ is clearly visible. The detailed micrograph of the kissing bond flaw with a typical angled direction and a zigzag shape near its end is evident in Figure 9e.

Around the SZ and TMAZ interface, the presence of the kissing bond appears to be the extension of the hooking defect in TMAZ, although the bond width is not as wide as it is for the hooking defect as stated by Threadgill et al. [34]. Furthermore, such bonds (kissing and hooking) are imperfections due to remnant oxide films on the original workpiece surfaces, which had not broken up due to the insufficient deformation during stirring. In fact, the distribution of the oxides is strongly connected to parameters such as welding speed, and tool rotational speed brings variations in the degree of stirring produced in the material [35,36].

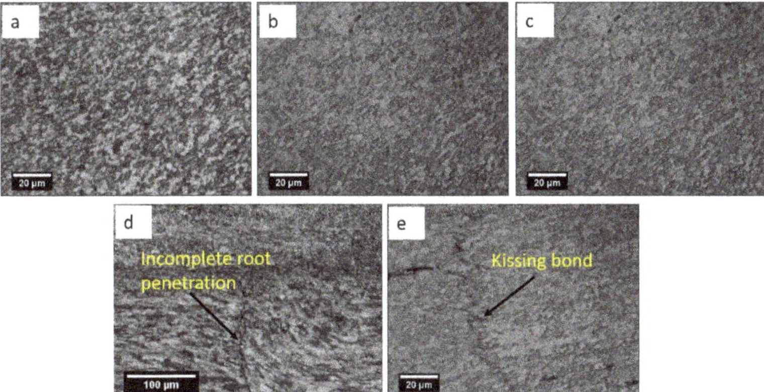

Figure 9. Scarf joint micrographs obtained for specimen produced by using tools of (**a**) cylindrical pin; (**b**) tapered cylindrical pin; (**c**) square pin; (**d**) hexagonal pin; (**e**) triangular pin.

3.3. Mechanical Testing

The effect of different pin profiles on tensile strength is explored for butt and scarf joints of Al 6063-T6. In the case of butt joints, the tapered cylindrical pin profile showed the maximum strength (162 MPa) and toughness (26 Joule) compared to the welds made applying other pin profiles, and this may be due to grain refinement and dynamic recrystallization (DRX) in the SZ [10]. Fine grain structure increases the strength of joints because the resistance in the movement of dislocation increases, as the number of grain boundaries. Furthermore, the cylindrical pin produces joints with better tensile properties than square, triangular and hexagonal. Joints produced using hexagonal and triangular pin profiles exhibited poor tensile strength due to non-homogeneous movement of materials around the pin. FS welded butt and scarf joints fabricated with cylindrical tool exhibited the highest elongation (11.16 and 7.75 mm, respectively), whereas joints with tapered cylindrical pins showed elongation 7.95 and 7.01 mm, respectively. This difference in elongation may be attributed to coarser grains from cylindrical tool weld, which aids in stretching by allowing the grain boundaries to deform more. Scarf joints with cylindrical pins showed maximum tensile strength (137 MPa) and toughness (21 Joule) compared to joints with other pin profiles.

Furthermore, in both joint configurations, the results obtained for square, triangular and hexagonal shaped tool pin profiles are found to be weak in strength. As pulsating action produced by pins with flat faces not only on depends on the number of flat faces in the pin but also on other parameters too. Thus, the FSW process parameter combinations appear to be unable to generate sufficient temperature and the rate of stirring in which the pulsating action could be effectively engaged.

It is reported that, due to inappropriate movement and insufficient plasticization of material during FS welding, the joints are prone to defects like kissing bonds, voids, tunnel defects, etc. [37,38]. It was observed that the FS Welded joints obtained by applying triangular and hexagonal pin profiles possessed tunneling defects, which resulted in the degradation of the tensile strength as shown in Tables 5 and 6. Similar results are also reported in literature [39].

Table 5. Butt joint configuration tensile and impact test results for different pin profiles.

Pin Profile	Peak Load (KN)	UTS (MPa)	Elongation (%)	Impact Strength (Joule)
Tapered Cylindrical	4.4	162	8	26
Cylindrical	4.1	160	11	24
Square	4.5	158	7	21
Hexagonal	3.5	117	5.3	22
Triangular	3.4	116	4.6	20

Table 6. Scarf joint configuration tensile and impact test results for different pin profiles.

Pin Profile	Peak Load (KN)	UTS (MPa)	Elongation (%)	Impact Strength (Joule)
Tapered Cylindrical	3.3	129	7	16
Cylindrical	4.4	137	8	21
Square	1.6	77	5	09
Hexagonal	4.4	121	8	11
Triangular	1.7	63	3	18

3.4. Fractography

To reveal how the fracture occurred during tensile testing, SEM observations of the fractured section were carried out. The fracture mechanism of the joints produced by testing different tool pin profiles and over two different joint configurations are described by the typical fractographs at the mid-section of the fractured tensile specimens in Figure 10.

The SEM fractographs reveal that a combination of deep and shallow micro-voids (less in number in case of butt joints) were present (Figure 10a), whereas micro voids in the case of scarf joint configurations were seen in large numbers that are relatively big in size compared to butt joints (Figure 10b). Furthermore, ductile striations spread over a large part are also visible. Furthermore, energy dissipation in plastic strains is evident from the deep dimples at certain places in the micrograph. Finely populated deep dimples are observed in butt joints (Figure 10a), which implies typical ductile mode of fractures observed in many aluminium alloys. The scarf joint shows shear type fracture with less number of deep dimples near the SZ (Figure 10b). The ductile fracture with void nucleation and coalescence is distinct in this micrograph.

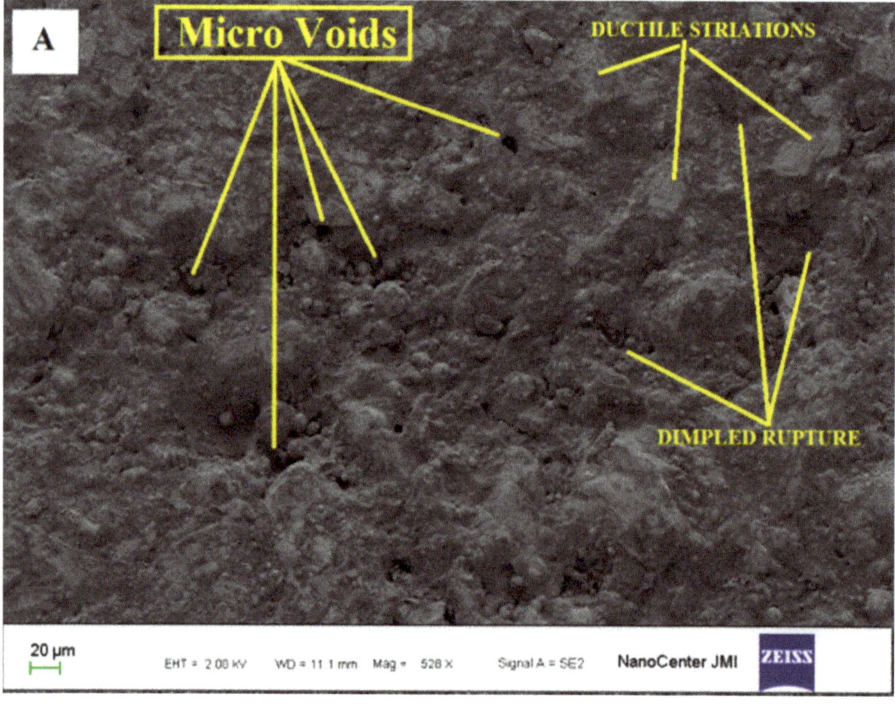

Figure 10. *Cont.*

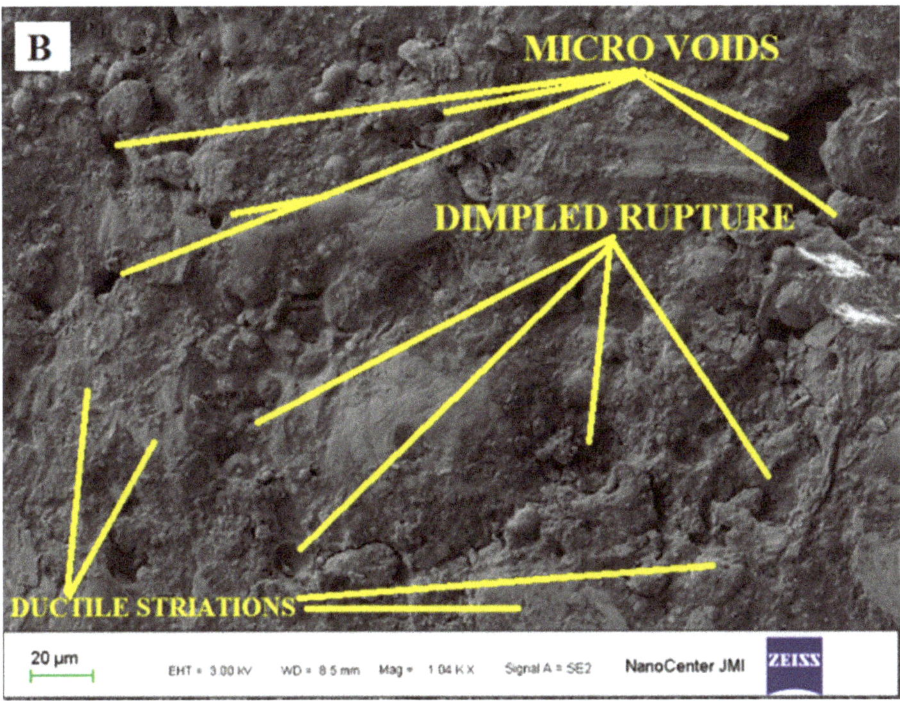

Figure 10. SEM fractography of the welded joints produced by (**A**) tapered cylindrical pin for buttjoint configuration and (**B**) cylindrical pin for scarf joint configuration.

3.5. Micro-Hardness Distribution

Micro-hardness distribution in the welded joints is performed to plot the topography of hardness variations across various regions of the welds. The micro-hardness distribution across the welded samples (transverse section) of the butt and scarf joints produced by different pin profiles are given in Figure 11a,b.

It is quite evident from Figure 11a that the micro-hardness profile is "W" shaped for all butt joint samples obtained by applying different pin profiles except for the square pin profile. The available literature supports the formation of "W" shaped profile for precipitation strengthened materials. In the stir zone, the average micro-hardness is 70 HV for all the samples except square pin profiles, and it is about 85 HV for square pin profile welded samples. Square pin profiles create a maximum pulsating action and provides a higher volumetric area (swept/static volume ratio is higher) for material mixing thus results in good material mixing and higher micro-hardness. The micro-hardness further increases at the TMAZ and suddenly drops at the HAZ of the retreating side of the joint obtained using a square pin profile. The sudden increment at the TMAZ may be attributed to the intermetallic compound (IMC) formation, and its decrement in HAZ is due to the grain coarsening and dissolution of the strengthening precipitates [40–43].

The micro-hardness profile at the stir zone for butt joining is stable (see Figure 11a); however, for all the scarf joints, there is a sudden drop in micro-hardness near the centerline as shown in Figure 11b. The basic material movement in FSW is advancing to the retreating side, but the complex nature of the scarf joint requires an effective vertical material movement, which is not fulfilled completely during FSW. Thus, the formation of heterogeneous mixture at the stir zone is the possible reason for the decrement of micro-hardness in the scarf joint for different pin profiles. However, the square joint creates the maximum micro-hardness among all the joints due to the effective pulsation, which can create better material stirring and mixing.

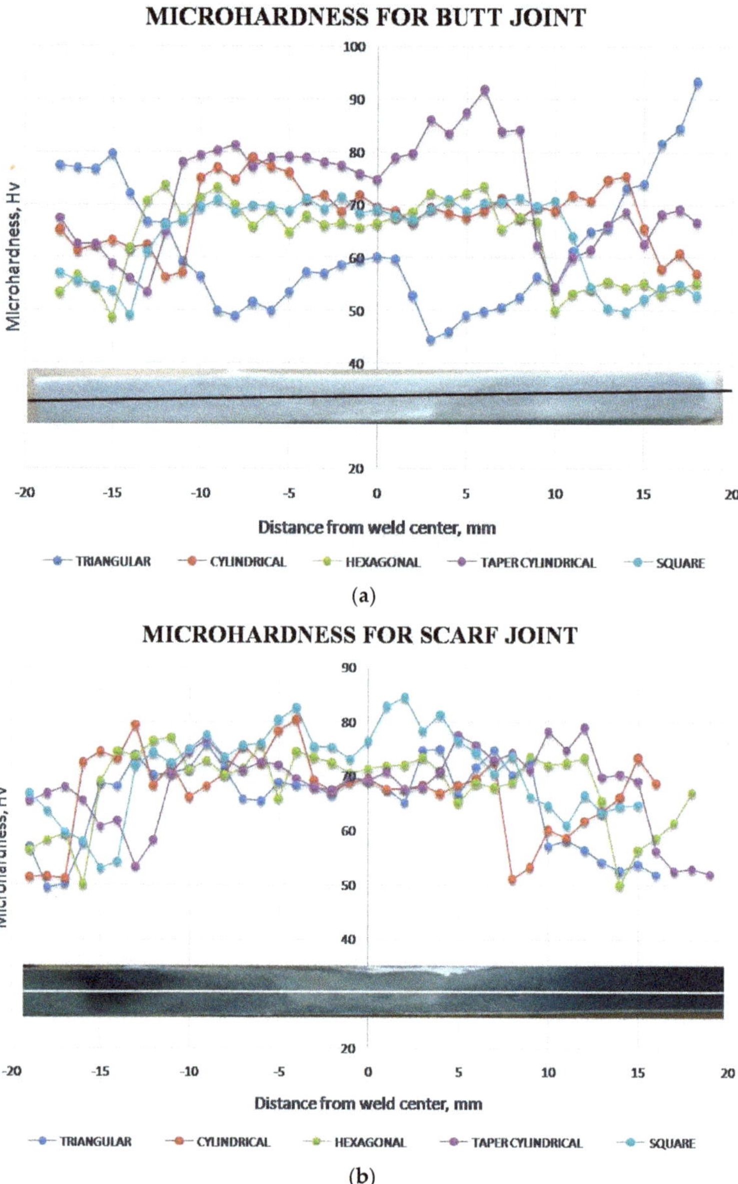

(a)

(b)

Figure 11. Micro hardness profiles of welded joints produced in (**a**) butt joint configuration and (**b**) scarf joint configuration.

4. Conclusions

FSW of Al 6063-T6in butt and scarf joint configuration using five different tool pin profiles (straight cylindrical, tapered cylindrical, threaded cylindrical, triangular, and square) has been successfully performed. Along with butt joint configuration, a completely new joint configuration i.e., scarf joint (lap-butt), is tried for FSW. Defects that form during the course of action were also reported and discussed, and feasible recommendations have been made to prevent their formation. Based on the results obtained, various conclusions are drawn below:

1. FS welded butt joints fabricated with Tapered Cylindrical tools exhibited the highest tensile strength (162 MPa), whereas triangular tools showed the lowest tensile strength (115.6 MPa).
2. Maximum impact strength of the FS welded butt joint is found to be 26 joules for Tapered Cylindrical tools.
3. The low strength obtained in the case of scarf joints is due to relatively new joint configuration and improper features of the joints such as inclination angle, plate positioning and improper plunge.
4. Tunnel defects are found on the advancing side of the butt joint fabricated using triangular pin profiles due to the improper flow of material and inadequate consolidation.
5. Hooking, kissing and zigzag line defects were observed in the weld zone of scarf joint configurations due to improper combination of process parameters employed for welding.
6. FSW on scarf joints has been performed on the parameter combinations, which were optimized for butt joints.

Acknowledgments: The authors are grateful to the Deanship of Scientific Research, King Saud University for funding through the Vice Deanship of Scientific Research Chairs.

Author Contributions: Arshad Noor Siddiquee and Mohd Azmal Hussain wrote the paper, Arshad Noor Siddiquee and Zahid A. Khan designed the experiment, Pankul Goel and Noor Zaman Khan performed the experiments, and Mustufa Haider Abidi and Abdulrahman Al-Ahmari analyzed the data.

Conflicts of Interest: The authors declare no conflict of interest

References

1. Arbegast, W.J. A flow-partitioned deformation zone model for defect formation during friction stir welding. *Scr. Mater.* **2008**, *58*, 372–376. [CrossRef]
2. Packer, S.M.; Matsunaga, M. Friction stir welding equipment and method for joining X65 pipe. In Proceedings of the Fourteenth International Offshore and Polar Engineering Conference, Toulon, France, 23–28 May 2004; International Society of Offshore and Polar Engineers: Mountain View, CA, USA, 2004.
3. Dawes, C.J.; Thomas, W.M. Friction stir process welds aluminium alloys. *Weld. J.* **1996**, *75*, 41–45.
4. Cabibbo, M.; McQueen, H.J.; Evangelista, E.; Spigarelli, S.; Di Paola, M.; Falchero, A. Microstructure and mechanical property studies of AA6056 friction stir welded plate. *Mater. Sci. Eng. A* **2007**, *460–461*, 86–94. [CrossRef]
5. Rhodes, C.G.; Mahoney, M.W.; Bingel, W.H.; Spurling, R.A.; Bampton, C.C. Effects of friction stir welding on microstructure of 7075 aluminum. *Scr. Mater.* **1997**, *36*, 69–75. [CrossRef]
6. Akinlabi, E.T.; Akinlabi, S.A. Friction stir welding process: A green technology. *World Acad. Sci. Eng. Technol. Int. J. Mech. Mechatron. Eng.* **2012**, *6*, 2514–2516.
7. Khan, N.Z.; Siddiquee, A.N.; Khan, Z.A. *Friction Stir Welding: Dissimilar Aluminum Alloys*; CRC Press: Boca Raton, FL, USA, 2017.
8. Vijayavel, P.; Balasubramanian, V.; Sundaram, S. Effect of shoulder diameter to pin diameter (D/d) ratio on tensile strength and ductility of friction stir processed LM25AA-5% SICp metal matrix composites. *Mater. Des.* **2014**, *57*, 1–9. [CrossRef]
9. Khan, N.Z.; Khan, Z.A.; Siddiquee, A.N. Effect of shoulder diameter to pin diameter (D/d) ratio on tensile strength of friction stir welded 6063 aluminium alloy. *Mater. Today Proc.* **2015**, *2*, 1450–1457. [CrossRef]
10. Ulysse, P. Three-dimensional modeling of the friction stir-welding process. *Int. J. Mach. Tools Manuf.* **2002**, *42*, 1549–1557. [CrossRef]
11. Grujicic, M.; Arakere, G.; Yalavarthy, H.V.; He, T.; Yen, C.-F.; Cheeseman, B.A. Modeling of AA5083 material-microstructure evolution during butt friction-stir welding. *J. Mater. Eng. Perform.* **2010**, *19*, 672–684. [CrossRef]
12. Mishra, R.S.; Ma, Z.Y. Friction stir welding and processing. *Mater. Sci. Eng. R Rep.* **2005**, *50*, 1–78. [CrossRef]

13. Elangovan, K.; Balasubramanian, V. Influences of tool pin profile and tool shoulder diameter on the formation of friction stir processing zone in AA6061 aluminium alloy. *Mater. Des.* **2008**, *29*, 362–373. [CrossRef]

14. Cavaliere, P.; Campanile, G.; Panella, F.; Squillace, A. Effect of welding parameters on mechanical and microstructural properties of AA6056 joints produced by friction stir welding. *J. Mater. Process. Technol.* **2006**, *180*, 263–270. [CrossRef]

15. Liu, G.; Murr, L.E.; Niou, C.S.; McClure, J.C.; Vega, F.R. Microstructural aspects of the friction-stir welding of 6061-T6 aluminum. *Scr. Mater.* **1997**, *37*, 355–361. [CrossRef]

16. Murr, L.E.; Liu, G.; McClure, J.C. A tem study of precipitation and related microstructures in friction-stir-welded 6061 aluminium. *J. Mater. Sci.* **1998**, *33*, 1243–1251. [CrossRef]

17. Sato, Y.S.; Kokawa, H.; Enomoto, M.; Jogan, S. Microstructural evolution of 6063 aluminum during friction-stir welding. *Metall. Mater. Trans. A* **1999**, *30*, 2429–2437. [CrossRef]

18. Sato, Y.S.; Urata, M.; Kokawa, H. Parameters controlling microstructure and hardness during friction-stir welding of precipitation-hardenable aluminum alloy 6063. *Metall. Mater. Trans. A* **2002**, *33*, 625–635. [CrossRef]

19. Svensson, L.E.; Karlsson, L.; Larsson, H.; Karlsson, B.; Fazzini, M.; Karlsson, J. Microstructure and mechanical properties of friction stir welded aluminium alloys with special reference to AA 5083 and AA 6082. *Sci. Technol. Weld. Join.* **2000**, *5*, 285–296. [CrossRef]

20. Dehghani, M.; Amadeh, A.; Akbari Mousavi, S.A.A. Investigations on the effects of friction stir welding parameters on intermetallic and defect formation in joining aluminum alloy to mild steel. *Mater. Des.* **2013**, *49*, 433–441. [CrossRef]

21. Khodir, S.A.; Shibayanagi, T. Friction stir welding of dissimilar AA2024 and AA7075 aluminum alloys. *Mater. Sci. Eng. B* **2008**, *148*, 82–87. [CrossRef]

22. Nakata, K.; Kim, Y.G.; Ushio, M.; Hashimoto, T.; Jyogan, S. Weldability of high strength aluminum alloys by friction stir welding. *ISIJ Int.* **2000**, *40*, S15–S19. [CrossRef]

23. Khan, N.Z.; Khan, Z.A.; Siddiquee, A.N.; Al-Ahmari, A.M.; Abidi, M.H. Analysis of defects in clean fabrication process of friction stir welding. *Trans. Nonferrous Met. Soc. China* **2017**, *27*, 1507–1516. [CrossRef]

24. Khan, N.Z.; Siddiquee, A.N.; Khan, Z.A.; Shihab, S.K. Investigations on tunneling and kissing bond defects in FSW joints for dissimilar aluminum alloys. *J. Alloys Compd.* **2015**, *648*, 360–367. [CrossRef]

25. Chen, Z.W.; Pasang, T.; Qi, Y. Shear flow and formation of Nugget zone during friction stir welding of aluminium alloy 5083-O. *Mater. Sci. Eng. A* **2008**, *474*, 312–316. [CrossRef]

26. Saeid, T.; Abdollah-Zadeh, A.; Assadi, H.; Ghaini, F.M. Effect of friction stir welding speed on the microstructure and mechanical properties of a duplex stainless steel. *Mater. Sci. Eng. A* **2008**, *496*, 262–268. [CrossRef]

27. Kim, Y.G.; Fujii, H.; Tsumura, T.; Komazaki, T.; Nakata, K. Three defect types in friction stir welding of aluminum die casting alloy. *Mater. Sci. Eng. A* **2006**, *415*, 250–254. [CrossRef]

28. Buffa, G.; Campanile, G.; Fratini, L.; Prisco, A. Friction stir welding of lap joints: Influence of process parameters on the metallurgical and mechanical properties. *Mater. Sci. Eng. A* **2009**, *519*, 19–26. [CrossRef]

29. Cui, L.; Yang, X.; Zhou, G.; Xu, X.; Shen, Z. Characteristics of defects and tensile behaviors on friction stir welded AA6061-T4 T-joints. *Mater. Sci. Eng. A* **2012**, *543*, 58–68. [CrossRef]

30. Iqbal, A.; Khan, N.Z.; Siddiquee, A.N. Friction stir welding of different joint configurations: A review. *J. Mater. Sci. Mech. Eng.* **2015**, *2*, 19–24.

31. Shirazi, H.; Kheirandish, S.; Safarkhanian, M.A. Effect of process parameters on the macrostructure and defect formation in friction stir lap welding of AA5456 aluminum alloy. *Measurement* **2015**, *76*, 62–69. [CrossRef]

32. Dubourg, L.; Merati, A.; Jahazi, M. Process optimisation and mechanical properties of friction stir lap welds of 7075-T6 stringers on 2024-T3 skin. *Mater. Des.* **2010**, *31*, 3324–3330. [CrossRef]

33. Cao, X.; Jahazi, M. Effect of tool rotational speed and probe length on lap joint quality of a friction stir welded magnesium alloy. *Mater. Des.* **2011**, *32*, 1–11. [CrossRef]

34. Threadgill, P.L.; Leonard, A.J.; Shercliff, H.R.; Withers, P.J. Friction stir welding of aluminium alloys. *Int. Mater. Rev.* **2009**, *54*, 49–93. [CrossRef]

35. Sato, Y.S.; Park, S.H.C.; Kokawa, H. Microstructural factors governing hardness in friction-stir welds of solid-solution-hardened Al alloys. *Metall. Mater. Trans. A* **2001**, *32*, 3033–3042. [CrossRef]

36. Field, D.P.; Nelson, T.W.; Hovanski, Y.; Jata, K.V. Heterogeneity of crystallographic texture in friction stir welds of aluminum. *Metall. Mater. Trans. A* **2001**, *32*, 2869–2877. [CrossRef]

37. Rai, R.; De, A.; Bhadeshia, H.K.D.H.; DebRoy, T. Review: Friction stir welding tools. *Sci. Technol. Weld. Join.* **2011**, *16*, 325–342. [CrossRef]

38. Zhang, Y.N.; Cao, X.; Larose, S.; Wanjara, P. Review of tools for friction stir welding and processing. *Can. Metall. Q.* **2012**, *51*, 250–261. [CrossRef]

39. Bayazid, S.M.; Farhangi, H.; Ghahramani, A. Effect of pin profile on defects of Friction Stir Welded 7075 Aluminum alloy. *Procedia Mater. Sci.* **2015**, *11*, 12–16. [CrossRef]

40. Bussu, G.; Irving, P.E. The role of residual stress and heat affected zone properties on fatigue crack propagation in friction stir welded 2024-T351 aluminium joints. *Int. J. Fatigue* **2003**, *25*, 77–88. [CrossRef]

41. John, R.; Jata, K.V.; Sadananda, K. Residual stress effects on near-threshold fatigue crack growth in friction stir welds in aerospace alloys. *Int. J. Fatigue* **2003**, *25*, 939–948. [CrossRef]

42. Prime, M.B.; Hill, M.R. Residual stress, stress relief, and inhomogeneity in aluminum plate. *Scr. Mater.* **2002**, *46*, 77–82. [CrossRef]

43. Liu, H.J.; Fujii, H.; Maeda, M.; Nogi, K. Mechanical properties of friction stir welded joints of 1050–H24 aluminium alloy. *Sci. Technol. Weld. Join.* **2003**, *8*, 450–454. [CrossRef]

Article

Mechanism of Solder Joint Cracks in Anisotropic Conductive Films Bonding and Solutions: Delaying Hot-Bar Lift-Up Time and Adding Silica Fillers

Shuye Zhang [1,*], Ming Yang [2], Mingliang Jin [3,4], Wen-Can Huang [3], Tiesong Lin [1], Peng He [1], Panpan Lin [1] and Kyung-Wook Paik [5]

[1]   State Key Laboratory of Advanced Welding and Joining, Harbin Institute of Technology, Harbin 150001, China; hitjoining@hit.edu.cn (T.L.); hithepeng@hit.edu.cn (P.H.); 13b309019@hit.edu.cn (P.L.)

[2]   Yik Shing Tat Industrial Co., Ltd., Shenzhen 518101, China; yangming.hitsz@gmail.com

[3]   Department of Chemical and Biomolecular Engineering, KAIST, 291 Daehak-ro, Yuseong-gu, Daejeon 305-338, Korea; 20104170@kaist.ac.kr (M.J.); moonchanhwang@kaist.ac.kr (W.-C.H.)

[4]   Department of Nano-Structured Materials Research, National NanoFab Center (NNFC), 291 Daehak-ro, Yuseong-gu, Daejeon 305-338, Korea

[5]   Department of Materials Science and Engineering, KAIST, 291 Daehak-ro, Yuseong-gu, Daejeon 305-338, Korea; kwpaik@kaist.ac.kr

*   Correspondence: syzhang@hit.edu.cn or syzhang@kaist.ac.kr

Received: 11 December 2017; Accepted: 3 January 2018; Published: 9 January 2018

Abstract: Micron sizes solder metallurgical joints have been applied in a thin film application of anisotropic conductive film and benefited three general advantages, such as lower joint resistance, higher power handling capability, and reliability, when compared with pressure based contact of metal conductor balls. Recently, flex-on-board interconnection has become more and more popular for mobile electronic applications. However, crack formation of the solder joint crack was occurred at low temperature curable acrylic polymer resins after bonding processes. In this study, the mechanism of SnBi58 solder joint crack at low temperature curable acrylic adhesive was investigated. In addition, SnBi58 solder joint cracks can be significantly removed by increasing the storage modulus of adhesives instead of coefficient of thermal expansion. The first approach of reducing the amount of polymer rebound can be achieved by using an ultrasonic bonding method to maintain a bonding pressure on the SnBi58 solder joints cooling to room temperature. The second approach is to increase storage modulus of adhesives by adding silica filler into acrylic polymer resins to prevent the solder joint from cracking. Finally, excellent acrylic based SnBi58 solder joints reliability were obtained after 1000 cycles thermal cycling test.

Keywords: adhesive thermos-mechanical property; SnBi58 solder joint morphology; flex-on-board assembly; thermal compression bonding; ultrasonic bonding

---

1. Introduction

Amid the current trend for wearable electronics assembly, flex-on-board (FOB) assembly is attracting a greater attention because of its important role in replacing conventional physical contacts-based socket type interconnections. In the case of conventional socket type interconnections, there are three main drawbacks, which include: physical contact, large package size, and low packaging density [1]. For reducing the package thickness from 4 mm to 0.1 mm, and fine pitch capability to less than 100 µm, FOB application, as shown in Figure 1, is an obvious choice in the replacement of conventional socket type interconnections [2].

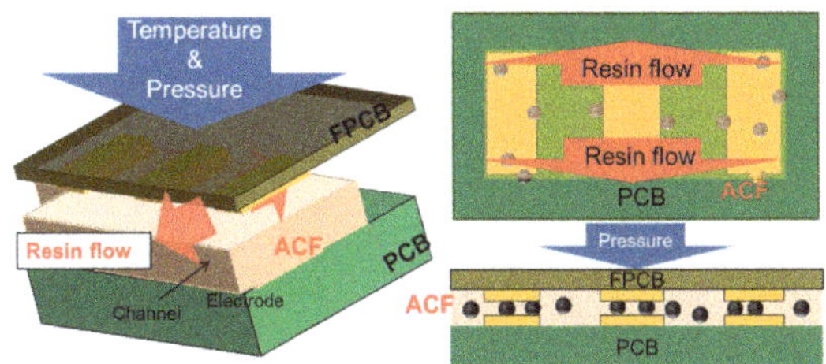

Figure 1. Bonding process and formed flex-on-board interconnection.

The anisotropic conductive films (ACFs) are the interconnection materials to assembly FOB applications. ACFs generally consist of thermosetting polymer resins and micron-sized conductive particles, with the cruel requirement in fine pitch applications of providing electrical paths only in the z-axis, and insulation property by polymer resins in the x-y plane. Physical contacts between the conductive particles and metal electrodes are the main electrical paths for micron-sized Au/Ni coated polymer ball joints and metal ball joints, but they will easily fail with the thermal expansion of polymer resin in high-power applications [3]. Therefore, micron-sized solder particles are added to polymer resin matrixes to replace micron-sized Au/Ni coated polymer ball joints and metal ball joints to form metallurgical joints for lower joint resistance, higher power-handling capability, and better reliability, in Figure 2.

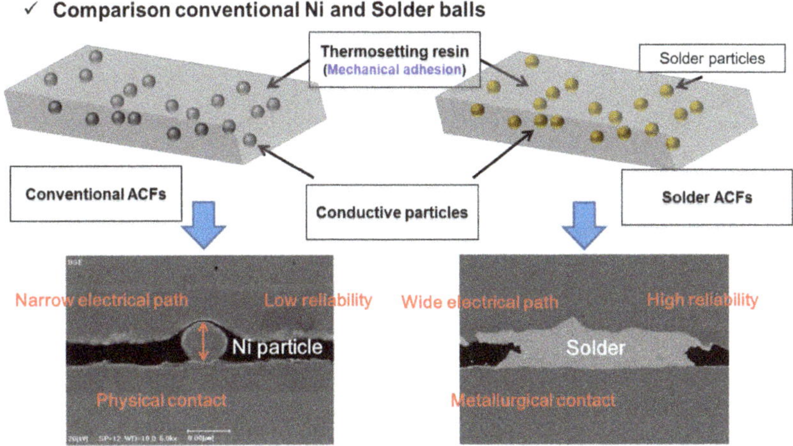

Figure 2. A comparison of the conventional Ni and Sn solder metallurgical anisotropic conductive films (ACFs) joints.

In these solder ACFs, solders are Sn-based Sn-58Bi or Sn-Ag-Cu (SAC) alloys, depending on their melting temperature. Sn-58Bi (139 °C) and Sn-3.0Ag-0.5Cu (221 °C) solder ACFs can be assembled at 200 °C and 250 °C bonding temperature, respectively. According to previous study [4], 200 °C was the optimized bonding temperature for Sn-58Bi solder ACFs joints in FOB applications. Below 200 °C, resin traps were always found at Sn-58Bi solder joints. Above 200 °C, a joint height larger than solder ball dimension was achieved and even open circuits were formed, due to ACFs faster curing property.

To remove the Sn solder oxide layer for enhanced performance, two bonding methods have been used. One method uses an ultrasonic (US) bonding [5], and the other method is to add flux materials in adhesives using a conventional thermo-compression (TC) bonding [6]. Figure 3 shows the difference on the final completed moment between the TC bonding and US bonding methods in Figure 3. At the end of the TC bonding, the bonding pressure will be automatically lifted-up, however, the lift-up time of the bonding pressure can be controlled in the cooling process by the US bonding.

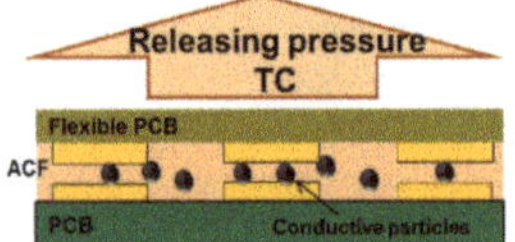

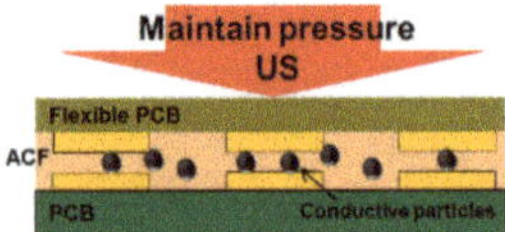

Figure 3. The comparison of the completed moment between thermo-compression (TC) and ultrasonic (US) bonding: Releasing pressure and maintaining pressure.

Because the polymer resins were still at a high temperature environment when the bonding pressure was removed, the potential polymer rebound without the bonding pressure protect might be issued for solder ACFs joints, due to viscoelastic property. The solder joint was molten and could not provide any protect against the polymer rebound, as a result, cracks of solder ACFs joints would occur. In terms of viscoelastic theory, the polymer resin shows a dominant elastic property instead of the viscos property, because the polymer rebound and the disappear of the bonding pressure were in a vertical direction and the polymer rebound was theoretically caused by the shape recover of polymer resins without pressure protect at a high temperature. In addition, there was not any change of the horizontal sheer forces in the polymer resin on the moment of releasing bonding pressures [7]. Therefore, the elastic property of polymer resins at a high temperature was very important for cracking solder ACFs joints. However, there is no research regarding this area, and no available method to solve the solder joint cracks. Figure 4 shows a Sn-58Bi solder joint crack after a FOB assembly.

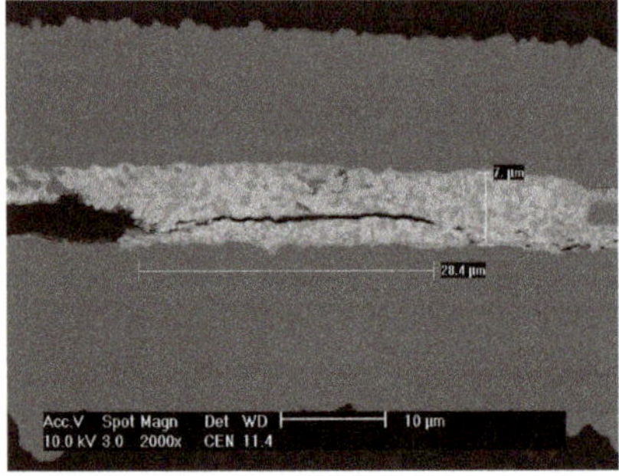

Figure 4. SnBi58 solder joint cracks after a thermo-compression bonding.

In this study, we used adhesives thermos-mechanical property to illustrate resin elastic property when the bonding pressures were removed, and the corresponding SnBi58 solder joint morphologies were observed using various adhesives materials and two bonding methods. In details, three typical adhesive resins were compared with different thermos-mechanical properties. It was found that solder joint cracks were always made at fast curable acrylic resins with a poor thermos-mechanical property. Its glass transition temperature (Tg) was too low at 45 °C. To solve the solder joint cracks at this 45 °C low Tg material, we developed the lift-up time of bonding pressure in the cooling process by an US bonding method. Using this, crack-free solder ACF joints were achieved when bonding pressure was lift-up in the cooling process until 70 s lower than polymer Tg. The second method to add 0.2 µm silica fillers to increase resin modulus using a conventional TC bonding. It was found that cracks at Sn-58Bi joints can be removed by adding to 10 wt % amount of SiO_2, where 1.38 MPa storage modulus was needed at 200 °C. Finally, −45/125 °C reliability was carried out to evaluate the improved solder

ACFs joints for low temperature curable acrylic resins by 1000 cycles. All of the joint crack experiments are conducted and lead by our research group.

2. Experiments

2.1. Test Vehicles and Materials

1-mm-thick FR-4 printed circuit board (PCB) and a 50-μm-thick polyimide based flexible printed circuit (FPCB) substrate which plotted 500-μm-pitch Cu patterns with electroless nickel immersion gold (ENIG) finish in Figure 5.

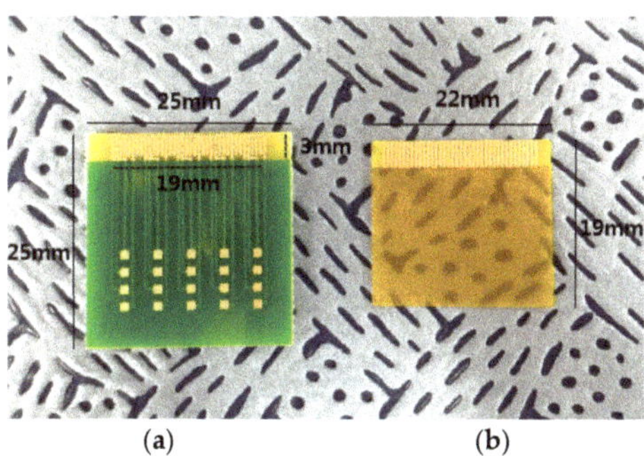

(a)   (b)

Figure 5. 500-μm-pitch flex-on-board (FOB) test vehicles. (**a**) printed circuit board (PCB), (**b**) polyimide based flexible printed circuit (FPCB).

For the ACFs, three kinds of thermosetting polymer resins were used, in name of acrylic base, imidazole base and cationic epoxy base. These products were bought from a commercial adhesive company. We added 5 wt % 8-μm-diameter Ni particles, 0.2 μm silica fillers, 30 wt % 25–32 μm diameters Sn58Bi particles, and 2 wt % flux material into the pure bought adhesives and then proceeded the film coating process. So, we got the anisotropic conductive films.

Ni particles were used to resist bonding pressures and maintain solder joint morphology during bonding in Figure 6. Silica fillers were directly added and mixed with the solution of polymer resins and conductive particles. Table 1 gives the specifications of the added materials, such as weight percentages of the pure polymer resins and the calculated volume percentages of the total ACFs materials with additives. Figure 7 shows the fresh ACFs were put on the PCB ENIG metal electrodes before bonding process [8].

Table 1. The specification of solder ACFs.

Solder ACFs	Weight Percentage				Calculated Volume Percentage			
	SnBi58 Solder (8.56 g/cm^3)	Polymer Resin (1.25 g/cm^3)	Ni Particle (8.9 g/cm^3)	Silica Filler (2.65 g/cm^3)	SnBi58 Solder (8.56 g/cm^3)	Polymer Resin (1.25 g/cm^3)	Ni Particles (8.9 g/cm^3)	Silica Filler (2.65 g/cm^3)
ACF 1	30%	1	5%	0%	6.25%	92.75%	1%	0%
ACF 2	30%	1	5%	5%	6.1%	89.7%	0.97%	3.23%
ACF 3	30%	1	5%	10%	5.9%	86.9%	0.93%	6.27%

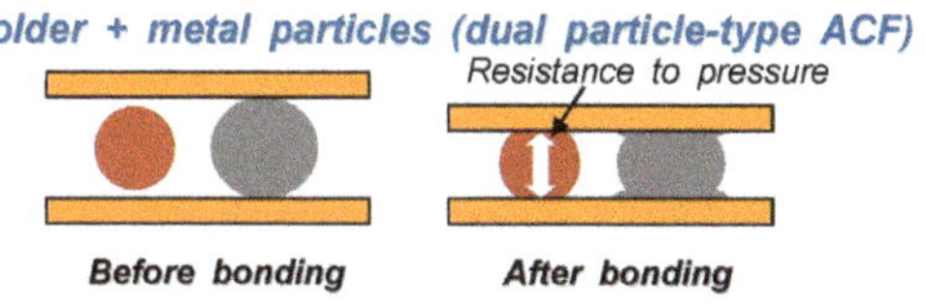

Figure 6. Role of Ni spacer at solder ACFs joints.

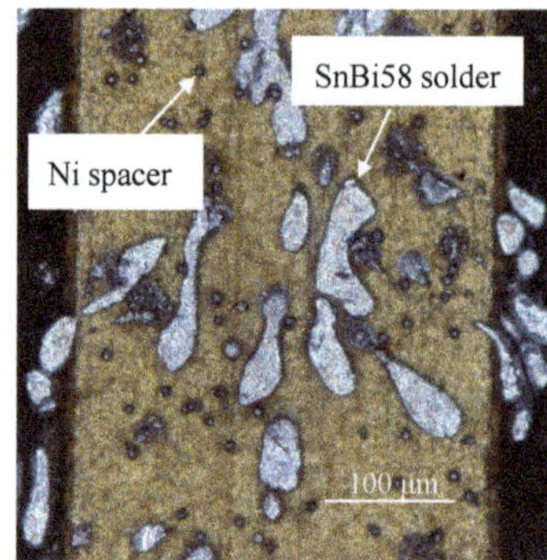

Figure 7. The laminated ACFs on the PCB electroless nickel immersion gold (ENIG) metal electrodes.

2.2. Two Bonding Methods

2.2.1. Thermo-Compression Bonding

The bonding parameters (peak temperature, time, and pressure) were set as 200 °C 10 s and 1 MPa. Figure 8 shows a schematic of TC bonding process. Polymer resin was cured by the heat conduction from the high temperature of hot bar. Figure 9 gives the in-situ temperature of solder ACFs joint by a TC bonding, which is measured by a K-type thermocouple every 0.1 s. In this case, the hot-bar was automatically lift-up on the peak temperature.

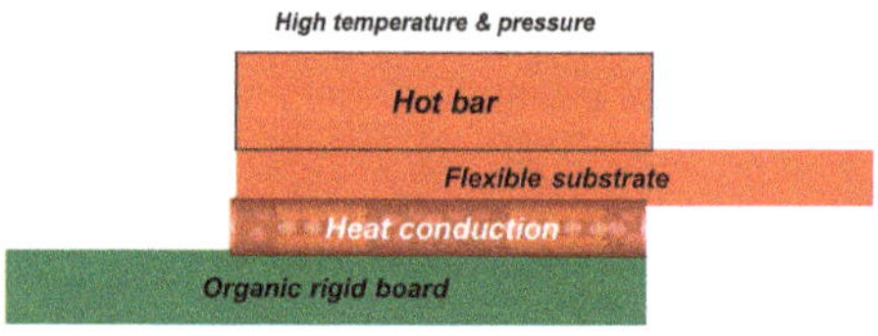

Figure 8. A schematic of TC bonding.

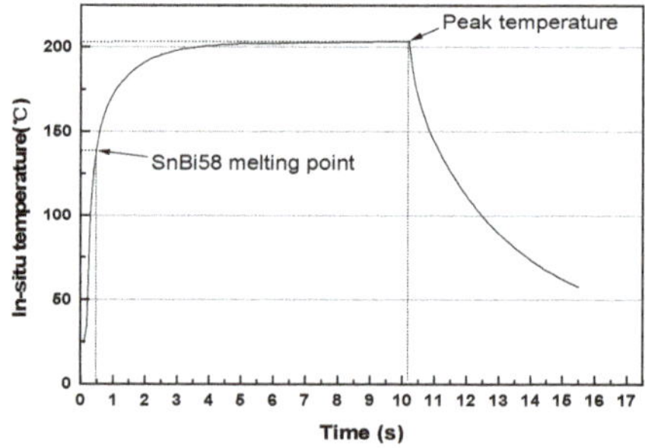

Figure 9. The in-situ temperature profile of solder ACFs joint by a TC bonding.

2.2.2. Ultrasonic Bonding

Figure 10 shows the schematic of ultrasonic bonding process. In this case, ultrasonic vibration was applied at room temperature and polymer resin was cured by the spontaneous ultrasonic vibration. Bonding pressure can be controlled as a function of time. So, Figure 11 gives the in-situ ACFs joint temperature profile and the delaying ultrasonic horn lift-up time using the ultrasonic bonding. In this case, ultrasonic vibration is applied for 10 s, and then the 1 MPa bonding pressure was maintained until 70 s to room temperature.

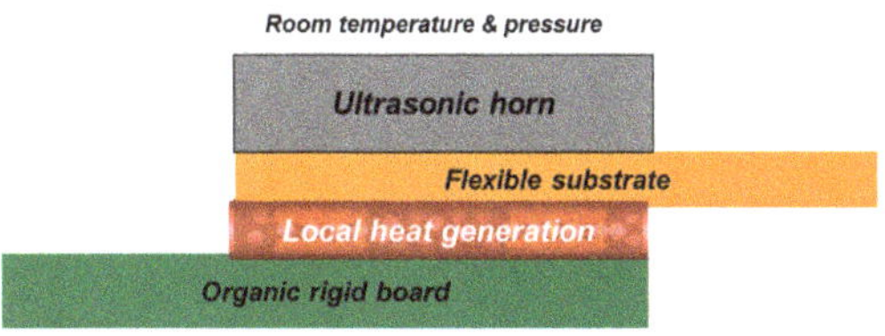

Figure 10. A schematic of ultrasonic bonding.

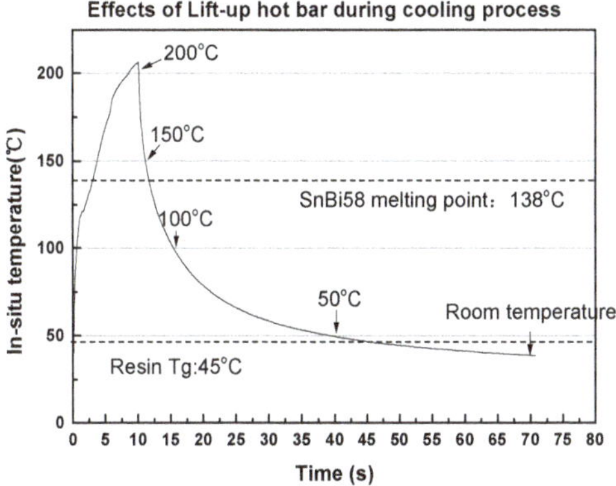

Figure 11. The in-situ temperature profile of solder ACFs joint by US bonding.

2.3. Differential Scanning Calorimetry (DSC)

For analysis of solder ACFs joint crack, DSC was performed. The heating rate was 20 °C/min from 30 °C to 300 °C and furnace cooled to room temperature at a nitrogen environment. Although a rate of 10 °C/min is traditional, a faster heating rate for the metallic solder materials does not affect much regarding solder melting temperatures. Polymer resin was not tested in this experiment, so the glass transition temperature of resin was not considered. At here, solder melting and solidifying behaviors can be adjusted. The test sample was 20 mg weight.

2.4. Thermomechanical Analysis

A thermomechanical analyzer (DMA) was used to measure the viscoelastic property of the cured polymer resin as a function of temperature. Figure 12 shows the DMA measurement for cured ACFs. A 10-mm-length, 2.5-mm-width, and 50-μm-thickness resin film was under a tensile force, where the static stress was 50 mN and the dynamic stress was 10 mN under 0.1 Hz sinusoidal load, and the temperature profile was fixed at a heating rate of 5 °C/min from 30 °C to 200 °C. For the elastic property in the z direction, the elastic strain was discussed. According to Equation (1), elastic strain is the elastic deformation of polymer resin driven by the sinusoidal tensile load, which is the ratio of length change (ΔL) over the initial ACF length (L). The higher elastic strain indicates larger plastic rebound of polymer resin. In addition, the coefficient of thermal expansion (CTE) of adhesives were also measured by a 50 mN constant tensile force as temperature increased.

$$Strain = \frac{\Delta L}{L} \tag{1}$$

Figure 12. DMA thermo-mechanical property measurement for cured ACFs [9].

2.5. Joint Resistance and Morphologies

After bonding process, ACFs joint resistance was measured using a kelvin 4-point-probe method. From Ohm's Law, we know that resistance is equal to voltage divided by current. The resistance of the desired part can be known if we apply the constant current going through it and measure the voltage dropped across it. The calculated ratio of overlapped voltage to current is regarded as the joint resistance [10].

In our design, delta mode measurement by KI 6220 nano-voltmeter from a Keithley company at America was used. The constant current was applied through both PCB and FPCB Cu lines and solder ACFs joints, and the voltage measurement was from another way and the overlapped grown part was the measured solder ACFs areas in Figure 13. 10 times were set up for the accurate measurement of one solder ACF joint area. Area size was 0.3 mm^2. 40 channels were measured for one kind of ACF joint.

In addition, solder ACFs joint images after the bonding process were observed by a scanning electron microscope (SEM). The backscattered electron mode and 2000 magnification were selected for an obvious comparison between Sn and Bi elements. Typical solder information, such as solder joint heights, shapes, morphologies, and cracks, was recorded.

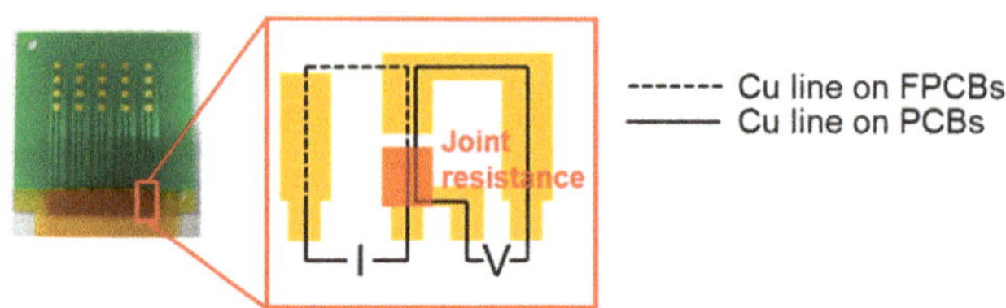

Figure 13. Electrical design for a four-point-probe measurement.

2.6. Reliability Evaluation

A $-45/125$ °C reliability was carried out to evaluate the improved solder ACFs joints at low Tg materials for 1000 cycles. The test samples for the reliability test were TC bonded using acrylic based solder ACFs with 0, 5, and 10 wt % of 0.2 μm silica fillers added. The reliability test was performed with a dwell time of 15 min from -40 °C to 125 °C for 1000 cycles. Joint contact resistances were measured every 200 cycles. After 1000 cycles, failure modes of ACF joints were detected by SEM analysis.

3. Results and Discussion

3.1. Elastic Property of Polymer Resin

Figure 14 shows a 50 mN tensile sinusoidal load with 10 mN amplitude and 0.1 Hz frequency in DMA. Figure 15 illustrates the measured strain of acrylic based SnBi58 solder ACFs in respect to the sinusoidal load in DMA as a function of temperature. Viscoelastic materials, such as elastomers, amorphous polymers, and semicrystalline polymers [11], were always in the polymer resins. For example, elastomers are usually thermosets as well as it may also be thermoplastic. The long polymer chains are cross-linked during a curing reaction. The molecular structure of elastomers can be imagined as Figure 16 and the dots represent cross-links. When the stress is removed, B configuration will return to the original A configuration [12]. This elasticity is derived from the long chains to reconfigure themselves to distribute an applied stress. This is the reason for the elastic behaviors of elastomers.

In Figure 15, it was found that the measured strain was increased at high temperature, especially above *Tg*. Actually, the measured strain is divided into two parts, the first is polymer relaxation under a tensile force and the second is elastic strain respond to sinusoidal mechanical load. Both the polymer relaxation and elastic strain were increased as the temperature increased. This is because elastomers that have cooled to a glassy or crystalline phase will have less mobile chains, and, consequentially, less elasticity than those manipulated higher than the glass transition temperature of the polymer [13]. So the polymer *Tg* is very important to reduce elasticity.

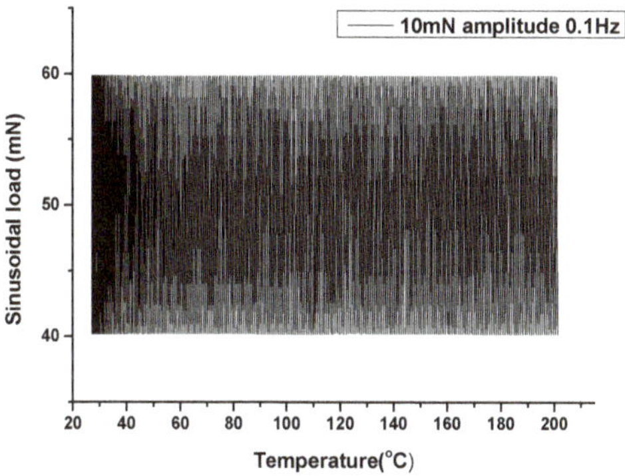

Figure 14. Mechanical load in a tensile mode by DMA.

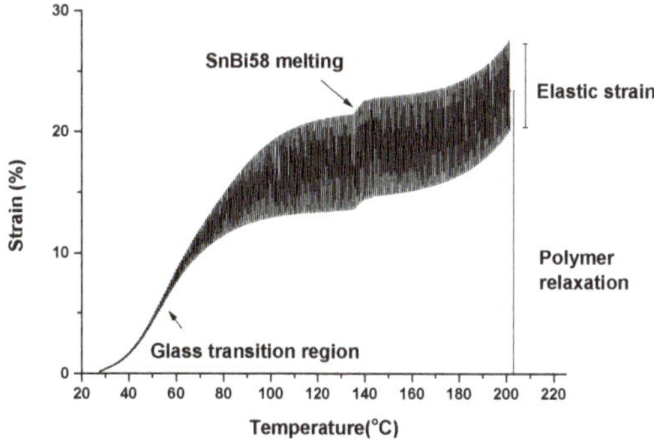

Figure 15. Strain of acrylic based SnBi58 solder ACFs in respect to a sinusoidal load in DMA as a function of temperature.

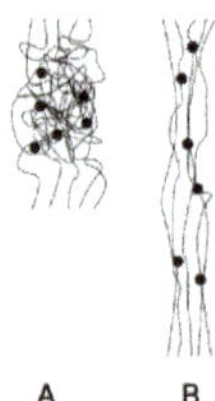

Figure 16. Elastomer structures: (**A**) is an unstressed polymer; (**B**) is the same polymer under stress.

Because Tg of acrylic resin is 45 °C, polymer transits from a rigid glass state to a soft rubber state, elastic strain in respect to same driving sinusoidal force is small below Tg and becomes large above Tg due to viscoelastic property. What is more, SnBi58 solder particles are made up of 30 wt % of total solder ACFs, and the strain measurement is in a tensile mode, so the dimension was stretched when the solder melts on 139 °C.

In this study, viscoelastic property was studied using a dynamic mechanical analysis method, which is by applying a small oscillatory stress and measuring the resulting strain. Complex dynamic modulus G can be used to represent the relations (2) to (4) between the oscillating stress and strain [14]:

$$G = G' + iG''$$

(2)

where $i^2 = -1$, G' is the storage modulus and G'' is the loss modulus:

$$G' = \frac{\sigma_0}{\varepsilon_0} \cos \delta$$

(3)

$$G'' = \frac{\sigma_0}{\varepsilon_0} \sin \delta$$

(4)

where σ_0 and ε_0 are the amplitudes of the applied stress and the measured elastic strain, respectively, and δ is the phase shift between them. Because G' and G'' only have phase shift difference, and both of them can represent ε_0. So, we use G' storage modulus to estimate resin elastic property. Figure 17 is the derived storage modulus of acrylic based polymer resin from DMA. Since acrylic resin is a low Tg material, the lower storage modulus indicates the larger elasticity of polymer resin.

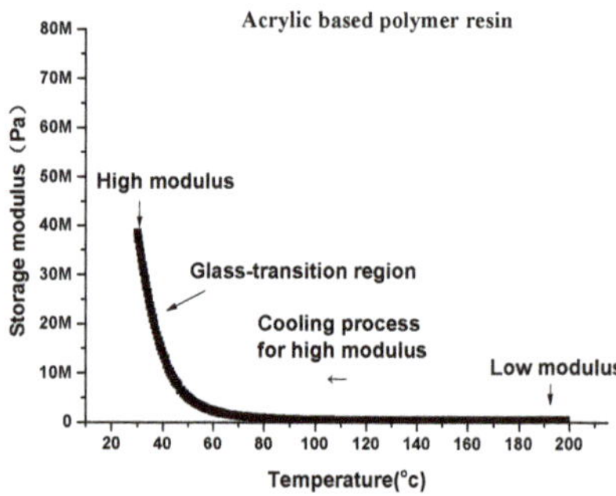

Figure 17. Storage modulus of acrylic based solder ACF as a function of temperature.

3.2. Effects of Resin Tg on Solder Joint Morphology

We prepared three types polymer resins with different Tg in Figure 18. When compared with acrylic adhesives by 45 °C Tg, the conventional epoxies cured by imidazole and cationic agent showed higher Tg (over 100 °C) and storage modulus. Especially at 200 °C when hot-bar was removed at the end of TC bonding, the cured acrylic adhesives only had 0.4 MPa storage modulus, but cationic epoxy and imidazole epoxy showed 7.1 MPa and 5.7 MPa, respectively.

Figure 19 shows SnBi58 solder ACFs joints morphologies using three types resins and a conventional TC bonding method. It only showed cracks at acrylic adhesives, regardless of bonding pressures. However, there is no solder joint cracks when using other two types adhesive materials under same bonding pressures. Therefore, our first conclusion is that solder joint crack can be solved by using higher Tg adhesives. In Figure 19, cracks are found mostly along the outer interface and sporadically inside the solder, probably due to the weak intermetallic compound of Ni_3Sn_4 formation at the interface after bonding. However, in Figure 19g, cracks are found in the solder, probably due to the crack initiation at the weak point of solder joints and then propagation into the bulk solder joints. All of the results can be replicate for the same specimens of ACFs recipes.

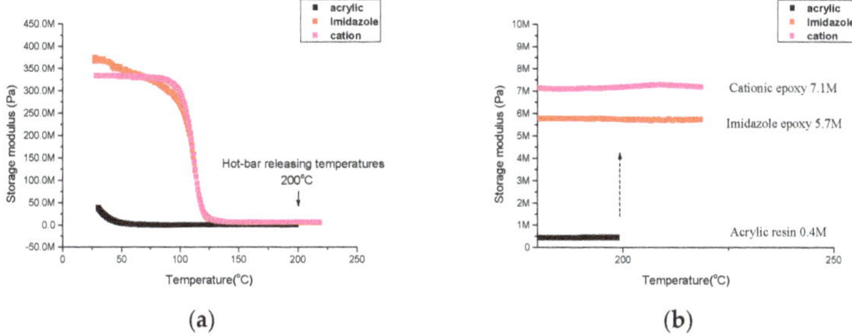

Figure 18. Thermal mechanical properties of typical adhesives films. (**a**) 30 °C–250 °C (**b**) magnified 200 °C–250 °C.

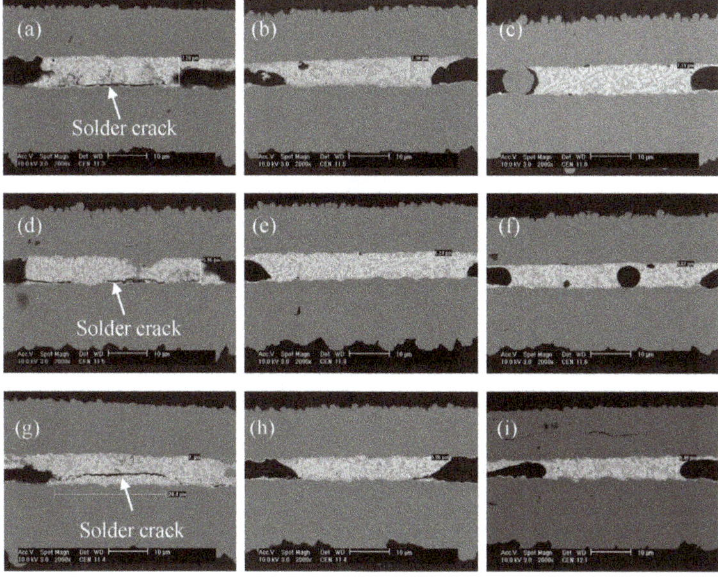

Figure 19. SEM images of SnBi58 solder ACFs joints by 200 °C 10 s TC bonding as a function of resin types and bonding pressures. (**a,d,g**) Acrylic base resins with 1, 2, and 3 MPa bonding pressures; (**b,e,h**) Imidazole epoxy with 1, 2, and 3 MPa bonding pressures; (**c,f,i**) Cationic epoxy with 1, 2, and 3 MPa bonding pressures.

Although we assumed that the molten solder joints were cracked under tensile forces by the polymer rebound when bonding pressures were removed at high temperature, however, there is another concern by CTE mismatch between shrink adhesives and solders during the cooling process. Therefore, we carried out the cooling analysis.

Figure 20 shows endothermic and exothermic behaviors of SnBi58 solder materials in a DSC analysis. In a heating period, SnBi58 solder melts at 139 °C and stays at a liquid state until cooling to 125 °C, and SnBi58 solders were totally solid below 90 °C. It means that the cooled adhesives are able to cause the solidified SnBi58 solder joints cracks above 90 °C, because the fully solid SnBi58 joint is with 30.9 Gpa young's modulus, which is a hard material. Figure 21 summarizes thermal expansion properties of three typical adhesives. Table 2 gives shrinkage percentages of typical adhesives films during cooling. 90 °C is lower than *Tg* of imidazole epoxy and cation epoxy, so there was a huge dimension shrinkage when cooling through glass transition region at 2 epoxies. From 200 °C cooling to 90 °C, acrylic resin, imidazole epoxy, and cation epoxy showed −11.2%, −13.2%, and −5.2% dimension shrinkage, however, there is no solder joint cracks at imidazole epoxy based SnBi58 solder ACFs joints in Figure 19. Thus, solder ACFs joint cracks were not related with higher compressive stress by adhesives shrinkages in the cooling process. The mechanism of solder ACFs joint cracks should be adhesive rebound, especially for lower modulus acrylic adhesives.

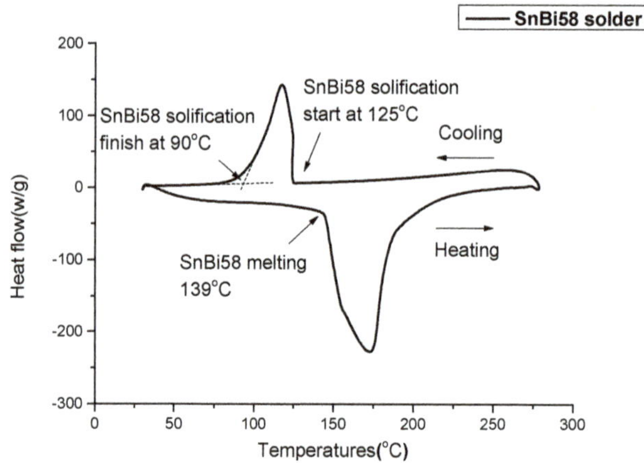

Figure 20. DSC behaviors of SnBi58 solder during heating and cooling process.

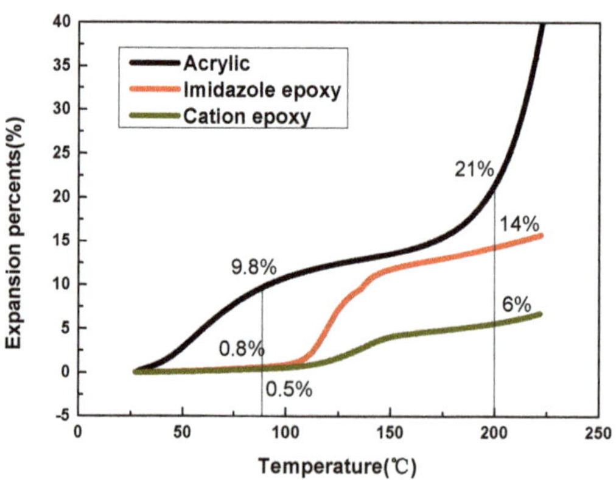

Figure 21. Thermal expansion properties of typical adhesives films.

Table 2. Shrinkage percentages of typical adhesives films during cooling.

Dimension Change	Acrylic	Imidazole Epoxy	Cation Epoxy
200 °C	21%	14%	6%
90 °C	9.8%	0.8%	0.8
200 cooling to 90 °C	−11.2%	−13.2%	−5.2%

3.3. Effects of Ultrasonic Bonding on Increasing Resin Modulus

Learning from that high modulus may help to solve the solder joint morphology; an ultrasonic bonding with maintaining pressures in the cooling process was tried. Table 3 summarized increased storage modulus of acrylic adhesives at lower temperature. Figure 22 shows the acrylic based SnBi58 solder ACFs joint morphologies during an US bonding method with different lift-up time of bonding tool. It was found that a crack-free SnBi58 solder ACFs joint can be obtained by maintaining pressure below 30 °C, due to higher storage modulus of the acrylic adhesives below its T_g (45 °C). At here, above 5 MPa storage modulus was needed to avoid solder joint cracks using an ultrasonic bonding. The enhanced process was done by 70 s, however, maintaining pressure time is too long for assembly over 30 s.

Table 3. Storage modulus of acrylic adhesive during cooling process.

Temperature	200 °C	150 °C	100 °C	50 °C	30 °C
Modulus	0.4 MPa	0.44 MPa	0.53 MPa	5.2 MPa	34 MPa

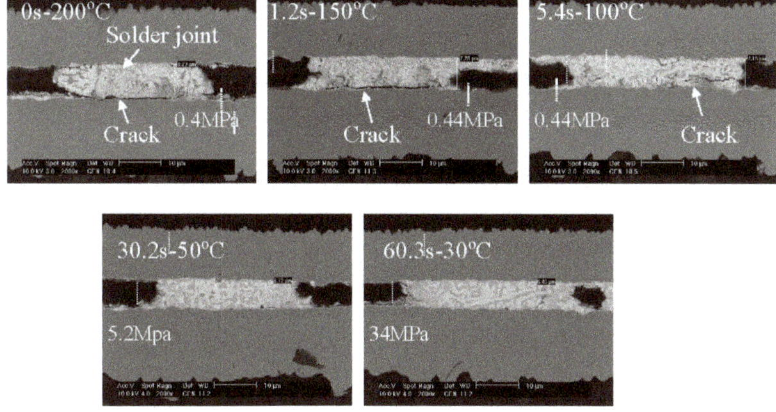

Figure 22. Effects of delaying ultrasonic horn lift-up times (0, 1.2, 5.4, 30.2, and 60.3 s) on the SnBi58 solder ACFs joint morphologies.

3.4. Effects of Silica Fillers on Increasing Modulus of Acrylic Resin

From the ultrasonic bonding experiment, we concluded that resin modulus is the key to determine solder joint cracks. In order to find out the specific modulus for TC bonding to prevent solder joint from cracking, the second approach is to increase resin storage modulus by directly adding silica fillers into acrylic polymer resins. Figure 23 shows that the measured strain was decreased of acrylic based SnBi58 solder ACFs by adding 0, 5, and 10 wt % of 0.2 μm silica fillers. Both the amount of the polymer expansion and the elastic strain of polymer resin were reduced by adding higher contents of silica fillers.

Storage modulus of acrylic based SnBi58 solder ACFs with addition of 0, 5, and 10 wt % 0.2 μm silica was shown in Figure 24. By adding 5 and 10 wt % 0.2 μm silica filler, the storage modulus 0.4 MPa, 0.9 MPa, and 1.3 MPa was prepared at 200 °C, respectively.

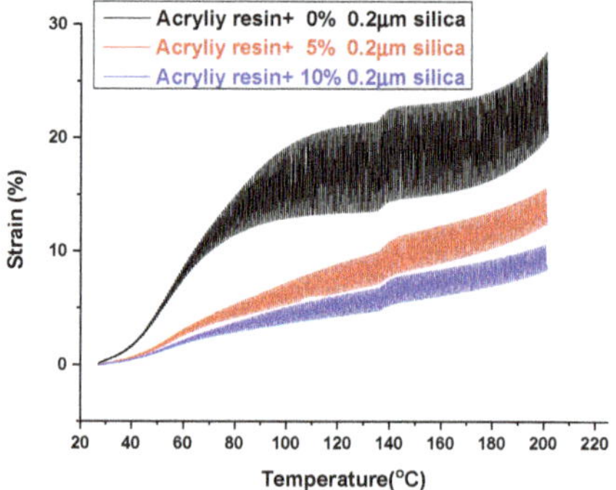

Figure 23. Strain of acrylic based SnBi58 solder ACFs with added silica fillers as a function of temperature.

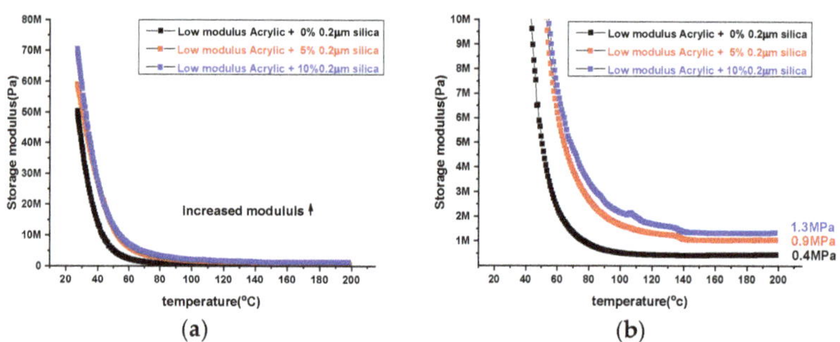

Figure 24. Elastic modulus of acrylic based SnBi58 solder ACFs by adding 0, 5, and 10 wt % 0.2 μm silica fillers at (**a**) 0–80 MPa and (**b**) 0–10 MPa ranges.

3.5. Reliability Evaluation for Modified Solder ACFs Joint

Solder joint morphologies of acrylic based SnBi58 solder ACFs added by 0, 5, and 10 wt % 0.2 μm silica filler before and after 1000 cycles reliability were listed in the Figure 25. According to the increment of elastic modulus by adding silica fillers from 0 wt % to 10 wt %, as shown in Figure 24b, solder joint crack was completely removed at acrylic based SnBi58 solder ACF joints during 200 °C TC bonding with a pressure release during a cooling process. The optimized storage modulus for adhesives in TC bonding was above 1.38 MPa at 200 °C for crack-free SnBi58 solder joints.

In terms of reliability evaluation, Figure 25 shows the joint morphology after reliability. For acrylic based SnBi58 solder ACFs without added silica fillers, the joint failure occurred at the interface between SnBi58 solder joints and Cu metal electrode after 1000 cycles reliability test, because initial solder joint cracks have already existed before the reliability test. If 0.2 μm silica filler was less than 5 wt %, then a small solder crack still remained and propagated at SnBi58 solder joints, resulting in unstable joint contact resistance during reliability. If 0.2 μm silica filler was more than 10 wt %, a complete solder joint morphology was obtained after the 1000 cycles reliability, because initial solder joint crack was perfectly removed. Figure 26 shows electrical performance in the 1000 cycles T/C reliability [15], and the most stable joint resistance was achieved by adding 10 wt % 0.2 μm silica fillers and crack-free solder joint.

TC bonding	Acrylic based SnBi58 solder ACF + 0% 0.2um silica filler	Acrylic based SnBi58 solder ACF + 5% 0.2um silica filler	Acrylic based SnBi58 solder ACF + 10% 0.2um silica filler
As bonded			
After 1000 cycles reliability			

Figure 25. Solder joint morphologies of acrylic based SnBi58 solder ACFs added by 0, 5, and 10 wt % 0.2 μm silica filler before and after 1000 cycles reliability test.

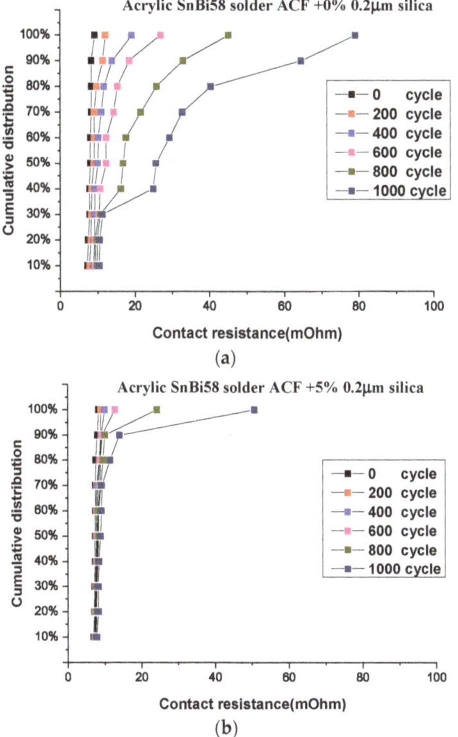

Figure 26. *Cont.*

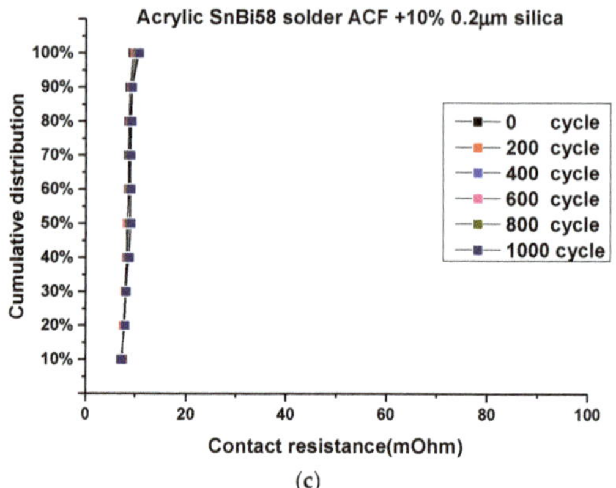

Figure 26. Solder joint contact resistances of acrylic based SnBi58 solder ACFs up to 1000 cycles T/C reliability as a function of (**a**) 0, (**b**) 5, and (**c**) 10 wt % 0.2 μm silica fillers addition.

The significance of this research is to present the low resin modulus is the determined factor to cause solder joint cracks, and we provided two methods to prevent solder joint from cracking by increasing resin modulus: by delaying hot-bar lift-up time and adding Silica Fillers. The other methods, which can solve the cracks in solder joints, will be investigated and considered in future.

4. Conclusions

In this study, we demonstrated the storage modulus of adhesives was the key factor for solder joint morphology. Crack at solder joints occurred due to low storage modulus. Two ways were suggested to increase resin modulus to solve the solder joint cracks. The first way was to maintain bonding pressures until cooling lower than its Tg (45 °C) during an ultrasonic bonding. The second way was to add 10 wt % 0.2 μm SiO_2 to the low modulus acrylic ACFs solder ACFs. For ultrasonic bonding, above 5 MPa of storage modulus was needed to prevent solder joint from cracking. However, the maintaining pressure time 70 s was too long for manufacturing factories over 30 s. Using a conventional TC bonding, it was found out that resin storage modulus above 1.38 MPa at 200 °C was needed for a crack-free SnBi58 solder joint morphology. Finally, the modified solder joint morphology was tested by a −45/125 °C reliability. After 1000 cycles, the modified solder joints showed the most stable electrical performance. Solder joint cracks can be reduced after reliability by adding 10 wt % 0.2 μm SiO_2 fillers.

Acknowledgments: The authors thank National Natural Science Foundation of China (Grant 51474081 and 61702317) for research funding support.

Author Contributions: Shuye Zhang and Kyung-Wook Paik conceived and designed the experiments; Shuye Zhang, Mingliang Jin, Wen-Can Huang, Tiesong Lin and Peng He performed the experiments; Shuye Zhang and Panpan Lin analyzed the data; Shuye Zhang contributed reagents/materials/analysis tools; Shuye Zhang wrote the paper.

Conflicts of Interest: The authors declare no conflict of interest.

References

1. Singh, P.; Viswanadham, P. *Failure Modes and Mechanisms in Electronic Packages*; Springer: Berlin, Germany, 2012.

2. Kiilunen, J.; Frisk, L. Reliability analysis of an ACA attached flex-on-board assembly for industrial application. *Solder. Surf. Mt. Technol.* **2014**, *26*, 62–70. [CrossRef]

3. Lee, K.; Saarinen, I.J.; Pykari, L.; Paik, K.W. High power and high reliability flex-on-board assembly using solder anisotropic conductive films combined with ultrasonic bonding technique. *IEEE Trans. Compon. Packag. Manag. Technol.* **2011**, *1*, 1901–1907. [CrossRef]

4. Zhang, S.; Lin, T.; He, P.; Paik, K.W. Effects of acrylic adhesives property and optimized bonding parameters on Sn 58Bi solder joint morphology for flex-on-board assembly. *Microelectron. Reliab.* **2017**, *78*, 181–189. [CrossRef]

5. Zhang, S.; Kim, S.H.; Kim, T.W.; Kim, Y.S.; Paik, K.W. A study on the solder ball size and content effects of solder ACFs for flex-on-board assembly applications using ultrasonic bonding. *IEEE Trans. Compon. Packag. Manag. Technol.* **2015**, *5*, 9–14. [CrossRef]

6. Kim, S.H.; Choi, Y.; Kim, Y.; Paik, K.W. Flux function added solder anisotropic conductive films (ACFs) for high power and fine pitch assemblies. In Proceedings of the 2013 IEEE 63rd on Electronic Components and Technology Conference (ECTC), Las Vegas, NV, USA, 28–31 May 2013; pp. 1713–1716.

7. Golla, D.F.; Hughes, P.C. Dynamics of viscoelastic structures—A time-domain, finite element formulation. *ASME J. Appl. Mech.* **1985**, *52*, 897–906. [CrossRef]

8. Zhang, S.; Paik, K.W. Effects of Cooling Processes and Silica Filler Contents of Solder ACFs (Anisotropic Conductive Films) on the Joints Reliability. In Proceedings of the 2016 IEEE 66th on Electronic Components and Technology Conference (ECTC), Las Vegas, NV, USA, 31 May–3 June 2016; pp. 737–742.

9. Ugural, A.C.; Fenster, S.K. *Advanced Strength and Applied Elasticity*; Pearson Education: London, UK, 2003.

10. Graffeuil, J.; Blasquez, G. Caractérisation des matériaux et des composants semiconducteurs au moyen de mesures de bruit de fond. *Acta Electron.* **1983**, *25*, 261–279.

11. Biswas, A.; Manivannan, M.; Srinivasan, M.A. Multiscale layered biomechanical model of the pacinian corpuscle. *IEEE Trans. Haptics* **2015**, *8*, 31–42. [CrossRef] [PubMed]

12. Urayama, K. An experimentalist's view of the physics of rubber elasticity. *J. Polym. Sci. Polym. Phys.* **2006**, *44*, 3440–3444. [CrossRef]

13. Van Krevelen, D.W.; Nijenhuis, K.T. *Properties of Polymers*; Their Correlation with Chemical Structure; Their Numerical Estimation and Prediction from Additive Group Contributions; Elsevier: Amsterdam, The Netherlands, 2009.

14. Vinogradov, G.V.; Yanovsky, Y.G.; Titkova, L.V.; Barancheeva, V.V.; Sergeenkov, S.I.; Borisenkova, E.K. Viscoelastic properties of linear polymers in the fluid state and their transition to the high-elastic state. *Polym. Eng. Sci.* **1980**, *20*, 1138–1146. [CrossRef]

15. Zhang, S.; Paik, K.W. A study on the failure mechanism and enhanced reliability of Sn58Bi solder anisotropic conductive film joints in a pressure cooker test due to polymer viscoelastic properties and hydroswelling. *IEEE Trans. Compon. Packag. Manaf. Technol.* **2016**, *6*, 216–223. [CrossRef]

MDPI

St. Alban-Anlage 66

4052 Basel

Switzerland

Tel. +41 61 683 77 34

Fax +41 61 302 89 18

www.mdpi.com

Metals Editorial Office

E-mail: metals@mdpi.com

www.mdpi.com/journal/metals

9 783039 289974